Troubleshooting
Process Operations

Fourth Edition

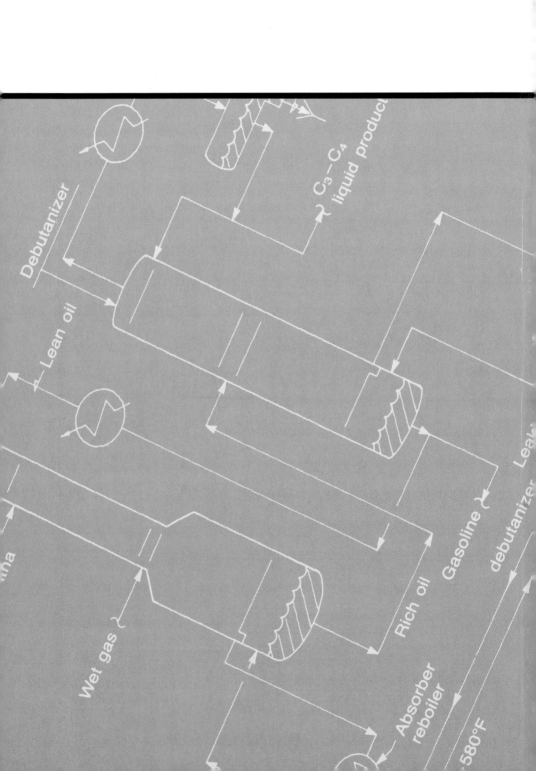

Troubleshooting
Process Operations
Fourth Edition

Norman P. Lieberman

PennWell®

Copyright © 2009 by
PennWell Corporation
1421 South Sheridan Road
Tulsa, Oklahoma 74112-6600 USA

800.752.9764
+1.918.831.9421
sales@pennwell.com
www.pennwellbooks.com
www.pennwell.com

Marketing: Jane Green
National Account Executive: Barbara McGee

Director: Mary McGee
Managing Editor: Stephen Hill
Production Manager: Sheila Brock
Production Editor: Tony Quinn
Book Designer: Susan E. Ormston
Cover Designer: Charles Thomas

Library of Congress Cataloging-in-Publication Data

Lieberman, Norman P.
 Troubleshooting process operations / Norman P. Lieberman. -- 4th ed.
    p. cm.
 Includes bibliographical references and index.
 ISBN 978-1-59370-176-5
 1. Petroleum refineries--Maintenance and repair--Handbooks, manuals, etc. I. Title.
 TP690.3.L53 2009
 665.5'30288--dc22   2009005758

Printed in the United States of America

7 8 9 10   22 21 20 19

Dedication

To the memory of my great grandfather, Fischel Lieberman, pioneer and first settler of the Russian Colony in the Rio Negro Province, Patagonia, Argentina. Born in South Russia in 1856, he died and was buried in the old Jewish cemetery in the town of General Roca, Argentina, in 1939. He founded the oldest and still existing Jewish temple in South America. Fischel turned the desert into an apple orchard. His spirit lives within me, and his orchard still blooms along the reedy banks of the Rio Negro River.

Other Books by Norm Lieberman

Troubleshooting Refinery Processes, 1980
Troubleshooting Natural Gas Processing, 1986
Process Design for Reliable Operations, 3rd Edition, 2006
A Working Guide to Process Equipment, 3rd Edition, 2007
Troubleshooting Process Plant Control, 2008
Process Engineering for a Small Planet, 2009

norm@lieberman-eng.com
1-504-887-7714

Contents

Section 3 Practical Problems

Section 4 Gas Drying and Compression

Section 5 The Process Engineer's Job

Preface to the Fourth Edition

My mother was cheap. She would never buy anything new. "Norman," she would say. "Use what you've got. We're poor people."

Actually we weren't all that poor. But Mom was cheap. She ate potato skins because she couldn't allow food to be wasted. She would walk a mile to save the dime for the trolley. No wonder she had her health until a month before she passed on. Too bad that my mother was not in charge of our little planet.

"People," she would have shouted. "Listen! We live on a poor little planet. We can't afford new things. Use what you've got. There won't be anything left for my grandchildren."

"Don't waste food or water or air or hydrocarbons. Listen dummies! Money doesn't grow on trees." She would have commanded.

I've spent my career trying to make my mother proud of me; to show her that I'm not a dummy. That's why I wrote this book. It's all about using what we have to make gasoline, jet fuel, and diesel oil without wasting hydrocarbons or electricity or combustion air or boiler feed water. We don't need new naphtha reformers or catalytic cracking units or sulfur recovery plants or new delayed cokers. We certainly don't need any new refineries. We have all the hardware we need; if we will just interface with the equipment in the field at a technical level.

Ladies and gentlemen, get out of your office. Forget about the meetings, the reports, the emails, and the other bureaucratic nonsense. "Get out of the house Norman!" Mom would yell. "Get some fresh air. It's a nice day outside. The sun is shining."

Okay Mom. We're all going. We're all going out to save our poor little green planet. We know that time is short. But we're no dummies. We know what we need to do and how to do it. We'll get it in the end.

Norm Lieberman
New Orleans, Louisiana
October 18, 2008
1-504-887-7714
norm@lieberman-eng.com

Preface to the Third Edition (1991)

Ten years ago, I wrote the first edition to this book. For the past seven years, I have operated my refinery troubleshooting business. During this period I have worked through enough process operating problems and equipment design errors to fill a dozen volumes. Much of my business continues to be conducted with the 15 or so major refiners in the United States and Canada. The tech service engineers, who work with me on clients' jobs, always ask what special techniques I employ to solve these problems. The procedure I use is the same one I used 10 years ago:

1. Discuss the problem with the shift operators.
2. Personally collect field data and carefully observe the operation of the unit.
3. Develop a theory as to the cause of the malfunction.

The error my clients often make is that they develop a theory, usually with process computer simulations, as to the cause of the malfunction. The theory is then reviewed with management and other technical personnel at a large meeting. If no one objects to the theory, it is accepted as the solution to the problem.

Typically, no one at the meeting has discussed the problem or the solution with the shift operators, nor has anyone personally observed the process deficiency in the field. Finally, the intended solution is not put to a plant test to see if it is consistent with the problem. This approach to solving refinery process problems by the major oil companies often results in wasting capital resources and engineering man-hours.

I would be pleased to discuss any problems or comments relating to the information imparted in this text as they may apply to specific operating or design problems.

Preface to the Second Edition (1985)

A process plant is more like a man than a machine, and the process engineer's trade is closer to the practice of medicine than to mechanical engineering. As a process plant troubleshooter, I find more in common with my family doctor than with my neighbor, a computer programmer. A process unit resembles man also in that, to function, it must be able to respond properly to a variety of circumstances that the designer never quite anticipated. I like to think that our creator built into mankind the same degree of flexibility that we find in a tried-and-true cat cracker.

Process plant engineering, especially troubleshooting, is different from most other branches of technology in another respect: It is not advancing very quickly. The principles of distillation, hydraulics, phase separation, and heat transfer, as they apply to process applications, have been well known for quite some time. The challenge in troubleshooting consists of untangling the influence that human error, mechanical failure, and corrosion have on these well-known principles. The aspect of the job that makes it so difficult is that most process problems are initiated by human error—a never-ending source of surprise.

This book is written for the practicing refinery operator or process engineer. It is based on my 20 years of trials and tribulations as a process plant operator, troubleshooter, and designer, as well as those of my friends and colleagues in the process industry. Recently, an acquaintance of mine, a young fellow pursuing a graduate degree in chemical engineering, read the manuscript for this book. He inquired as to what percentage of process troubleshooting technology he had mastered as a result.

"Not much," I responded, "but I hope my book will give you a proper respect for the magnitude and complexity of process plants in general and troubleshooting assignments in particular."

Preface to the
First Edition (1980)

Rummaging through the attic, I happened upon my old college textbooks. Much to my dismay, these treasured volumes, elucidating the principles of mass transfer, fluid flow, and differential calculus, seemed slightly incomprehensible and rather irrelevant. After 15 years of applying the fundamentals of chemical engineering in a dozen refineries, I still did not feel ready for that final exam in advanced thermodynamics. I can, however, diagnose and repair any and all basement sump pump problems. The operation and maintenance of my home air conditioner is no mystery. I can even adjust the air registers on my furnace and remove the accumulated deposits from the water heater.

The years dedicated to operating, designing, and troubleshooting petroleum processes have taught me these skills. Long, humid nights spent listening to the roar of giant steam turbines under a South Texas sky has schooled me in the challenges facing refinery shift workers. Such are the experiences I wish to relate in this book. It is not only the facts, but the feelings of working in a petroleum refinery that I hope to pass along.

Introduction

One warm spring day in 1977, the telephone rang in my Chicago home. Once again, one of the Amoco refineries had a problem. Thus began a typical troubleshooting assignment.

This refinery depended on twin plants to recover sulfur removed from the crude oil. The Environmental Protection Agency (EPA) regulates missions in refineries, and the ability to recover sulfur can and has limited refinery throughput.

On this day the refinery was in trouble. Due to a boiler-tube failure, brought on by a combination of bad luck and poor judgment, one of the two sulfur recovery plants had suddenly shut down. With only one plant operable, refinery crude run had been reduced by 25%. Possibly as a political gesture to corporate headquarters—more likely for lack of anything better to do in a desperate situation—I was called upon to help.

To help! But to help do what? Was I supposed to advise on repair and start-up of the plant that was out of service? Was I supposed to investigate the cause of the boiler failure? Maybe they wanted me to devise a method to squeeze more capacity out of the plant that was still operating. The novice troubleshooter should realize at the start of his career that people asking for assistance usually have only the vaguest idea what they want done. More often than not, they need help in diagnosing the problems and not with implementing the solution.

By Saturday afternoon I had arrived at the plant site. Needless to say, most refinery failures occur on weekends. As I had suspected, based on prior experience, no data had been assembled for my review, no meetings had been set up to solicit my advice, and no instructions had been left to define my task. Actually, the professional staff had all forgotten I was coming and had gone home.

This was just as well. The troubleshooter should begin by talking to the unit shift operators. These people run the plant 24 hours a day and, although they don't always know why something happened, they can often tell you what really transpired.

After a ritual exchange of pleasantries, I sat down with the shift foreman. We discussed the situation; rather, he talked and I concentrated on suppressing my impatience. There is something about being in a refinery on a Saturday afternoon that makes one want to get on with the job and get home.

It soon became apparent that the refinery really was in a difficult spot. The remaining sulfur recovery plant was limited to 100 tons per day (T/D) of sulfur. The refinery normally made 130 T/D. Consequently, crude run had been cut by 40,000 barrels per day (B/D) to avoid emitting sulfur pollutants to the atmosphere.

Limited! As soon as any operator uses the word limited, the troubleshooter should respond, "Which piece of process equipment is limiting plant capacity?" After some evasion, the foreman referred me to the chief operator, who would be able "to answer my question in more detail." (The psychology of dealing with chief operators is a subject unto itself, as mentioned in chapter 37. Suffice it to say that utmost diplomacy is always warranted.) The chief came right to the point. The sulfur recovery plant was limited by front-end pressure. The hydrogen-sulfide feed gas (H_2S) would spill to the flare whenever the pressure in the feed drum exceeded 10 pounds per square inch gauge (psig).

Figure I–1 illustrates the setup. The control valve, upstream of the feed drum, was used to hold the pressure in the drum below 10 psig. As the flow through the plant was raised by opening this valve, the pressure drop in the plant increased. As a result, pressure in the feed drum also rose.

The plant superintendent had left instructions to avoid spilling hydrogen sulfide to the flare. The operators were merely conforming to these instructions. Raising the feed gas charge above the 100 T/D rate would cause the feed drum to exceed 10 psig.

But why was the pressure control valve, which allowed feed gas to spill to the flare, set at 10 psig? The feed drum was designed to withstand much higher pressures. The chief operator informed me that they held to the 10-psig limit because "We always do it this way." For a recital of the historical circumstances supporting this limit, I was referred to the senior shift operator—Mr. Leroy Jackson.

Demonstrating the cooperativeness that refinery workers display when they really know what they are talking about, he explained the problem. "When we raise gas flow to the feed drum, its pressure goes up," said Jackson. "This doesn't hurt anything. Except at about 10 psig, the feed flow recorder reaches the end of the chart. Then we don't

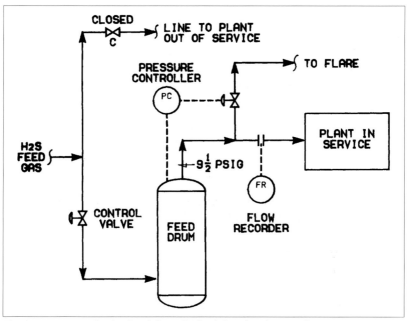

Fig. I–1. Pressure in the feed drum limited the plant's capacity.

know the plant feed rate. Of course, this doesn't really matter, but it's sort of convenient to be able to read this flow."

He was referring to the flow recorder shown in Figure I–1. When I explained how important it was in terms of dollars per day to increase sulfur plant feed, Mr. Jackson registered surprise. "Why, if someone had told me what was up, I'm sure we could have done something to get that extra sulfur through the plant."

Well, we did do something; we installed a new range tube on the flow recorder. A range tube is part of an orifice plate flow recorder transmitter. The longer the range tube, the greater the flow that can be measured by a flow recorder. We doubled the length of the range tube. This increased the maximum feed gas flow that could be recorded by 41% (i.e., the square root of two). Changing a range tube takes about 30 minutes and about as many dollars.

The next step was to increase the setting on the pressure control valve, which spilled feed gas to the flare. Mr. Jackson increased this setting from 10 psig to 15 psig. The task involved climbing several ladders and turning a dial inside the local pressure controller box.

The hardest part in any troubleshooting job always involves the human element. In this case, I had to convince the foreman to try something new. It does very little good to say, "There is now no rational reason for you not to increase the feed gas rate." The trick is to allow the man making the critical decisions to think it is his idea to increase the charge rate.

With some apprehension the foreman issued instructions to the chief operator to increase feed ever so slightly while making sure that the plant did not slide into oblivion as a consequence. By the end of the next day, the refinery had reestablished normal crude runs; sulfur production had risen to 125 T/D.

In retrospect, it may seem ludicrous to have cut back production because of a limited range on an unimportant flow recorder. Both the problem and solution seem to be the sort of thing local supervision should have handled. Perhaps troubleshooting of this type is considered too trivial for the trained chemical engineer.

Most refinery difficulties have a simple origin. However, this simple origin is clouded by false data, misconceptions, superficial observations, and third-hand reports. If the answer was obvious, you would not have been called upon. Your technical training is one tool you take into the field to reveal the underlying problem, but confining your investigation to technical areas will severely limit your chances of success.

The capacity limit of this sulfur plant was, in a sense, not due to the small range of the flow recorder. A communications breakdown between the unit operators and first-line supervision had resulted in an artificial limitation. The troubleshooting engineer is most effective when he overcomes this failure to communicate. This is best done by personally gathering data, making direct field observations, and most importantly, soliciting the opinions of the shift operators. This type of activity, when joined with sound technical training, makes a powerful combination with which to tackle refinery problems.

Even the most competent and experienced operating superintendent can become ineffective when given an incomplete account of a problem. He is often too involved in administrative matters to find the time to go out into the field and get the straight story. This is a weakness common to all large organizations.

In the tale just recited, my contribution was to go to the most important source of data available: the shift operator. One should cut through all the layers of supervision and ask the people who turn the valves. To summarize, the answers are in the field—not in the office.

section **1**

Specific Processes

I search for something once well known but long since forgotten.

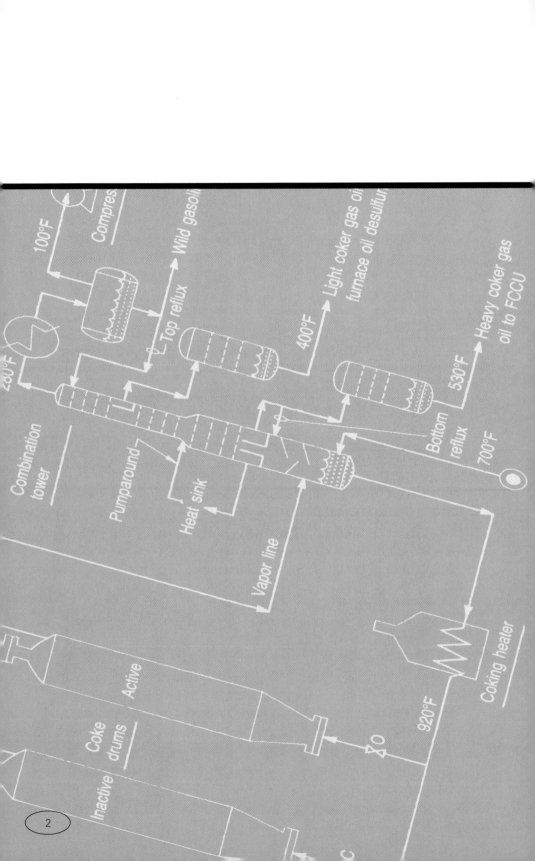

Crude Distillation

On January 23, 1983, the Good Hope Refinery was shut down. Our problem was simple: We had run out of money. As the technical manager, I was faced with a choice. I could become unemployed, or I could become a consultant. I easily concluded that the latter choice would come across smoother when I told my mother about the latest development in my career.

Now the first thing a consultant needs is clients; that was going to be a problem because I did not have any. Fortunately, an old school chum, who was working for AXECO, heard I was in financial difficulties and offered to help.

He suggested that AXECO would retain my services for one day to review the operation of its crude distillation unit. Evidently, crude runs had fallen from 195,000 barrels per stream day (B/SD) to 192,000 B/SD on the No. 2 crude unit. My colleague went on to explain that AXECO's technical staff were tied up in planning reviews and therefore management had decided to hire an outside consultant to troubleshoot the problem.

Figure 1–1 summarizes the process flowsheet of No. 2 crude unit. I began my investigation by asking the operators why they were limited to only 192,000 B/SD of crude charge. They responded that the color of the furnace oil (see figure 1–1) was the limiting factor on crude charge. Whenever the crude rate was increased above 192,000 B/SD, they reported, the furnace oil color would take on an unacceptable brown tinge.

These observations surprised me, because if the furnace oil is dark, the fluid catalytic cracking unit (FCCU) feed cut (i.e., the next lowest product draw) must be even darker. However, when I inquired as to the color of the FCCU feed drawoff, the operators responded that this product was right on spec. That is, it was black.

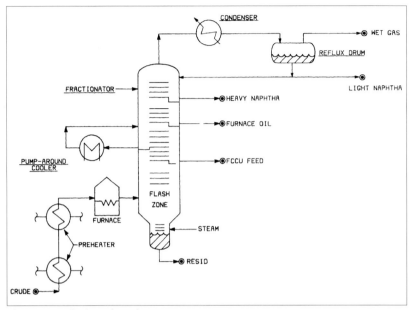

Fig. 1–1. Typical crude unit.

"No, that is not right," I explained, "FCCU feed is not supposed to be black. Gas oil (the FCCU feed is often referred to as atmospheric gas oil) is supposed to be a translucent greenish-purple. Black gas oil means that there is crude oil residual components in FCCU feed."

Crude residual components (called resid) contain metals such as nickel and vanadium. These metals, especially the nickel, accumulate on the cracking plant zeolite catalyst. The nickel promotes hydrothermal reactions in the FCCU (or "Cat"). Such reactions preferentially produce low-value fuel gas and catalytic coke, consequently reducing the production of more valuable diesel oil and gasoline. Hence, black gas oil downgrades a refinery's ability to produce motor fuels.

I explained all of this to the operators, but they disagreed. They explained that their FCCU feed was right on spec; it was black, it had always been black, and it was supposed to be black.

Does Anyone Know the Question?

I now understood something important about No. 2 crude unit. The problem was not that throughput had slipped a few thousand

B/SD. The real problem with this distillation unit was that it was not making a clean split between the first side cut (FCCU feed) and the bottom's product (resid). And certainly, I have noted this phenomenon only too often. That management tends to concentrate on throughput rather than on fractionation efficiency.

Whenever one encounters a fractionation problem, the first logical troubleshooting step is to perform a pressure-drop survey across the tower. If the ΔP (pressure drop) is too low, it means the distillation trays are dumping or weeping. If the ΔP is too high, it means the trays are flooded or entraining. Either flooding or dumping will promote entrainment of black crude from the flash zone (see figure 1–1) into the FCCU feed. Hence the three trays above the flash zone in the No. 2 crude unit, commonly called wash oil trays, were suspect.

Interpreting pressure drop data

By using a differential pressure indicator, it is possible to measure the pressure drop across a trayed section in a distillation tower with an accuracy of ± 0.03 pounds per square inch (psi). The measured pressure drop across the three-sieve wash oil trays was 0.54 psi or 0.18 psi per tray.

To convert from pressure drop in psi to pressure drop in inches of water, multiply by 27.7:

$$(0.18\,\text{psi}) \cdot \frac{(27.7\,\text{in. water})}{\text{psi}} = 5.0\,\text{in. water}$$

However, the liquid flowing across the wash oil trays is not cold water with a 1.0 specific gravity (sp gr); it is hot gas oil with a 0.74 sp gr. To convert from inches of water pressure drop per tray to inches of pressure drop of the actual liquid flowing across the tray, multiply the ΔP by the ratio of the specific gravities:

$$(5.0\,\text{in. water}) \cdot \frac{(1.00)}{0.74} = 6.74\,\text{in. liquid}$$

This means that the average pressure drop per tray in the wash oil section was 6.74 inches of hot gas oil. Now the tray spacing (i.e., the vertical dimension between the tray decks) for the wash oil trays was 24 inches The pressure drop per tray as a percentage of tray spacing is an important measure of the capacity of a sieve tray (or common valve tray). To calculate this value, divide the observed pressure drop by the tray spacing:

$$\frac{(6.74 \text{ in. gas oil})}{(24 \text{ in. tray spacing})} = 28\%$$

I have found, for most refinery services, when the above percentage exceeds 20% to 25%, a further increment in tray loading will promote tray flooding and entrainment. In the case of the No. 2 crude unit, the 28% value calculated in the equation above explained why the FCCU feed was black: The wash oil trays were flooding and severe entrainment of black resid was contaminating the gas oil product.

The solution is an aspect of the problem

A quick calculation indicated that the wash oil trays were operating well below their design rating. Evidently, though, they were flooding sufficiently so that not only was resid being entrained into the FCCU feed, resid entrainment was being carried seven trays higher and discoloring the furnace oil product as well. But why were the wash oil trays overloaded at a crude rate well below their design capacity?

Continuing the pressure survey, next I measured the pressure drop across the bottom three trays used to steam strip light gas oil from resid. I found that the ΔP was 0.81 psi. The tray spacing was 18 inches and the resid had a 0.79 sp gr. Therefore:

$$\frac{(0.81 \text{ psi}) \cdot (27.7 \text{ in. per psi})}{(18 \text{ in.}) \cdot (0.79) \cdot (3 \text{ trays})} = 53\%$$

The 53% factor for the bottom three stripping section trays proved that these lower trays were flooding much worse than the upper wash oil trays. When lower trays are badly flooded, the excessive entrainment carried up the column may cause overloading and flooding of the upper trays. This is true even if the higher trays are operating below their rated capacity.

The cause of the crude capacity limit now started to come into focus. The dark furnace oil limited crude run. The black FCCU feed caused the furnace oil to discolor. The flooding of the wash oil trays was the direct cause of the black FCCU feed. The severe flooding of the bottom steam stripping trays was causing the wash oil trays to flood. But what was causing the stripping trays to flood?

The capacity of stripping trays, such as those shown in figure 1–2, is largely a function of the stripping steam rate. The trays were designed to handle 4,000 lb/hr of steam. The operators were using about 4,000 lb/hr of steam because the unit engineer had instructed them to run the stripper section at design conditions. However, the 4,000 lb/hr of steam was causing the bottom trays to flood, so I reduced the steam rate to 2,000 lb/hr.

Many engineers believe that problems and solutions exist independently. Not so. The solution to a problem is simply one subtle aspect of the problem and flows naturally from the problem itself. Arbitrary solutions, created in an environment isolated from the problem, will not work. To reveal the solution hidden inside each problem, we use two tools:

- Direct field measurements
- Discussions with unit operating personnel

Eliminating the black gas oil

After the steam rate was cut, the ΔP across the stripping trays gradually declined from 0.81 psi to 0.28 psi. An hour later, the FCCU feed was translucent green and the furnace oil was clear. The gas oil content of the resid had not increased due to the reduction in stripping steam. After all, trays which are flooding will not bring the oil and steam into intimate contact. Hence, an increment of stripping steam which causes trays to flood simply wastes steam and reduces the stripping tray efficiency.

The crude rate was then increased by 3,000 B/SD to 195,000 B/SD. More importantly, the FCCU feed was no longer black. I asked the operators how long the gas oil had been black. One of the older men responded that the problem had started after a turnaround eight years ago. It was during this turnaround that the bottom of the tower had been modified. The three stripping trays had been replaced to increase capacity. Ever since, he concluded, the FCCU feed had been black.

A Tray Construction Error

The sun was setting over the AXECO plant when the No. 2 crude unit engineer, having completed his meeting schedule, came out to the unit to check on my progress. While he was pleased that the crude rate was up and that the gas oil was no longer black, he was concerned about the low steam rate to the stripping section.

I suggested to him that the stripping section was suffering from some unknown form of mechanical damage and that the next time they opened the tower during a turnaround, they should correct this damage. Then, they could go back to using the design steam rate.

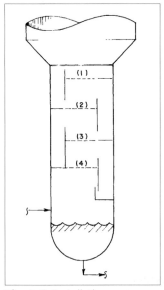

Two years later, I was working for ARCO on a vacuum tower problem in Ferndale, Washington, when my colleague at AXECO called again. He related that the tower on No. 2 crude unit had been opened for inspection and that they wanted me to fly down to their plant to inspect the stripping trays. Figure 1–2 shows what I found when I entered the column.

The downcomer from tray 3 was resting on tray deck 4. On closer inspection, I saw that the trays had not been damaged at all. The boilermakers who had installed the new trays 10 years before had fitted up the downcomer improperly.

Fig. 1–2. Installation error causes premature flooding.

With no clearance between the bottom of tray 3's downcomer and tray deck 4, the resid could not flow through the downcomer. Hence, it had to drain through the sieve holes. At 4,000 lb/hr of stripping steam, a large static head of liquid was needed to overcome the force of the upflowing steam. This static head of resid filled the space between the tray decks and caused excessive re-entrainment of resid into the flash zone which then flooded the wash oil trays.

When I reduced the steam rate to 2,000 lb/hr, the stripping trays drained down through the sieve holes to a more normal level, and the flooding subsided.

Correcting the tray problem

A 2-inch metal strip was cut off the bottom of the faulty downcomer. The bottom edge of the downcomer was kept ¼ inch below the top edge of the outlet weir on tray 4. This ¼ inch is called a "downcomer seal." When the tower was returned to service, the operators reported that the stripping steam could be raised to 4,000 lb/hr without turning the FCCU feed black. The increased stripping stream rate reduced furnace oil losses to resid by about 1,300 B/SD. The incremental value of this recovered furnace oil over downgrading to crude tower bottoms was $3.50/bbl or $1,700,000/yr.

Is it really possible for a world-class refinery to lose almost $2 million a year for 10 years because a downcomer was accidentally made 2 inches too long? Apparently, yes.

Typical Troubleshooting Problems

Even though a crude unit is relatively simple, the opportunities for improving its operations are comparable to those of more exotic refining processes. Some of the problems that the troubleshooter is likely to encounter are:

- Off-color naphtha
- Low furnace oil production
- High distillate end point
- Excessive light ends in cat cracker feed
- Condenser tube failures
- Poor operation of pumparound circuits

Decreased Fractionation

Accounting for a loss in fractionation is a common trouble-shooting assignment. For example, crude unit operators find that they can no longer meet furnace oil end-point specs unless they sacrifice furnace oil yield. On one unit, furnace oil production had dropped from 7,000 B/SD to 4,000 B/SD. Possible explanations for this type of problem are tray flooding, improper heat balance, and tray damage.

Trays can flood even below design loads because of fouling of the tray decks. Flooded trays lose fractionating efficiency (see chapter 19).

Check the pressure drop across the suspect trays. If the pressure drop per tray (in feet of liquid hydrocarbon) exceeds 25% of the tray spacing, flooding is likely.

Upset tray decks

Although trays can corrode through, a more common cause of damage is unit upsets. A high liquid level, above the flash zone, will cause the trays to be bumped by the up-flowing vapors. Slugs of water can dislodge tray decks when the water suddenly flashes.

It seems that many tray upsets occur during short unit outages. At such times steam may back into the fractionator and form pockets of water. Also, any operating malfunction that causes a pressure surge in the overhead or flash zone (check the pressure recorder) can disrupt tray decks. If the operating engineer sees a loss in fractionation following such incidents, he should proceed as follows:

Quantify the reduced fractionation. This is done by measuring the 5%–95% gap. For example, the gap between heavy naphtha and furnace oil is calculated by subtracting the naphtha, 95 vol% American Society for Testing Materials (ASTM) temperature, from the furnace oil, 5 vol% ASTM temperature. Compare this gap against historical data. A reduction of 20°F to 30°F in the gap is significant.

Check the pressure drop across the trays. Low pressure drop is symptomatic of broken tray decks. If the pressure drop per tray (in feet of liquid) is less than 10% of the tray spacing, tray damage and/or dry trays are quite likely.

X-ray the tower. On one 11-foot diameter fractionator, up-ended tray decks could distinctly be seen on the x-ray film.

Recommending to management that a crude unit be shut down to repair trays can be a milestone in an operating engineer's career. If the trays are found intact when the tower is opened, the recommendation can be an albatross around one's neck. X-raying is a good way to avoid this risk.

Improper heat balance

Hot vapors, flowing up the fractionator from the flash zone, are partially condensed by contact with cooler pumparound liquid. The heat absorbed by the pumparound stream is used to preheat crude.

As the pumparound circulation rate is increased, both heat removal from the fractionator and from crude preheat increase. This saves furnace fuel.

Higher pumparound rates will reduce fractionation. This is shown in figure 1–3. In this sketch, trays 3, 4, and 5 provide the fractionation between the furnace oil and FCCU feed. Trays 6 and 7 are the pumparound trays and, hence, contribute little toward separating the two products.

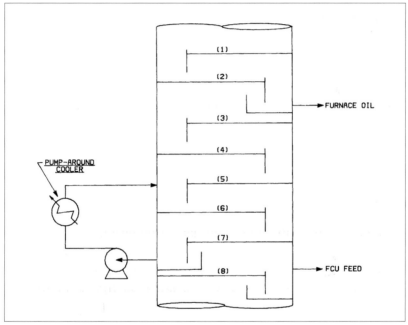

Fig. 1–3. Pumparound duty affects fractionation.

By increasing the pumparound heat removal below the furnace oil draw (tray 2), less liquid is left to be condensed above the furnace oil drawoff. This, in turn, reduces the amount of liquid that spills over tray 2 and down to trays 3, 4, and 5. The decrease in liquid (i.e., internal reflux) flowing across these trays impairs the separation between furnace oil and FCCU feed.

Under extreme circumstances, the trays below the furnace oil draw may have no liquid at all on them. Dry trays do not fractionate. When this happens, the furnace oil end point will skyrocket. Reducing the pumparound duty will correct this problem.

Raise pumparound to save energy

Operators are usually more interested in making on-spec products than in saving energy. Hence, in the field one will often find pumparound rates cut back to 50% of their proper level. Try increasing pumparound circulation until the product separation is adversely affected. Perhaps an end-point spec can be extended. Up to the point at which the trays flood, increasing pumparound flow will save energy.

On one unit, preheat was increased by 8°F simply by opening up the two discharge valves on the pumparound circuit. The operators could not recall when or for what reason these valves had been pinched back.

Note that as the pumparound duty is decreased, the vapor and liquid loads on the trays above the pumparound return tray will increase. This should ordinarily enhance fractionation. However, the reduction in pumparound duty could cause trays 3, 4, and 5 shown in figure 1–3 to flood. This would reduce the separation efficiency between FCCU feed and furnace oil.

The question then is, as an operator reduces pumparound duty, how can he tell if fractionation between two adjacent products is getting better (due to increased internal reflux on the trays) or worse (due to tray deck flooding and entrainment)? The simple answer is to observe the difference in draw temperatures (ΔT), between the adjacent cuts.

An increase in this ΔT between FCCU feed and furnace oil accompanying a reduction in pumparound duty indicates enhanced fractionation. A reduction in this ΔT indicates trays 3, 4, and 5 in figure 1–3 are flooding.

Light-naphtha end point

Operators may attempt to maximize heavy naphtha production at the expense of light naphtha by increasing the fractionator top reflux rate, which drops the tower top temperature. The water vapor in the overhead hydrocarbon vapors begins condensing at its dew point. If the tower top temperature is too low, water will condense on the top trays. This water is corrosive and will eat holes in the tray decks. Over a period of months, the degree of separation between light and heavy naphtha will consequently deteriorate.

If the light-naphtha end point is high and the tower has been operating within 10°F of its calculated water dew point, corrosion of the top trays is indicated. The procedure to calculate water dew point is given in the Appendix.

Replacing tray decks and downcomers with Monel steel (high nickel content) will help prevent this corrosion. The area underneath the downcomers is especially prone to this type of leakage due to hydrochloric acid attack.

Dirty naphtha

Naphtha products may suddenly assume a yellowish cast and then return to a normal water-white condition. If this proves to be a recurring problem, the difficulty is probably water in the top reflux.

First, obtain a sample of the reflux naphtha. Does it contain free water? If so, manually drain the water from the reflux drum until hydrocarbons blow out of the drain. About 30 minutes of water-free refluxing should clear up the dirty naphtha. A malfunctioning reflux drum water–naphtha interface level control valve is often the culprit. Perhaps the steam stripping rate has been increased, and the water drain line cannot handle the increased flow.

Steam Stripping Cat Cracker Feed

A side-stream product drawn from a crude tower is only half fractionated. Refluxing below the drawoff tray controls the product end point. Front-end fractionation also takes place in the side-stream stripper.

A typical stripper configuration is shown in figure 1–4. Steam enters the bottom of the stripper and reduces the hydrocarbon partial pressure. The product partially vaporizes to reestablish vapor-liquid equilibrium. The heat for this vaporization comes from the product itself, not from the stripping steam.

On many crude units, steam stripping of naphtha, kerosene, and furnace oil appears adequate. On the other hand, stripping of FCCU feed is frequently poor. Figure 1–5 shows the effect of stripping. Hydrocarbons boiling below 500°F belong in naphtha reformer feed or kerosene, not in the FCCU charge.

Obtain a sample of FCCU feed produced at the crude unit. Run an ASTM atmospheric distillation. Ask the lab technician to stop the distillation at 680°F; this avoids thermal cracking. If more than a few percent is distilled off below 500°F, examine the operation of the FCCU feed stripper.

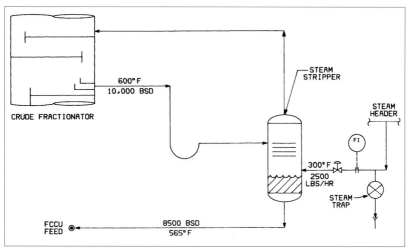

Fig. 1–4. Fractionation depends on stripping.

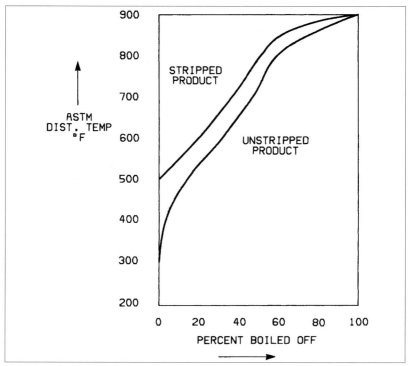

Fig. 1–5. The effect of steam stripping.

Causes of inadequate stripping

Several reasons have been observed for inefficient stripping. Operators note correctly that preheat is improved and steam is saved by minimizing stripping steam. However, far more energy is wasted by unnecessarily running naphtha and kerosene through the FCCU.

Perhaps someone has retrofitted the stripper feed to preheat crude before it flows to the stripper. This is a good energy-saving scheme for the crude unit; however, a steam stripper will not work on subcooled feed. The feed must be at its bubble point (i.e., the tower drawoff temperature). Otherwise, the steam will not be an effective stripping agent.

Using stripping steam upsets the stripper bottoms level, causes erratic fractionator operation, and produces excessively wet products. Consequently, the operators have stopped using it. Such symptoms are indicative of slugs of water in the stripping stream. Check that the steam line has a steam trap. Also, the takeoff for the stripping steam line from the steam header should come off the top of the header pipe.

Increasing steam flow interferes with the ability to withdraw product from the fractionator. The stripper vapor line may be too small or partially plugged. Check the increase in stripper top pressure at various steam rates. This pressure should not increase more than two or three psi over the full range of steam flows. Also, the stripper feed line should have a loop seal, as shown in figure 1–4. This loop prevents steam from backing up the feed line.

Steam is ineffective because of upset trays. At the design steam rate, check the pressure drop across the trays. For a typical four-tray stripper, the pressure drop should be roughly 1 foot of water. If the pressure drop is very much less than this, the stripping trays are likely damaged.

Steam Stripping Summary

Not long ago, I was retained by one of the major oil companies to assist in upgrading their operation of a 110,000-B/SD crude unit. This older crude unit was part of a large, fully integrated Gulf Coast refinery.

The chief process engineer's plan to achieve this objective was to implement an "expert system." This system is a form of artificial intelligence, utilizing a computer. Basically, the expert system generates a series of questions for the unit operators to answer so as to guide them in optimizing crude unit operations. My part in this project was to provide a technical basis to permit the computer programmers to formulate the software for the expert system. As Lloyd, the chief process engineer phrased it, "Mr. Lieberman's knowledge of crude unit operations will be converted into artificial intelligence, which will then be presented to the crude unit shift operators in the form of our expert system."

Defining the project scope

To initiate the expert system, I planned to spend several days on the unit. I wanted to discuss current unit operating problems with the shift operators and familiarize myself with the equipment layout. Also, I planned to observe how the unit responded (see figure 1–6) to various step-changes in stripping steam rates and pumparound flows. My primary objective was to develop a set of operating parameters to maximize production of jet fuel.

Lloyd readily agreed that maximizing jet fuel production was the correct goal for our expert system. He suggested we embark immediately on the software formulation phase of the project. "All crude units are pretty much the same," he said, "and as you're a crude unit expert, we shouldn't have to waste our time at the plant site."

At this juncture, I decided this particular consulting assignment was not my cup of tea. Direct field observations are the core of all process engineering projects, and I felt my client would be better served by a different consultant. However, as we exchanged views, Lloyd raised a decisive point: "You realize, Mr. Lieberman, that there is a substantial consulting fee at stake." I decided to compromise and settled for a half-day of on-site field observations. After all, Lloyd was the client's representative.

Stripping steam rates

Referring again to figure 1–6, we can see that jet fuel can be downgraded to lower value products as follows:

Heavy virgin naphtha. Poor stripping of jet fuel will cause a low flash point of the jet fuel product. To raise the flash point, the end point of the heavy virgin naphtha product is increased by pulling harder on the heavy naphtha draw. In effect, lighter boiling range jet fuel components are downgraded to heavy virgin naphtha.

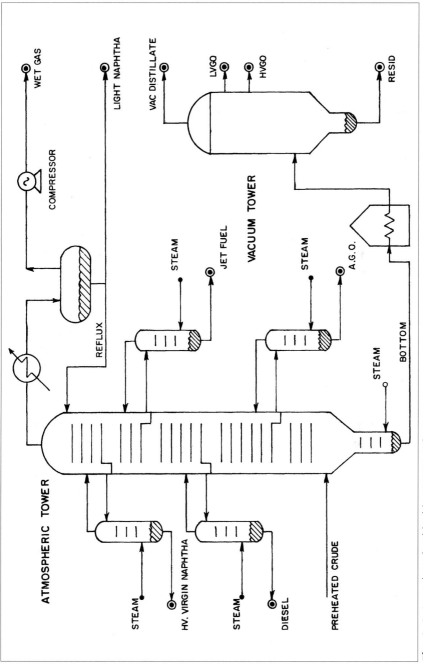

Fig. 1–6. A crude unit with side-stream strippers.

Diesel oil. Poor steam stripping of diesel will increase the jet fuel composition of the diesel product by 10% to 15%.

Atmospheric gas oil (AGO). Heavy boiling range jet fuel components will be downgraded to fluid cracker feed due to poor stripping of the AGO product.

Vacuum distillate. This stream was directed to the refinery's recovered oil system (i.e., slop) and eventually reprocessed at the delayed coker. The jet fuel content in the slop would hence be commingled with low-value light coker gas oil. Poor steam stripping efficiency in the bottom of the atmospheric fractionator could leave up to 1,000 B/SD of jet fuel in the vacuum distillate product.

As jet fuel production was only 65% of the volume indicated in the crude analysis, we agreed that proper operation of these four strippers would be the first objective of our expert system.

Gathering the artificial intelligence

Regrettably, neither Lloyd nor any of his staff was available to join me for the field observation. The engineering staff was scheduled to attend an Integrated Evaluation seminar that morning. However, Mr. Zipperany, the senior shift operator, was assigned to answer questions.

I explained to Zip that we would manually optimize stripping steam rates and then integrate the techniques we used in the expert system. Zip noted that this was unnecessary because all parameters on the unit had been optimized long ago. Zip explained, "We've been running at these stripping steam rates because they are the rates we always run at, because they are the correct rates, which we know because we always run at these rates."

"How much steam are you using on the atmospheric gas oil stripper?" I asked.

"Our standard 1,400 lb/hr-we always use this much steam in the AGO stripper," Zip explained.

Using 1,400 lb/hr of steam to strip 7,000 B/SD of gas oil should produce a temperature drop of roughly 20°F between the gas oil feed and the stripper bottoms. This results from the heat of vaporization in the stripper originating from the gas oil itself. The steam's function is to reduce the hydrocarbon partial pressure across the stripping trays. The steam normally does not contribute much heat to the oil. Therefore, there is a linear relationship between the percent of gas oil feed vaporized and the temperature drop of the gas oil in the stripper. A 20°F temperature drop corresponds to about 12 wt% vaporized; a 2°F temperature drop equates to little, if any, vaporization.

I checked the inlet and outlet temperature indicators (TIs) which were 622°F and 621°F, respectively. Apparently, the stripping steam was ineffective. Why? Possible explanations were:

- Trays in the stripper became flooded.
- Trays in the stripper were damaged.
- The actual steam rate was much lower than 1,400 lb/hr.
- The gas oil entering the stripper was sub-cooled below its boiling point.

I asked Zip if he was sure that the steam flow meter was accurate. He responded that all meters were checked routinely. Puzzled, I decided to check the flow meter myself. I found that the steam flow to the AGO stripper was blocked in and actually zero.

Slowly opening the steam inlet, I adjusted the steam flow until the AGO stripper bottoms temperature dropped to about 600°F, which is an indication of reasonable stripping efficiency.

Diesel oil stripping

About 12,000 B/SD of diesel oil product was being stripped with 3,600 lb/hr of steam. Once again, the stripper inlet and outlet temperatures were almost identical. Perhaps the stripping trays were flooded and thus being rendered ineffective.

The most common cause of tray deck flooding in any fractionator is high liquid level. When the liquid level rises above the steam inlet, the stripper will flood. I asked Zip if they were maintaining a normal liquid level in the bottom of the diesel oil stripper. In response, he indicated a circular chart. A pen inscribed a perfect red circle on the chart at the 65% level. The fact that the indicator level did not vary made me suspicious.

"Does the level always hold so steady?" I inquired.

"Yep," Zip answered, "it's been holding at 65% for years-just as you see it."

I measured the differential pressure across the stripper between the top of the gauge glass assembly and the vent connection on the vapor outlet line. The indicated pressure differential was 4.5 psi. To convert from ΔP to feet of liquid the following formula is used:

$$\text{Ft of liquid head} = \frac{(\Delta p) \cdot (2.31\ \text{ft})}{\text{sp gr}}$$

where

　　sp gr = the specific gravity of the liquid at the stripper bottoms temperature.

Evidently, the pressure differential across the stripper equaled 16 feet of liquid head. As the stripper was only 19 feet high, the tower was essentially full of liquid. The gauge glass confirmed my theory: it was full of liquid with no visible interface.

Zip manually pulled the liquid level down in the diesel oil stripper until the interface appeared in the gauge glass. The stripper bottoms temperature began to fall, indicating that stripping efficiency was being restored. Diesel production dropped by about 1,000 B/SD, and the jet fuel product rate increased by a corresponding amount.

Looking over the control panel, I felt very pleased. Zip, however, was upset. "Look at the level indicator chart," he fumed. "It's a mess."

The pen was rapidly moving between 20% and 40% on the chart. Of course, the 65% indicated level was really 100% of the true bottoms level.

Jet fuel stripper

One of the important functions of a crude unit is to meet the flashpoint specification for the jet fuel product. This is best done by varying the steam rate to the jet fuel stripper. However, the operators on this crude unit were adjusting the heavy virgin naphtha draw rate to meet the jet fuel flash spec.

Zip informed me that adjusting the steam to the jet fuel stripper did not influence its flash point. That is, the steam was ineffective in removing lighter hydrocarbons (i.e. naphtha) from the stripper feed. This was odd, because the stripper draw temperature was 435°F and the stripper bottoms temperature was 395°F. Typically, a 40°F ΔT indicates good stripping efficiency. However, when I inspected the stripper column, I saw that both the feed line and the stripper vessel's shell were entirely without insulation. The 40°F temperature reduction was primarily due to the ambient heat loss.

For the steam to a stripper to be effective, the oil must enter the stripper close to its bubble point. If the feed is sub-cooled, the stripping steam cannot reduce the hydrocarbon partial pressure enough to promote vaporization of the oil. This renders the steam ineffective. The concept is easily understood if we remember that the heat of vaporization for stripping comes from the oil itself and not from the steam.

Pulling lighter jet fuel components into heavy virgin naphtha to meet the jet fuel flash spec not only reduces jet fuel production, it also downgrades the quality of the naphtha reformer feed.

Bottom's stripper

About 8,000 lb/hr of steam was consumed in the atmospheric tower bottom's stripper. This rate had been optimized many years ago to minimize the production of vacuum distillate (see figure 1–6).

Regardless, I tried cutting the steam back to 5,000 lb/hr to see how much the vacuum distillate rate would increase. Surprisingly, the flow of vacuum distillate remained constant. The only notable result on the crude unit was a drop in flow of wet gas from the reflux drum. The 3,000 lb/hr cut in the bottom's stripping steam had unloaded the overhead condenser. This caused the reflux drum temperature to drop and hence reduced the flow of wet gas to the off-gas compressor.

"This is okay," Zip noted, "That compressor is bottlenecking our unit. Now that it's unloading, we can rev up the crude rate."

What was wrong with the bottom's stripping trays? Possibly, over the last few years they had been damaged. Perhaps water had entered the crude column on start-up and blown the tray decks out? As I could find no measurable ΔP across the trays, I was fairly certain that the trays were no longer intact.

The wrap-up meeting

Later that day, Lloyd and his staff assembled for a progress report. Lloyd explained his plans to implement artificial intelligence on the crude unit.

"Mr. Lieberman, who will now speak," he said, "will provide the technical basis for the expert system."

I suppose I might have been more diplomatic in my opening comments. "Your problem on the crude unit is not lack of artificial intelligence. What you all lack is technical service support for the shift operators on a day-to-day basis. What you need is not artificial intelligence, but ordinary intelligence."

Lloyd finally did agree with my initial analysis and retained the services of a different consultant to implement the expert system.

Leaking Drawoff Trays

Leaking product drawoff trays can result in wasting large quantities of energy. This is a major problem in towers equipped with valve trays, as compared to those towers still using bubble-cap trays. The troubleshooter can recognize a drawoff tray with excessive leakage when the following two conditions are satisfied: All valves on the product drawoff line are wide open, and the product has a lower-than-desired end point.

Be careful! On one unit, the kerosene drawoff tray was thought to be leaking, based on the above criteria. The refinery lived with the problem for two years. Finally, out of curiosity, a newly assigned engineer climbed up in the tower. He found the kerosene drawoff control valve stuck halfway open. He forced open the control valve with the handjack, and a $2,000/day problem was solved. (The control valve position indicated in the control room is not necessarily the actual control valve position.)

Welded trapout pans

Having concluded that a product trapout tray is leaking, the operating engineer must formulate an effective and permanent repair plan. The prerequisite to accomplish this is to determine the accident that initiated tray leakage.

Figure 1–7 details a product drawoff facility. If a large percentage of the liquid leaving the trapout tray is to be withdrawn at the drawoff pan, both the tray and pan must be reasonably leakproof. The calculated volume of liquid leaking through a valve tray deck is quite small at reasonably high (30+% of design) vapor rates. Unfortunately, this calculation is often irrelevant in regard to a commercial crude tower. Frequently, the workmen who install a tray deck do not do a perfect job. This results in a poor seal between the tray deck sections themselves and also between the tray ring and the tray.

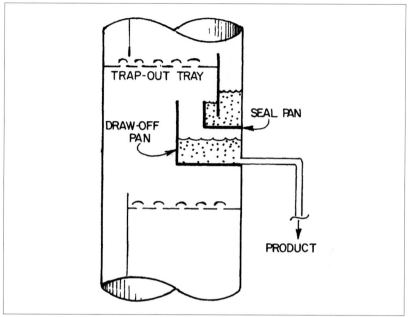

Fig. 1–7. Always use both seal and draw-off pans.

Valve trays cannot be tested for leaks. One cannot simply look at a tray installation and conclude it is tight. To measure leakage, a water level must be established on the tray deck with the rate of leakage actually measured, and the rate of leakage cannot be determined with a valve tray.

One refiner who had to continue using a valve tray in trapout service made the following modifications, which successfully stopped excessive tray leakage. First, the tray deck sections after being assembled inside the tower were seal-welded together. Second, to stop leakage at the tray ring, a strip was welded onto the periphery of the tray. The strip was set back a few inches from the tray ring; its height was the same as the overflow weir. A wiper ring was installed above the new peripheral strip (see figure 1–8).

A common alternate to the wiper ring and seal strip is to use a 3-inch-by-3-inch expansion ring. This is just a thin wall section of angle iron bent to conform to the vessel wall. The top edge is welded to the vessel I.D., and the bottom edge is welded to the chimney tray. One needs only to note on the spec sheet, "Expansion Ring Required," for the tray vendors to provide the requisite hardware. An expansion ring installation is shown in figure 3–5.

Even a perfectly assembled valve tray may become leaky onstream. Usually, a small slug of water enters the fractionator, flashes, and produces a pressure surge. The trays are bumped but not upset. If a product drawoff has lost capacity after a brief unit outage (such as a short power failure), water is the probable culprit.

A bolted-together drawoff pan (see figure 1–7) is also a source of leakage. Find out if the pan was leak-tested during the last unit turnaround. On one fractionator, too many drain holes were drilled in the pan. One strategically located drain hole is sufficient.

The operating superintendent of a medium-sized crude unit obliterated all of the above difficulties. He tore out the valve trapout tray and drawoff pan. A new, all-welded pan was installed. Several large chimneys allowed for vapor passage. The pan was seal-welded to the tray ring and the problem was corrected. (Note, in large-diameter towers, a seal weld can fail due to thermal expansion. Use of an expansion ring as described above is recommended.)

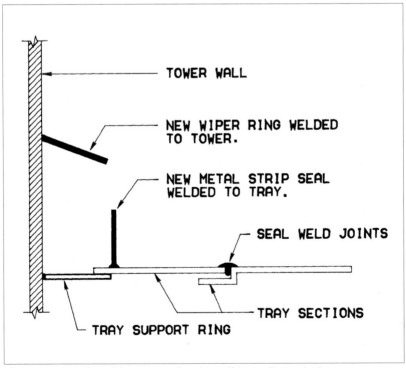

Fig. 1–8. Strategic welding on a valve drawoff tray will stop leaks.

Overhead Condenser Corrosion

Crude oil contains various chloride salts. Some of these salts decompose in the furnace to produce hydrochloric acid. The HCl boils overhead in the fractionator and dissolves in the water as it precipitates in the condenser. The resulting low pH water is very corrosive, and frequent overhead condenser tube failures can occur. If such failures become a chronic problem, check the following:

1. The desalter should remove 90+% of the salts in crude. If it does not, check the desalter temperature (usually 270+°F). Steam condensate should be used for wash water. Is the proper dosage of desalter chemical being injected? Is the voltage up to design?

2. Many refiners inject a small amount of caustic in the heat-exchange train downstream of their desalter. The caustic neutralizes the HCl evolved in the furnace. This is a fine way to stop condenser corrosion, but it also promotes fouling in the heat exchangers downstream of the caustic injection point.

3. Check the pH of the water withdrawn from the reflux drum. If it is low (5–6), the operators may not be adding sufficient NH_3 or neutralizing chemicals to the overhead system.

Many engineers rely on their desalter and corrosion-inhibitor chemical vendor to formulate a corrosion protection plan for their unit. This is usually a good idea. However, one needs to know enough about this aspect of the unit to determine if the vendor representative is competent. One useful method is to ask competing chemical vendors to come up with alternative programs.

Only the $MgCl_2$ and $CaCl_2$ salts in crude oil decompose to HCl at temperatures above 350°F. This HCl is very hydroscopic. That is, the first drop of water that condenses in the overhead condensers will absorb all the HCl it contacts. This leads to a very low localized pH. The operators see the effect of the hydroscopic nature of HCl as extremely accelerated localized corrosion at the inlets to the overhead crude condensers.

While the use of caustic injection to the preheat exchanger train, improved desalting, and proper use of neutralizing and filming chemicals is important, the key to success in stopping overhead corrosion is water recirculation.

Water from the reflux drum is recirculated into the overhead system just before the point at which water condensation begins. Where there are parallel condensing trains, the water must be uniformly distributed through spargers with a 10-psi to 20-psi pressure drop. Enough wash water must be recirculated to bring the overhead vapors to their calculated water dew-point temperature (see appendix). Note that the vapor from the overhead of the crude tower is already at its hydrocarbon dew point.

The recirculated water will dilute the HCl as it condenses and prevent the formation of low localized pH solutions. Injection of water-soluble corrosion inhibitors into the recirculated water is also a good idea.

Exchanger Train Fouling

Heat transfer in crude exchangers often declines because of tube-side fouling. The difference in crude preheat may be 50°F for a dirty versus a clean exchanger train.

A sudden drop in furnace feed temperature is commonly caused by a slug of bottoms sediment and water (BS&W) in crude charge. This happens when a tank in the crude supply system has been cleaned. The silt and waxy dirt were flushed down the crude line to the refinery. A floating suction in the refinery's crude charge tank minimizes the effects of these incidents.

Certain low-sulfur waxy crudes rapidly foul heat exchanger tubes. An easy method that has worked in some services to restore lost heat-transfer capacity partially is as follows:

1. Block in and bypass the tube side of the fouled exchanger.
2. Continue shell-side flow. This heats the tubes. Waxy deposits are melted off. Harder deposits are spalled off due to thermal shock.
3. After 20 minutes, restore normal flow.

For one set of exchangers that preheated crude to 250°F, this method doubled the observed heat-transfer coefficients.

Low crude-side velocity also promotes fouling. If the calculated velocity is less than 2–3 ft/sec, consider doubling the number of tube-side passes. Inadequate desalter operation or excessive use of caustic will salt up exchangers.

Anti-foulant chemicals may have a noticeable—but not a major—effect in reducing exchanger fouling. High velocities, exclusion of tank-washing bottoms from the refinery crude, and good desalting are the paramount factors.

Excessive tube-side pressure drop due to fouling can cause the channel head pass partition baffle to fail. (Consult the TEMA[1] data book for exchanger details.) This pass partition baffle prevents the crude from bypassing the tube bundle. However, as the tube-side ΔP rises, this baffle will eventually fail and lead to a sudden loss in the preheat exchanger duty. The maximum allowable pressure difference across this baffle should be listed on the exchanger detailed drawings.

Preflash Towers Save Energy

A refinery designed to run a 30° American Petroleum Institute (API) crude must unexpectedly switch to a 40° API crude. Sometimes, large amounts of natural gas condensate (butane and naphtha) are commingled with the crude supply. Such changes can lead to a reduction in crude running capacity. This happens because:

- The fractionator tower becomes overloaded and the trays flood.
- Increased vaporization of the lighter crude through the preheat exchanger train and furnace increases pressure drop, which reduces the capacity of the crude charge pump.
- The furnace duty must increase to vaporize a higher percentage of the lighter crude.
- The fractionator condensers cannot handle the increased load; consequently, the reflux drum runs hotter and excessive wet gas production results.

The answer to these problems is to install a preflash tower. If the unit has a flash drum upstream of its furnace, this drum can be retrofitted as a preflash tower. Figure 1–9 illustrates a preflash tower arrangement used with success at several refineries. Significant features of this arrangement are:

- The preflash tower is located midway in the preheat exchanger train, and its pressure "floats" on the fractionator pressure.

- Only one barrel of reflux for each four or five barrels of net overhead liquid is required. This low reflux ratio is permissible because the preflash overhead liquid product is charged to the crude fractionator, several trays above the heavy naphtha drawoff. This effectively redistills the poorly fractionated preflash liquid.
- Preflash overhead vapors are cooled by preheating crude.

Field tests have shown that the changes depicted in figure 1–9 result in no measurable loss in separation efficiency between light and heavy naphtha. For a typical low-sulfur crude (35°API), 6–8 vol% of the crude is taken overhead in the preflash tower. The observed reduction in furnace duty was 5%–10%.

Energy savings

A method to calculate energy savings resulting from retrofitting a flash drum as a preflash tower is:

$$\text{Energy saved} = SH \cdot M \left(T_{FZ} - T_{PF}\right) - H_{RF}$$

where:

SH = Specific heat of the hydrocarbon vapor in the preflash tower (about 0.5–0.6)

M = Total pounds per hour of the preflash tower net overhead (liquid plus vapor) product

T_{FZ} = Temperature of the fractionator flash zone

T_{PF} = Preflash tower feed temperature

H_{RF} = Preflash tower reflux heat duty (i.e., lb/hr of reflux multiplied by about 170 BTU/lb)

This calculation is approximate and applies when the heat removal from the tower pumparound sections cannot be increased because of equipment limitations. In order to calculate the heat savings quantitatively, a tower computer simulation is required. This approximate method has, however, been found satisfactory for most situations.

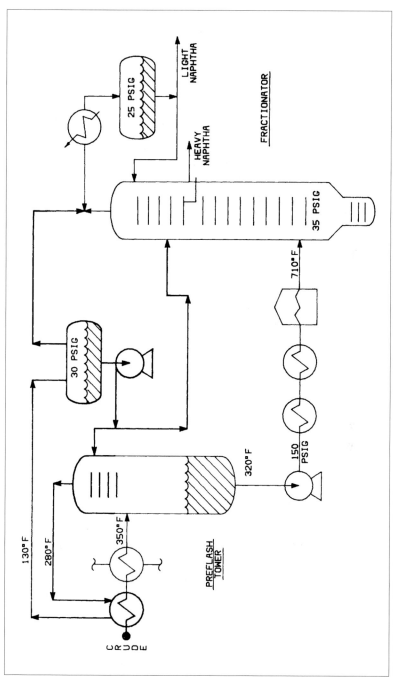

Fig. 1–9. Retrofitting for a preflash tower.

Preflash tower foaming

A common example of foam formation in the bottom of a fractionator inducing flooding occurs in a crude preflash tower. In this case, stable foam accumulates in the bottom of the column as a consequence of the "flow improver" chemicals added to crude oil. These chemicals reduce pressure drop in the crude pipelines. Once the foam level rises to the feed inlet nozzle, the trays flood and black distillate is produced.

Rising Energy Index

As the world crude oil supply dictates, most refiners are cutting crude runs. Too often, one result is an increase in the BTUs required to process a barrel through the crude unit. For the average crude unit, 90% of its energy input is provided by furnaces. Theoretically, both heat exchanger and furnace efficiency increase at lower loads. Why, then, does overall heat economy suffer at lower charge rates? A few items that the troubleshooter may ponder are:

- On one 4,000-BSD crude unit, an energy survey showed 45% of furnace duty was dissipated to ambient heat losses from hot uninsulated lines, manways, and so on.

- Cold air can be sucked into the furnace convective section through holes in the furnace skin. This reduces the efficiency of heat recovery from the hot flue gas. At lower crude rates, flue gas flow drops, but cold air in-leakage remains constant. Thus, at lower crude rates, holes in the furnace exterior will hurt efficiency more than at higher throughputs. A roll of aluminum tape can go a long way toward correcting this problem.

- A leaking FCCU feed trapout tray (see figure 1–1) wastes energy when the rate of leakage exceeds the desired internal reflux rate. The rate of leakage is not reduced as throughput decreases. Therefore, the BTUs wasted per barrel of crude charge increase as crude run drops. The engineer can observe this effect by noting the temperature difference: flash-zone temperature minus FCCU feed drum temperature. This difference is roughly 15°F. A larger temperature difference indicates a leaking drawoff tray. Especially on large crude units running at reduced charge rates, this particular problem has caused great energy losses.

- Low velocities through heat exchange and furnace tubes promote fouling and loss of heat transfer. To counter this increased fouling rate, an accelerated program of onstream exchanger and furnace cleaning is logical. However, many refiners tend to reduce this type of maintenance because their furnace capacity is not being pushed at the lower crude rates.

Do not forget to take advantage of lower crude rates by reducing the fractionator operating pressure. Very substantial energy savings will result if the tower flash zone pressure can be cut by 5 or 10 psi. Chapter 9 discusses a number of ideas along these lines.

Safety Note

The leading cause of fires and explosions in refineries is failure of carbon steel piping spool pieces in chrome process piping. Above 550°F and 2% sulfur, chrome piping is required. The problem is that 9% chrome steel, when new, appears very similar to carbon steel (C.S.).

The corrosion rate for C.S. piping is 10 times that of chrome. If a single 4 foot C.S. piping section is inadvertently flanged-up in 800 feet of 9% chrome line, the probability of locating the rapidly thinning spool piece is remote.

As piping ages, I can visually detect a C.S. section— unless it is insulated, as it normally is. Also, C.S. and 9% chrome piping are both magnetic. Thus, in reality only the plant inspection department has the electronic equipment to discriminate between chrome and carbon steel piping spool pieces.

Troubleshooting Checklist for Crude Unit

Decreased Fractionation
Reduced ASTM gap
Flooding
Upset tray decks
Pumparound rate too high
Top trays corroded
Water in reflux
Downcomer installation

Inadequate Steam Stripping
Long front end on product
ASTM
Operators saving steam
Cold stripper feed
Water in steam
Plugged vapor outlet
Stripper trays upset

Energy Wasters
Leaking trapout tray
Pumparound rate too low
BS&W in crude
Low exchanger velocities
Caustic injection in exchangers
Air leaks in furnace skin

Overhead Corrosion
Desalter efficiency
Caustic injection
Reflux drum water pH
Corrosion control chemicals

Preflash Tower
Redistill preflash naphtha
Reduce furnace duty
Increase unit capacity
Foaming

Reference

1. *TEMA Data Book*. Published by the Tubular Exchanger Manufacturers Association. Fourth Edition, 1959.

Delayed Coking Cycles

Coking is an old process but one that is becoming more important as the quality of the world's crude supply deteriorates. As the sulfur, metals, and Conradson carbon contents of crudes increase, coking the bottom of the barrel is looking better to many refiners. The greater part of the barrel of resid produced from a crude unit can be converted to gasoline, distillate, and gas oil in a coker. Thus, in the current economic environment, the coker is frequently the most important unit in the refinery.

By far the largest amount of coking capacity is represented by the delayed coking process. While delayed coking is simple enough, the cyclic nature of the process gives ample scope to the talents of the troubleshooter.

Process Description

Figure 2-1 illustrates a simplified flow diagram of a delayed coker. Hot resid feed flows to the bottom of the combination tower. The combination or fractionator tower bottom section acts as surge drum from which the coking heater is charged.

The heater raises the resid temperature to 900°F. The resid then flows into the bottom of one of a pair of coke drums, where it thermally cracks to gas, gasoline, gas oil, and solid coke. The coke gradually fills the drum—usually over a period of 12 to 24 hours— while the lighter products pass on as a vapor to the combination tower. Since the coking reaction is endothermic, the vapors leaving the top of the coke drum are roughly 110°F colder than the heater outlet temperature.

In the combination tower, the coke drum vapors are condensed and fractionated into four products: gas, wild gasoline, furnace oil, and heavy gas oil.

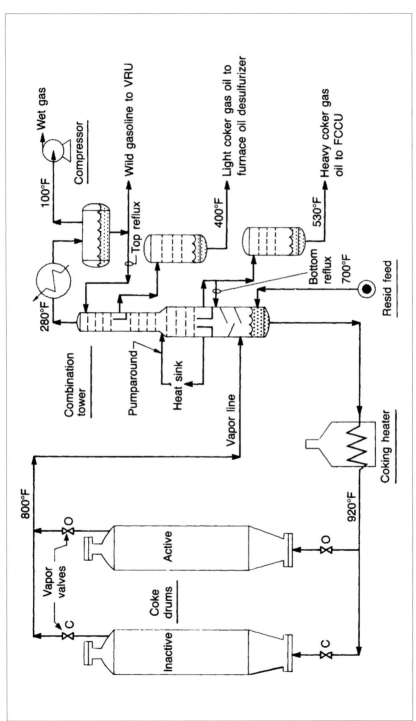

Fig. 2–1. A simplified process flow diagram of a delayed coker.

The Coking Cycles

While the continuous aspect of delayed coking is straightforward, problems arise in the batchwise filling and emptying of coke drums. Table 2–1 summarizes a typical cycle. While one drum of the pair is filling, the other drum is either steaming, quenching with water, hydraulically decoking, or warming with hot vapors.

The most commonly encountered difficulties in this batchwise operation are:

- **Foamovers.** Partially coked resid is carried over the top of the coke drum.
- **Soft coke.** The volatile combustible matter (VCM) of the coke is too high.
- **Shot coke.** Coke is made in small balls instead of the usual sponge structure. This coke takes its name from its buckshot appearance.
- **Plugged steam-water inlet.** Coke plugs the water inlet nozzle and slows down the coking cycle.
- **Excessive cycle length.** The capacity of a coke drum varies inversely with the cycle length.
- **Warm-up condensate.** This often mishandled stream can represent one-half volume percent on crude.

From an economic viewpoint, a foamover is the most serious of these problems.

Foamovers

A large carry-over of coke or partially coked resid from the coke drum is called a foamover. Preventing such foamovers is vital to continued operation of a delayed coker. Depending on the volume of material carried over, the effect on unit operability will range from bad to disastrous. Several problems are initiated by foamovers.

- **Coke lay-down in the coke drum overhead vapor lines.** This will cause an increase in the coke drum operating pressure. Reduced unit capacity may result.
- **Partial plugging of the combination tower bottom's screen.** This screen keeps large pieces of coke out of the suction of the coking heater charge pump. Figure 2–2

Table 2–1. Summary of a 24-hour coking cycle.

Step	Elapsed time, hour
1. *Fill* empty coke drum.	24.0
2. *Switch* feed from full drum into empty drum.	0.5
3. *Little steam*—Use about 2,000 lb/hr steam to strip lighter hydrocarbons out of the full drum to the combination tower.	1.0
4. *Big steam*—Line up full drum overhead to blowdown system and use about 20,000 lb/hr steam to cool the coke partially in the bottom head.	1.0
5. *Little water*—Slowly start a small amount of water into the bottom of the drum while reducing steam flow. Water flow is maximized consistent with not overpressuring the drum, usually about 50 gpm.	1.0
6. *Big water*—Open up the main water line and fill the coke drum with water, typically at 1,000 gpm.	3.0
7. *Drain water*—After the water in the drum stops boiling, coke is cooled and water is drained out the bottom.	2.0
8. *Remove heads*—The large heads on the top and bottom of the drum are unbolted and removed by the decoking crew.	0.5
9. *Drill pilot hole*—Using high-pressure water, a 4-ft diameter hole is cut from top to bottom through the coke.	0.5
10. *Cut* coke also using high-pressure (3,000) psig) water. Starting from the top, the coke is cut and falls through the pilot hole.	4.0
11. *Replace heads*.	1.0
12. *Steam test and purge* the drum to ensure their heads do not leak and that air is removed.	1.0
13. *Warm* up the empty drum by backing hot vapors from the active drum through the top of the inactive drum. Condensed liquid is pumped to the FCCU: feed and vapors are vented back to the combination tower.	6.0
14. *Open* the vapor valve of the empty drum all the way and prepare to switch feed.	1.0
15. *Idle* time allowed for slippage of individual steps.	1.5
Total	48 hours

Note: Commercial coking cycles range from 12 hours or less to 24 hours, with 24-hour cycles being most typical.

shows the effect of a carryover on the screen. A minor foamover will plug its bottom section. A symptom of this is the need to raise the combination tower bottom's level to secure a steady flow of liquid to the heater's charge pump.

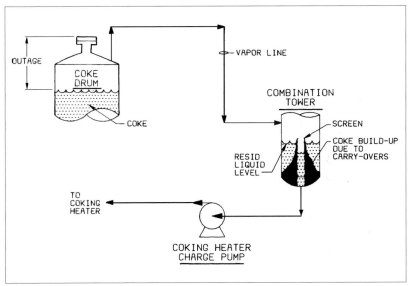

Fig. 2–2. The effect of coke drum carryover.

- **Complete plugging of the combination tower bottom's screen.** The unit must now be shut down to clean the screen. In some units, this screen is 25 feet high, but a serious foamover can plug it in one day.
- **Coking heater tubes are coked.** A sudden loss in heater feed will lead to coking of the tubes. If the firing rate is rapidly reduced, severe tube damage may be avoided. You can count on the need to steam-air decoke or pig-out the tubes after the coker is forced down because of a foamover.
- **A massive foamover will carry coke well up into the combination or fractionator tower.** If the coke gets up into the trayed section, many days will be required to clean the tower before it can be returned to service.
- **Plugging the blowdown system.** After the drums are switched, the full drum is vented to a blowdown system. If the drum then foams over, the entire blowdown system

can be plugged. As several pairs of coke drums will often share a common blowdown system, this type of foamover can shut down an entire coking complex. This unfortunate scenario actually happened to one six-drum unit, in Texas City in 1978.

Operating errors cause foamovers

There are no valid reasons for foamovers. They are all due to human error and are avoidable. The process operating engineer can help eliminate future foamovers by clearly defining the cause of a current foamover. As with all such incidents, it is difficult to determine, after the fact, what actually transpired. The guilty parties remain prudently silent. The following hints will give the operating engineer a few leads in his investigation:

Liquid hydrocarbons in a coke drum will cause a foamover when hot resid is first switched into the drum. The liquid is condensed when a drum is warmed and should be completely drained by the end of the warm-up period. Before the switch, the drain valve shown in Figure 2–3 is closed. An experienced operator can hear, as he closes this valve, if vapor is blowing freely through it. Also, a low temperature (450°F to 600°F) at the bottom of the drum indicates that the drain line is plugged.

The operating engineer should suspect that a foamover was precipitated by switching into an incompletely drained coke drum if the drum had not reached a normal temperature after a long warm-up period. Operators must be schooled never to switch into a drum containing liquid hydrocarbons. If the empty drum cannot be drained, it should be put on circulation.

Cool *coke drums* will have a marked tendency to carry over. The large volume of partially coked liquid in the drum will form a very high foam front. If the overhead vapor temperature of a coke drum is 30°F to 40°F below normal at the time of a foamover, you can conclude that the cause of the foamover was the low coke drum temperature.

Erroneous level indication has caused many a foamover. If the operators are not sure of the height of the foam front level, they will not know when corrective action is necessary. A properly functioning gamma-ray or neutron level indicator will always record a foamover. If a foamover is not recorded, it shows that the gamma-ray level indicator system is not working.

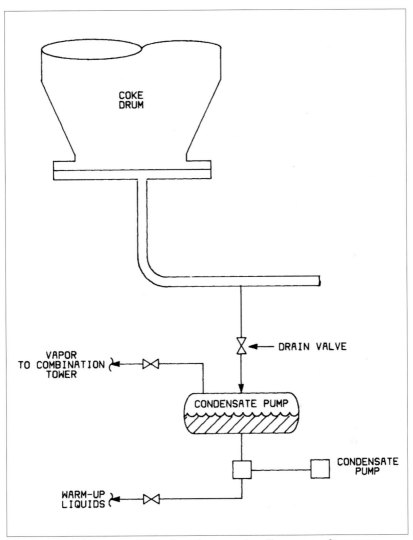

Fig. 2–3. Draining warm-up condensate properly will prevent a foamover.

A typical pair of coke drums will have six radiation level indicators (three for each drum). Some refiners will only have a single recorder for all six points. An operator must manually select which level indicator he wants to see recorded. An inexperienced person can become confused by this setup. The preferred method is to dedicate a separate pen recorder for each level indicator. Each coke drum will then have its own strip chart and three different colored pens.

A backup to the above system is a handheld Geiger counter. An operator can find the foam level in the coke drum by climbing up the side of the drum and checking for an increase in the level of radiation. This method cannot be used for neutron back-scatter (i.e., "K-ray") level indicators.

Silicone antifoam addition is an effective way to reduce the height of the drum foam front. Coke drums typically have a level indicator (as measured from the top flange) at 10, 20, and 30 feet. Silicone addition is started at the 20-foot level. If the foam front is not immediately knocked back below the 20-foot level, something is wrong.

The antifoam addition point can be plugged, the silicone mix pot may be full of diluents without any of the silicone chemical, or the antifoam charge pump might be inoperable. Regardless of the cause, one should switch out of a coke drum (or put the unit on circulation) once it has been established that silicone addition is not controlling the foam level.

When the foam front rises above the 20-foot level for a second time, double the silicone antifoam addition rate and prepare to switch out of the drum. If one waits until the foam front reaches the 10-foot level to initiate action, it may be too late to stop a foamover.

Overfilling coke drums implies operator negligence. If, after a foamover, the measured height of coke below the top head (i.e., the outage) is less than 10 ft, the cause of the foamover is obviously overfilling.

Changing the quality of resid charge to a coker from 13°API to 7°API may increase the coke make per barrel by 50%. The consequent rapid filling of a coke drum will fool operators who are not watching their job. Figure 2–4 quantifies the increase in coke yield with decreasing resid gravity.[1] Because the foam front in the coke drum is about 10 feet high, any outage of less than 10 feet means that some partially coked resid has been carried over into the bottom of the combination tower.

One factor that contributes to a high foam front is caustic. In one refinery, severe coke drum foaming was caused by the presence of caustic in the resid feed. Elimination of caustic injection at the crude unit desalter eliminated the caustic in the coker charge and reduced the coke drum foaming.[2]

Pressure surges on switchover or during drum filling may initiate a foamover. On one unit a faulty fractionator tower level controller allowed the resid level to rise above the vapor inlet nozzle. As

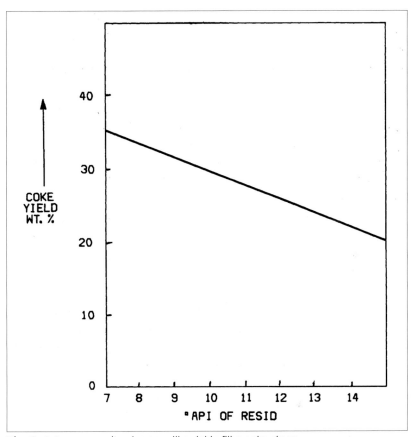

Fig. 2–4. Lower gravity charge will quickly fill a coke drum.

this nozzle was submerged, back pressure against the coke drum increased. After some time the control room operator noticed the high level. He rapidly pulled the resid level down. The sudden drop in coke drum pressure, as the resid dropped past the vapor inlet nozzle, started a coke drum foamover. This phenomenon is identical to that observed when opening a warm bottle of beer.

When switching feed out of a full coke drum into an empty drum, the pressure on the full drum will drop. Experienced operators carefully observe the pressure on the full drum during a switch. Allowing the pressure to drop too rapidly will start a foamover. A drop of 1 or 2 psi every five minutes is about right. The top radiation level indicator will show when a foamover starts during the feed switchover. To stop this foamover, the vapor valve on the full drum should be pinched closed until the drum pressure rises a few psi.

This all shows the importance of examining the coke drum pressure recorder chart when troubleshooting a foamover. This chart is analogous to the fingerprints left at the scene of a crime. The process engineer should be able to tell from the pressure recorder the portion of the drum cycle that was taking place at a given time.

During steaming, a full drum may suddenly start to carry over. At this point in the cycle, the coke drum overhead is usually lined up to the blowdown or quench system and not to the combination or fractionator tower. Since several pairs of coke drums may be served by a single blowdown system, a carryover into this system can incapacitate an entire coking complex.

A full coke drum that has been produced at an overhead vapor temperature of 770°F or less will have 5 or 10 feet of partially coked resid lying on top of the solidified coke. This semiliquid mass is easily carried over during coke drum steaming. The troubleshooter can verify this by obtaining a piece of coke recovered from the plugged blowdown lines; it will look like black glass.

If a coke drum begins foaming over during steaming, it will not help to reduce the steam rate because this also drops the coke drum pressure. The sudden drop in pressure accelerates the foamover. Rather, one should begin pinching on the overhead vapor valve to increase the coke drum pressure. Rising coke drum pressure, along with continued silicone antifoam addition, will stop this type of foamover if corrective action is taken quickly.

To avoid putting a coke drum in such a vulnerable position, an overhead temperature of 780°F or higher should prevail when the drum is almost full. This will eliminate the potential for carrying over an amorphous mass of partially coked reside. The black "glass" found in plugged lines is solidified amorphous coke.

Sometimes a coke drum is too full. Sometimes a coke drum is too cool. Even under such adverse circumstances, my experience teaches that maintaining an ascending coke drum pressure on the drum one is switching out of, will suppress coke drum foamovers. An early indication that a drum is starting to foam over is that the third level detector from the top starts to show a drop in density. Quick action to partly close the vapor valve from this drum will typically suppress the foamover. Even the smallest rate of the ascending pressure profile is quite effective in suppressing foamovers.

What to Do if a Coke Drum Is Not Ready

Between equipment failures and human fallibility, it sometimes happens that an empty coke drum is not ready to be switched into when the other drum is almost full. When this occurs, the unit must be put on circulation. Resid charge is stopped, and the liquid level in the bottom of the combination tower is held by pumping heavy coker gas-oil product below the drawoff pan. The coke drum is bypassed by opening a valve between the coking heater outlet and the combination tower vapor inlet. After awhile, to maintain liquid levels, outside gas oil is charged to the combination tower.

On occasion, coke drums have been overfilled and the unit shut down because the shift operators either did not know how or did not want to bother putting the coker on circulation. Another time, a coke drum was overfilled because the chief operator did not know if he had the authority to decide when to put the coker on circulation. A serious program to stop coke drum carryover must train operators to put the unit on circulation rapidly. From the moment the decision is made until the full coke drum is bypassed, about 30 minutes will elapse.

Coke Quality

The four aspects of coke quality that affect price are sulfur, metals, hardness (volatile combustible matter [VCM]), and physical structure. The sulfur content of coke will be 40% to 50% higher than that of resid feed. All the metals (mostly vanadium) in resid will concentrate in the coke. The operating engineer has no control over these properties, but he can have a major influence on the hardness and %VCM of the coke.

Coke VCM

Petroleum coke sold for fuel may have a high VCM; coke sold for calcining must have a low (10%–13%) VCM. Making coke too soft (high VCM) leads to potential foamover problems and increased coke yields at the expense of more valuable gas oil. Hard coke (low VCM) takes longer to cut.

The VCM of coke is primarily a function of coke drum overhead vapor temperature. Many refiners attempt to control coke VCM by monitoring heater-outlet (i.e., transfer-line) temperature. This is incorrect because the temperature at which the coking reaction takes place is a function of the vapor-line temperature. A very high transfer-line temperature (930°F) will not make hard coke if the drum insulation has deteriorated and the vapor line is only running at 770°F.

A one-unit (1%) change in VCM can be effected by changing the overhead vapor temperature by 7°F. This rule of thumb holds in the range of 10%–15% coke VCM. The vapor-line temperature is increased by dropping heavy gas oil into the heater charge or by raising the heater outlet temperature.

The coke drum and its contents represent a large heat sink. If an empty drum has not been properly warmed (650+°F), the vapor-line temperature is suppressed and softer coke is made. When a drum is almost full, an increase in the transfer-line temperature will take one or two hours to effect the critical vapor-line temperature.

One cause of soft coke is poor drum insulation. A few hours into the filling cycle, the temperature difference between the transfer line and vapor line ought not to be more than 110°F. A greater difference indicates inadequate drum insulation. Water may have penetrated underneath the weatherproofing and soaked the insulation, thus reducing its effectiveness.

The thermocouple used to measure the overhead vapor-line temperature should be contained in a polished stainless steel thermocouple well. The well should be cleaned after each cycle.

Time, pressure, and steam

If coke is left in a drum too long, it can become so hard as to be almost impossible to cut. On the other hand, reducing coking cycles will increase coke VCM. A rough rule of thumb is that a six-hour decrease in coke drum filling time will raise coke VCM by 1%–2%.

Lowering coking pressure appears to have a small effect toward making harder coke. More importantly, lower coking pressures significantly increase the production of valuable gas oil while cutting coke and gas yields. For example, a decrease in coke drum pressure of 10 psi will reduce the amount of coke made by several percent.[3]

Many refiners try to reduce coke VCM by steaming longer with superheated steam. This does not appear to do much good. Adding 50 gpm of water strips out as much hydrocarbon as does 25,000 lb/hr of 600°F steam. The water flashes to steam and is heated to 700°F by the time it reaches the soft coke lying in the top of the coke drum.

Physical structure

Coke that is to be calcined and formed into anodes is more valuable if it has certain physical properties. Foremost among these is the coefficient of thermal expansion. Some cokers produce a valuable product called needle coke (named for its crystalline structure). These properties are a function of the resid feed composition. The coke specification that the operations engineer can usually influence is VCM, and this is done by closely monitoring the coke drum overhead vapor temperature. Higher recycle rates and pressure also improve coke quality.

Shot coke.

Normally, coke cut out of a drum has a porous, sponge-like structure. Occasionally, coke is made in the form of hard, round balls the size of buckshot. This is called shot coke. The gravity of the coking heater charge is the overriding factor affecting the tendency to form shot coke:

- 7°API or less—shot coke formation likely
- 8°–9°API—shot coke formation possible
- 10° API or more—shot coke formation unlikely

Much has been said in the industry as to how to stop shot coke from forming. Other than raising the gravity of the resid by the addition of gas oil, I can only vouch for the following:

1. Add 10%, 12°API low-sulfur resid to 8°API high-sulfur reside.
2. Add 5% decanted (slurry) oil from an FCCU to a 6°–8°API high-sulfur resid.

One refiner reports that the "tendency to form shot coke can be eliminated by the injection of silicone." However, this operation was found to be prohibitively expensive and was discontinued.[4] Raising pressure also suppresses shot coke formation.

From the operator's viewpoint, shot coke is bad news: It slows down the rate at which the coke is quenched. Cooling water, entering the bottom of the drum, spreads radially in sponge coke at a rate of about 8 ft/hr. This rate is much slower in a drum full of shot coke. Thus, portions of the coke may remain unquenched.

Later, when the cutting water hits the hot coke, an eruption of steam and coke balls issues forth from the open top head. If this happens, allow the cooling water to remain in the drum an extra hour before draining.

Shot coke is also prone to packing down in the bottom of the coke drum. In this way, the flow of cooling water into the bottom of the drum is restricted. Instead of taking three hours to fill the drum with water, eight hours may be required. The time to drain the water out of the drum is also increased.

Thus, shot coke has the potential to extend the time of the coking cycle and, hence, may reduce coking capacity. The solution to this and other problems that extend coking cycles is discussed in the next section.

Common Coke Quality Questions

There is a very real relationship between shot, needle, and sponge coke. This relationship is influenced by coke drum operating conditions, crude type, and target VCM. The following 20 questions and answers attempt to clarify this complex relationship:

1. What is shot coke?

Shot coke is composed of precipitated asphaltenes. Asphaltenes are large hydrocarbon molecules that are dispersed in the lighter aromatic and paraffinic oils in vacuum residuum.

When these lighter molecules thermally crack and vaporize in the coke drum, the asphaltenes agglomerate and precipitate out of solution. The coke is called shot coke because the precipitated asphaltenes take the form of hard, spherical particles.

2. What is needle coke?

Needle coke is formed from condensed aromatic compounds. High boiling aromatic compounds (typically 950°F and higher) react in the coke drum by cross-linking the aromatic rings.

The resulting crystalline, silver-gray colored coke is called needle coke because of its erratic, but clearly noticeable structure resembling broken needles.

3. What is ordinary sponge coke?

Sponge coke is a mixture of needle and shot coke. The coke formed is spongelike in appearance, and has a dull, black cast. It is porous, with no structure. When the ratio of needle coke to shot coke falls below a certain point, the sponge coke will become less porous.

4. How can the component of coke yield caused by precipitating asphaltenes be reduced?

The yield of precipitated asphaltenes cannot be reduced because the asphaltenes in the coker feed will produce coke regardless of coke drum operating parameters.

5. Can the component of coke yield from cross-linked aromatics be reduced?

The formation of cross-linked aromatic rings can be reduced by lowering the coke drum pressure, raising the coke drum outlet temperature, improving the vacuum tower operation, and by adding steam to the coke drum.

6. Is there any needle coke in shot coke?

The outside surface of shot coke spheres is coated with a layer of needle coke. The coating gives shot coke its polished-surface appearance.

7. When coke is cut from a drum, sponge coke is found along the walls, and shot coke is found in the middle. What causes this segregation?

The segregation is caused by poor drum insulation. Poor insulation promotes the 950°F and higher-boiling-range aromatic hydrocarbons to cross link rather than vaporize.

8. Why is the coke in the top of a coke drum mostly sponge coke, and the coke in the middle and bottom of the drum mostly shot coke?

During coking, the top of the drum runs cooler than the middle or bottom of the drum. Therefore, the coke formed in the top of the drum contains a higher percentage of needle coke mixed with shot coke.

9. Why, at constant heater outlet temperature, does the coke drum vapor outlet run hotter when shot coke is produced compared to when needle coke is produced?

The precipitation of asphaltenes to form shot coke is a physical change involving no heat of reaction. Therefore, there is no temperature drop across the coke drum.

In contrast, needle coke formation is an endothermic reaction. This is shown by the 50°F–80°F drop in temperature between the coke drum inlet and outlet during needle coke production.

10. Is there any shot coke in needle coke?

Needle-grade coke contains no shot coke. Commercial needle coke is typically produced by charging FCCU slurry oil to the delayed coker.

All of the components in the slurry oil coker charge are produced as a vapor from the FCCU reactor fresh feed. Even if vacuum tower bottoms are charged to the FCCU, none of the asphaltenes in the resid will vaporize, and therefore, the asphaltenes will not be in the coker charge.

11. The coker feed nozzle at the bottom of the coke drum can plug with shot coke. How can this be prevented?

Steam can be started into the coke drum, at the rate of 100 lb/hr/ft of coke drum diameter, before switching out the drum.

12. Although shot coke formation can be tolerated in some cases, it occasionally pours out of the drum when the bottom head is removed. How can this be prevented?

The heater outlet temperature should be run about 10°F cooler during the first two hours of the coking cycle. This will promote the formation of sponge coke in the coke drum. The procedure will also aid in keeping the feed nozzle open during the quenching and draining period of the coking cycle.

13. Can shot coke be used as anode material?

The coefficient of thermal expansion of calcined shot coke is too high, making it unsuitable for high-quality anode production.

14. What is the percent VCM of shot coke?

The VCM content of shot coke is similar to that of sponge coke, in the range of 9%–12%.

15. Why is it important to reduce the sulfur content of needle coke, and how can it be controlled?

Sulfur causes puffing of the calcined needle coke. Increasing the delayed coker recycle ratio by cutting the heavy coker gas oil product end point will make more needle coke and reduce the sulfur content of it.

Other ways to improve needle coke quality include minimizing steam in the heater passes, reducing the gravity of the recycle gas oil, and raising the coke drum pressure.

16. Why does some sponge coke look denser and less porous than usual sponge coke?

The reason for the less porous appearance is that shot coke, in this case, is just beginning to form. This is sometimes called incipient shot coke.

17. Whenever it is desired to reduce the production of low-value coke to make more gas oil, the coke drum generates almost all shot coke. How can coke production be reduced without causing shot coke?

A reduction in coke production cannot be made without producing more shot coke. A certain amount of cross-linked aromatic coke, mixed with the asphaltic coke, is necessary to turn shot coke into sponge coke. This increases total coke production.

18. Why does the amount of shot coke in the drum vary, even though the drum temperature and pressure, and the recycle ratio and refinery crude composition are kept constant?

The formation of free radicals in the coker feed is favored by exposure of the crude or resid to oxygen. Particulate matter in the resid also accelerates shot coke formation. The particulates can be controlled by improving the operation of the crude unit desalter.

19. Does increasing the recycle of light coker gas oil promote shot coke formation?

Yes. Shot coke formation is increased because recycling light gas oil raises the coke drum vapor outlet temperature and reduces the production of cross-linked aromatic coke. This concentrates the asphaltic coke.

20. Increasing the recycle ratio is supposed to stop shot coke formation. However, in some cases, increased recycle ratio has no effect. Why?

Whether or not increased recycle ratio reduces shot coke formation is dependent on the composition of the incremental recycle.

Recycling $950°F–1,000°F$ coker gas oil will increase the formation of needle coke.

Recycling $600°F–700°F$ coker gas oil will reduce the formation of needle coke by raising the coke drum outlet vapor temperature. Heavy recycle is necessary to effectively suppress shot coke formation.

An additional area of concern relates to the quality of the coke produced from hydrotreated vacuum resid feeds. For a variety of reasons, premature furnace fouling, inclusion of semi-coked liquids in the quenched coke drum, and carryover of coke fines into the fractionator, the delayed coking of 100% hydrotreated feed has not been routinely successfully achieved in commercial operations. Attempts to do so have resulted in the production of high VCM coke and/or commingling of hydrotreated and virgin coker feedstocks.

Fighting Long Coking Cycles

How long should it take to turn a coke drum around? The elapsed time from when one switches out of a full drum until it is empty and ready to receive resid again is called the cycle time. Most delayed coking units are designed for an 18- to 24-hour cycle time. Shorter cycles can be used in refineries where coke drum volume limits capacity. Each of the steps detailed in the Troubleshooting Checklist at the end of this chapter can be contracted to save time.

The decided trend toward poor-quality crudes means that the tons of coke produced per barrel of resid are, on the average, increasing. Coke-drum size is therefore a more frequent bottleneck than is the capacity of the heater, fractionation tower, or wet gas compressor (see figure 2–1). With this in mind, a common assignment for an engineer is to determine how the operating crew can reduce cycle time.

As a basis for discussion, the details of operating on a 12-hour coke drum cycle will be reviewed. The discussion that follows applies to a 20-foot diameter coke drum producing about 1,000 T/D of anode-grade coke. The volatile combustible matter averages 12%. In really large-diameter coke drums (28-plus feet in diameter) producing hard coke with 8%–10% VCM, it is unlikely that a 12-hour coke drum filling cycle can be obtained. However, 14- to 16-hour cycles in 27-foot drums have become a common industry practice.

Human factor

Turning around a coke drum is a cooperative effort between the process operators and the decoking crew. It is labor intensive and normally accomplished without the intervention of supervisors. The work itself is best characterized as dirty, difficult, and if not done with care, dangerous.

To get a 12-hour cycle, every facet of the complex task of turning around a drum must be carried out at maximum speed. This means that both the decoking crew and the operators must be personally committed to the objective of short cycles.

One demonstrated method of obtaining this commitment is to put the decoking crew on a piece-work basis. Then the decoking crew cuts one coke drum and goes home. Typically, when the 12-hour cycle is being achieved, the decoking crew will spend four hours on the job site.

But it is somewhat unsettling to pay a person for eight hours of work and only have him on the job for half that time. In one refinery, management attempted to rectify this matter by assigning the decokers to general maintenance work after they completed cutting a full coke drum. Suddenly, the time to cut a drum of coke increased from four to seven hours.

Motivating operating personnel can be difficult. Adding labor-saving devices such as air operators on the large vapor valves will demonstrate to the process operators that refinery management is placing great emphasis on short cycles. A careful and honest review by first-line supervision explaining how increased coke production will enhance refinery profitability is also helpful.

Time-saving details

Figure 2–5 shows the valves used during the coking cycle. Prior to switching out of a coke drum, the process operators must be sure that the empty coke drum is truly empty. Trying to speed up the cycle by cutting hot feed (900°F) into a coke drum that is incompletely drained will cause a foamover.

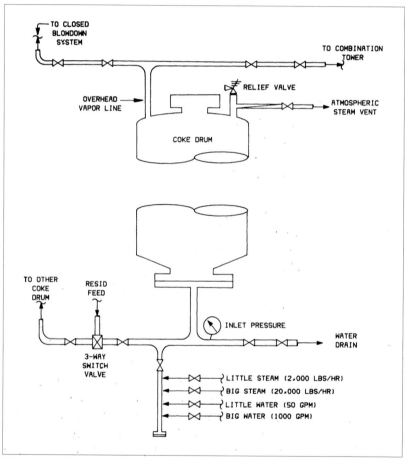

Fig. 2–5. Valves used during the coking cycle.

The quick way to be sure that a coke drum is clear of liquid is to use a "tailpipe" thermocouple (figure 2–6). This temperature point is located between the coke drum bottom head and the condensate collection drum. When the coke drum is draining freely, the tailpipe temperature will be close to the coke drum temperature.

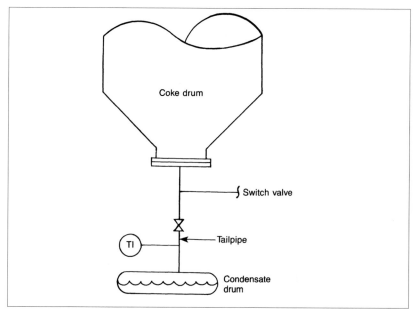

Fig. 2–6. Thermocouple verifies drainage (courtesy *Oil & Gas Journal*).

A cold tailpipe indicates a coke drum with liquid accumulated in the bottom head. About 30 minutes before switching time, the operating crew should verify that the tailpipe thermocouple is reading at least 370°F to 400°F.

Making the switch

The objective in switching the feed between coke drums is to avoid foamovers. I have witnessed several coke drum foamovers initiated by improper switching technique. Here is what transpired:

1. A soft drum of coke (i.e., average VCM of 14%) is produced due to inadequate coke drum temperature.

2. Operating personnel swing over the Wilson-Snyder switch valve too rapidly. This causes the full coke drum to depressure.

3. As indicated by the radiation coke drum level detector, the full coke drum's liquid level suddenly rises.

4. Operating personnel at the switch valve are advised from the control room that the full coke drum is rapidly overfilling. They then immediately complete the switch in the hope of minimizing the carryover of resid into the

coke drum overhead vapor line. This further reduces the pressure in the full drum, and a major carryover of partially coked resid ensues.

The proper method to control drum foamovers during a rapid switchover is to control coke drum pressure by pinching back on the outlet vapor valve to maintain the pressure in the full coke drum. While the operator at the switch valve is diverting the feed to the empty drum, he observes the full coke drum pressure. As this pressure falls during the switch, he closes off the vapor outlet valve from the full drum to hold the pressure within 5 psig of normal operating pressure. Using this technique, the switchover can be accomplished in about 20 minutes.

The larger the pressure difference between the coke drum and the fractionator, the greater the hazard of foamover during switching. Coke buildup in the overhead vapor line is the usual cause of the increased ΔP in the overhead vapor line.

Vapor valve operation

Coke drum overhead vapor valves are massive affairs. Switching a pair of 20-foot diameter drums may require the manipulation of four 18-inch and four 12-inch valves. Size aside, the vapor valves are difficult to open and close because of coke deposits on valve internal parts. Removing the insulation on piping upstream of the vapor valves or injecting a small liquid quench (typically 3 vol%) into the vapor line at the first 90-degree turn will help.

To minimize the amount of hard physical labor required to switch drums every 12 hours, the vapor valves should be fitted with locally controlled air operators. Such a device is little more than an air gun permanently mounted to the stem of each valve. The speed of the valve movement is controlled with a ¾-inch air valve.

Keep drain free

One of the frequent causes of extended coke drum cycles is the difficulty in filling the drum with cooling water. Once feed is switched out of a coke drum, the coke tends to settle back and partially plug the feed nozzle in the bottom head. This effect is especially pronounced when shot (i.e., BB-size pellet) coke is produced because of charging a low-gravity feed. Not only is the time required to fill the drum with water extended, but it becomes difficult to drain the drum.

To keep the feed nozzle open during a switch, a small amount of steam must be cracked into the drum that is being taken off-line. By the time the feed has been diverted from a full drum, the flow of steam should be several thousand pounds per hour. In practice, the steam flow is adjusted to maintain the pressure in the feed line several psi above the pressure in the bottom of the coke drum.

Water filling

Once the switch has been completed and a small flow of steam (4,000–5,000 lb/hr for a 20-foot diameter drum) has been established, the vapor valve on the full drum will be slowly reopened. Meanwhile, the operators should be lining up the cooling water to the coke drum. For short coke drum cycles to be maintained, it is necessary to skip the standard "big steam" portion of the cycle. Within 30 minutes of the switch, the flow of water to the coke drum should be established. The flow of water must be controlled automatically as follows:

1. The water rate is increased according to a predetermined time versus flow curve, as shown in Figure 2–6.
2. If the pressure of the drum being cooled increases too fast, the flow of water is restricted.

Figure 2–7 shows the required controls. A plastic sheet cut in the shape shown in Figure 2–8 to serve as a cam-type operator suffices to guide the water flow valve. An excessive increase in coke drum pressure delays the progress along this curve. Careful control of the coke drum pressure during quenching is a major factor in preserving drum mechanical integrity at reduced cycles. The automated device described does an effective job in controlling this pressure. (*Author's note: This section, while still valid, was written before computers were available on process units.*)

For the first 45 minutes, the flow of water will increase from 25 to 75 gpm. The entire flow will flash to steam and create a substantial back pressure in the coke drum. During this period, the coke-drum overhead should be diverted from the fractionator to the blowdown system.

Note that 1–2 wt% of the coke will be stripped out as gas-oil vapors. Once the flow of water markedly increases, the coke drum is no longer hot enough to turn all the water to steam and the amount of hydrocarbons being removed from the coke is greatly reduced.

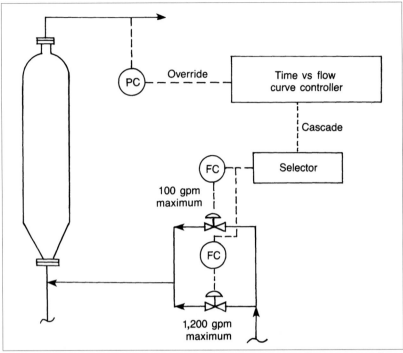

Fig. 2–7. Controls for automated water addition (courtesy *Oil & Gas Journal*).

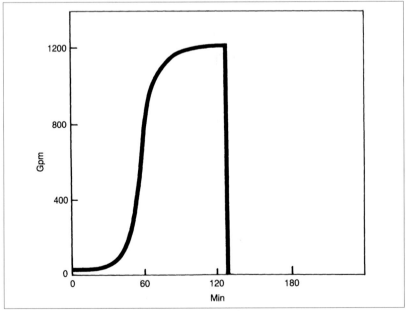

Fig. 2–8. Curve for adding water to coke drum (courtesy *Oil & Gas Journal*).

The pump that supplies cooling water to the drum should be sized for 1,500 gpm for a 20-foot diameter drum. Coke drums venting to enclosed blowdown or quench systems are not overflowed. A water level is established 10 feet above the coke level in the drum.

Using automatic water addition and an adequately sized pump permits a proper water level to be reached in two hours. The water in the drum is boiling, so occasional water addition may be required to hold a proper level. The drum's radiation instrumentation is used to follow the water level. The water spreads out from the center of the drum at a rate that has been measured at 8–10 ft/hr. For a 20-foot diameter drum, the water must sit in the drum for an hour before it is drained. For the last 10–20 minutes of this period, the coke drum overhead should be diverted from the enclosed blowdown system to a large-diameter (six inches for a 20-foot drum) atmospheric exhaust. This will minimize the coke drum pressure when the top head is unbolted.

Sometimes the radiation-level detector (especially when the more sensitive neutron back-scatter devices are used) will falsely indicate that the water level in the coke drum has risen to the correct level. Apparently, the level detector is fooled by the increasing density of steam as the coke cools.

You can ascertain when a coke drum has been drained without adequate water quenching by noting the appearance of the drain water. If 10 minutes after draining is initiated, the water comes out steaming hot, and 20 minutes later, steam and water blow out the drain line together, you can be sure that the drum was never fully quenched. The usual reason for this is that the bed of coke was never completely covered by water. The drained water should get steaming hot (i.e. about 160°F) just before draining is completed—not shortly after draining is initiated.

Decoking crew

Some refineries use one person to decoke a drum, while others use four. Having seen cokers operate on a 12-hour cycle with both a two- and a four-person crew, I conclude that the most effective crew size for short cycles is three decokers. During part of this decoking period, only two crewmembers will be involved in the job at hand while the third person is greasing bolts, lubricating the air-operated stem hoist, sweeping the cutting deck, and so on. Only when the coke drum is being unheaded (i.e., the top and bottom heads removed) and headed-up are all three required to work simultaneously.

Unheading

When the process operators begin draining down the coke drum, the decoking crew's job begins. Its first task is to remove the top head after ascertaining that the coke drum is essentially depressured. This is done by opening a bleeder on the top head.

One method to hasten unheading and heading-up is to tack-weld one of the two nuts used with each bolt onto the heads. For the top head, the nut underneath the coke drum flange is tack-welded; for the bottom head, the nut on top of the coke drum flange is attached. This eliminates the need for a person to hold a backup wrench when unbolting the drum with an air gun. Handling the top head can be greatly simplified by installing a minihead (3-foot diameter) atop the top head. This minihead is removed during decoking, while the large head is left in place.

While two crewmembers are raising the top head, the third member of the team should verify that water is draining freely from the coke drum. Sometimes, due to formation of hot coke or the late addition of steam to the bottom nozzle after the switch, the flow of water from the bottom nozzle is restricted. This problem can best be observed by using the "dump line" shown in Figure 2–9.

This line is also useful for draining the last few barrels of hot water out of the coke drum prior to dropping the bottom head. For a coker to operate on 12-hour cycles, it is necessary to drain the drum in one and a half hours. This period must not be wasted by the decoking crew. In addition to removing the top head, they should prepare the bottom head for removal.

Bottom-head removal can be expedited by installing a track-mounted air gun for unbolting the bottom head. The track is attached to the circumference of the coke drum cone as shown in Figure 2–10, and the gun is slid along the track. This feature, combined with tack-welding the nut to the bottom flange, permits one person to unbolt the bottom head easily.

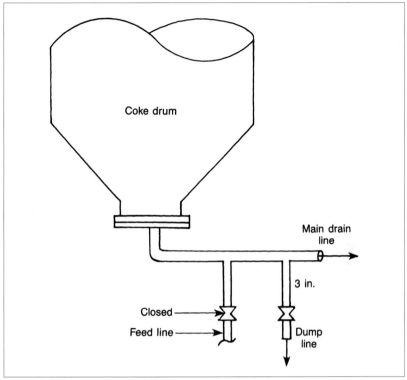

Fig. 2–9. Dump line speeds draining (courtesy *Oil & Gas Journal*).

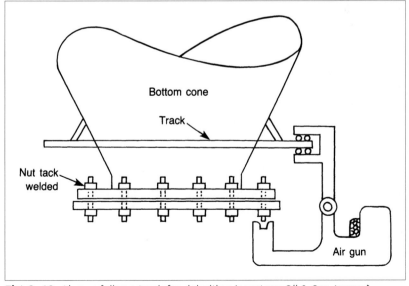

Fig. 2–10. Air gun follows track for debolting (courtesy *Oil & Gas Journal*).

While the drum is still draining, the bottom carriage, which is used to slide the bottom head out from under the drum, should be set in place underneath the bottom head. The carriage must be designed so that the resid feed line does not interfere with the movement of the carriage. That is, it should not be necessary to disconnect the resid feed line from the bottom head to set the carriage in place.

A unique device to lower the bottom head safely was developed in 1988 by Champlin Oil in Corpus Christi, Texas. The bottom head is attached by a hinge (see figure 2–11) to the bottom of the coke drum. The hinged bottom head is lowered by means of hydraulically powered pistons. A safety clamp keeps the hinged head firmly in place after the head is unbolted in the conventional way. (Since 1988, there have been great advances in automated bottom head removal not detailed in this text.)

I witnessed a drum full of shot coke unheaded with this device. The coke started to rollout in a most alarming fashion. However, the operator controlling the piston from a remote location simply closed the head partially until the shot coke stopped rolling out.

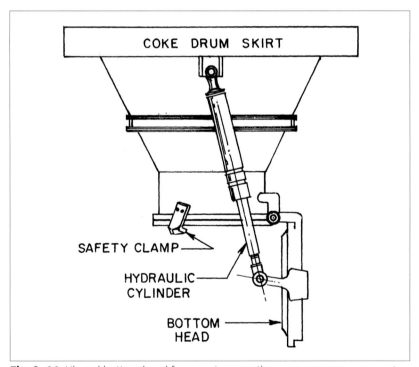

Fig. 2–11. Hinged bottom head for remote operation.

I also learned from this incident how to prevent massive amounts of shot coke from pouring out of the bottom of an unheaded coke drum. Simply be careful to drain the drum fully prior to unheading. It is the last few hundred gallons of quench water left in the drum that "wash" the shot coke out.

Resid line to bottom head

Many cokers have a removable spool piece between the bottom head of the coke drum and the resid feed line. This section of piping must be disconnected every cycle. Usually, it is not possible to drop the bottom head without going through this exercise. By designing the bottom-head removal carriage properly, this spool piece can be left permanently attached to the bottom head. This feature saves 10 minutes during the unheading operation and 20 minutes when the bottom head is replaced. A modern unheading device avoids this complication.

Bottom-head removal

The decoking crew should begin unbolting the bottom head while the drum is still draining. Leaving every third bolt in place, the drum depressured and mostly drained, keeps the head secured. As soon as the worker unbolting the head sees the flow of water diminish from the dump line, shown in figure 2–9, he should finish unbolting the head while a coworker disconnects the resid feed line. Using this procedure, the bottom head can be removed within 20 minutes after the water stops draining from the drum.

Next, the skirt, or telescoping chute, connecting the choke chute to the drum's bottom flange is hoisted into place and bolted up. While this step—which should take 10 minutes—is proceeding, the third member of the decoking crew should be lining up the cutting water system and lowering the drill stem with the pilot head attached into the top of the drum. The coke is now ready to cut, and the full decoking crew should assemble on the cutting deck.

Drainage problems

If the drum is not draining properly, the decoking crew should proceed as follows:

1. Wait as long as possible without jeopardizing the cycle.

2. Set the bottom carriage—which is used to slide the bottom head out from under the coke drum—up against the head.

3. Remove all but four bolts. Then loosen the remaining four bolts so that the bottom head is cracked open.

4. As the drum drains, continue to loosen the bolts to facilitate drainage.

In defense of this unorthodox procedure, I can only say that I have seen it done repeatedly and it is the only way to get a drum with a plugged drain line rapidly drained. Before applying this remedy, the decoking crew should ascertain that the drum is completely depressured and that the water level is well below the lowest radiation-level detection point. One way to improve drainage is to install a distributor at the feed nozzle to prevent the nozzle from plugging.

Draining can also be reestablished by blowing steam back into the bottom head for a few minutes. Make sure to block in the drain line prior to reinitiating steam flow.

Coke cutting

Cutting time varies with the hardness of the coke, the condition of the hydraulic cutting nozzles, and the available cutting water pressure. Of equal significance is the skill and experience of the person cutting the coke.

Initially, a 4-foot diameter hole is hydraulically drilled down the center of the coke drum using the pilot cutting head. Once the hole is completed and coke is cut out of the bottom cone, the stem is retracted and the cutting head installed. Now, starting 6 feet below the top of the coke level in the drum, slices of coke are cut out. The chunks of coke fall through the hole and out through the bottom of the coke drum.

How can this standard procedure be expedited? First, avoid producing very hard or soft coke. The soft coke (i.e., 14% VCM and higher) makes a lot of coke fines, which contaminate the cutting water and bog down the coke dewatering equipment. Hard coke, 10% VCM and lower, takes longer to cut. Second, a cutting water pressure-at the pump discharge-of 120 psi per foot of drum diameter is necessary. For large-diameter drums (24 foot and greater), specially designed nozzles on the cutting head are required to focus the jet of water near the walls of the coke drum. For 26-foot, 0-inch inside diameter (1D) drums the optimum conditions at the cutting head are 950 gpm at 2,900 psig.[5]

The cutting nozzles, which screw into the pilot and cutting heads, must be kept in first-class condition. This is simply a matter of calipering the nozzle diameters several times a week and changing out the worn nozzles. The rate of wear on these nozzles varies with the concentration of coke fines in the cutting water. A worn nozzle cuts like a dull knife. The jet of water is more diffuse and hence does not cut the coke cleanly. Rather, the coke is ground out instead of being cut. This produces additional fines to increase the wear in the nozzles. The self-defeating cycle is best controlled by adequate clarification of the cutting water.

Improper cutting techniques can also produce excessive coke fines. Interestingly, the same procedures that speed coke cutting also minimize coke fines. Cutting coke quickly makes less fines. By undercutting the coke 5–6 feet, chunks of coke are produced that will not jam the stem or plug the bottom outlet of the coke drum. Many workers cutting coke prefer to undercut the coke by 2–3 feet and avoid the possibility of encountering a chunk of coke big enough to jam the stem or plug the bottom hole. The amount of work to correct either of these difficulties is appreciable. Unfortunately, undercutting coke by 2 feet instead of 5 feet doubles cutting time and coke fines.

How long, then, should it take to cut a drum of coke? Assuming a typical size drum (20-foot diameter and 70-foot tangent), the following time intervals have been observed:

- Drill pilot hole   20 minutes
- Retract stem and change from pilot head
  to cutting head   10 minutes
- Cut coke   75 minutes

Cutting a drum of coke in one and three quarters hours is no mean trick. It can only be done by an experienced person determined to make every minute count. For example, he must be able to decide when a section of coke is undercut by the sound of water hitting the sides of the drum. He must have a "feel" for the stem at the maximum safe speed; he must take care not to leave sizable chunks of coke adhering to the sides of the drum. Any coke left in a drum goes through a second coke cycle and becomes superhard the next time the drum is opened for decoking.

Dual-purpose cutting head

This coke-cutting procedure involves the use of a combination coke-cutting/drilling tool made by Pacific Pumps (Dresser Industries), as shown in Figure 2–12. This cutting head is in use in a large number of U.S. cokers.

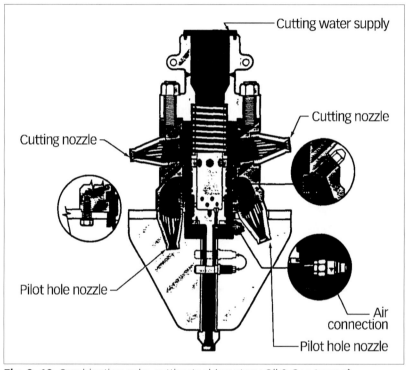

Fig. 2–12. Combination coke-cutting tool (courtesy *Oil & Gas Journal*).

The head permits switching from the pilot bit mode to the cutting mode, using the same head. Switching is a five-minute manual operation.

The tool design has jet nozzles that create a tighter stream of water that maintains its energy at the wall of large-diameter coke drums. The nozzle concentrates the power from the jet pump and provides a sharper hydraulic jet of water to more effectively cut the coke.

One operator reported that the combination head reduced the cutting time (time from completion of pilot drilling to the time the jet pump is stopped) from 2.5 hours using another bit design, to about 1.6 hours with the combination bit.

When cutting soft, fuel-grade coke or shot coke from a large-diameter drum, and utilizing a high-discharge-pressure jet pump, the main impediment to rapid coke cutting is coke cave-in which can bind the drill stem.

The following drilling and cutting procedure is designed to be used with the combination cutting head to help reduce the cave-in problem and increase the speed with which the coke can be removed from the drum:

1. Drill the pilot hole while gradually lowering the stem at a rate of approximately 10 ft/min. A maximum drill stem rotational speed of 15–20 rpm is recommended to create a pilot hole with a diameter of about 2.5 feet at the top of the coke bed. Do not cut coke out of the bottom cone.

2. After the pilot hole is completed, stop the drilling unit and oil the pneumatic motor.

3. Withdraw the drill stem with the flow of cutting water stopped. Remember that the top 5–10 feet of the coke bed can be like sand and can easily collapse, jamming the drill stem. This can be prevented by careful removal of the stem and by avoiding undercutting the soft coke layer. Undercutting can be prevented by carefully enlarging the pilot hole by running the cutting head down again.

4. Switch the combination cutting tool from the pilot drilling mode to the cutting mode by applying air pressure to a fitting on the combination tool and observing the position of a steel pin.

5. Open a 3- or 4-inch water connection located under the drum-relief valve to allow quench water to suppress the evolution of steam from the top head of the drum. This will improve the operator's visibility when the tool mode is switched.

6. Adjust the drill stem motor to rotate at a rate of 8–10 rpm and begin coke cutting.

7. As cutting progresses, observe the size of the coke lumps leaving the drum. Control the vertical speed of the drill stem so that the lumps leaving the drum are about fist size. Increase the vertical speed if the lump size increases, and decrease the speed if the lumps get smaller. Controlling the lump size is important to avoid cave-ins that result in bottom-head plugging.

8. The first pass of the cutting head should enlarge the pilot hole to 4–5 feet in diameter at the top of the coke bed (with water flowing on both the downstroke and the upstroke).

9. Make two more passes with the drill stem turning at 8–10 rpm. After the third pass, the level of coke in the drum will be reduced by about 20 feet. This means that, to save time, the cutting head should not be raised to the original starting point after each pass. Gauging the residual level of coke in the drum can be done by observing the evolution of steam from the top head. If no coke is being cut and no coke is falling through the drum, steam evolution will increase. Experienced operators can also note the sound of water striking the metal sides of the drum, indicating coke has been removed from that level in the drum.

10. The fourth and subsequent passes should be made with the drill stem readjusted to rotate at 6–8 rpm. Reducing the head rotation speed keeps the tangential velocity of the water jet at the coke interface at a relatively constant value as the coke wall approaches the drum wall.

11. In a typical operation, four to six more passes will be required to fully remove the remaining coke. Each of these remaining passes will take about 10–12 minutes.

12. After the last cutting pass, a tough ring of coke, several feet high and several feet above the bottom of the drum, will adhere to the drum wall. This can be removed using the high-energy jets of the combination tool. This generally takes only a few minutes. With other head types, this layer has to be removed by grinding.

The procedure depends on a good source of clarified water supply to the cutting head and frequent inspection of the cutting nozzles for wear.

The procedure will reduce cutting time for large coke drums making primarily fuel-grade coke. Field experience indicates that the procedure can decoke drums containing up to 1,000 tons of green coke in about two hours.

Heading up

As soon as the cutting water pump is shutdown, the decoking crew immediately begins replacing the top and bottom heads. Both heads must be worked on simultaneously. Flange surfaces are prepared while the coke cutting is in progress. On this basis, heading up the coke drum and reconnecting the resid feed line takes about 30 minutes, assuming the last section of resid feed-line piping is integral with the bottom head.

Hot coke

If a refiner is to operate on a short coke drum cycle, he can expect to encounter pockets of hot coke that the drum cooling water has not had time to quench. During the coke-cutting portion of the decoking cycle, hot coke is a very real hazard. A jet of cutting water hitting 700°F coke will cause a mini-eruption of fist-size coke from the top head. To prevent injury to personnel, a heavy-duty screen should be bolted across the top head. A 1-ft diameter opening in the center of the screen is sufficient to allow for horizontal displacement of the stem. For further personnel protection, a shatterproof, transparent shield should be placed between the cutting stem controls and the top head.

Steam test

Once the coke drum is headed-up, the next step is to free the vessel of air. This is done by blowing steam up from the bottom of the drum and venting it out the top. For a 20-foot diameter drum, a steaming rate of 30,000 lb/hr for 20 minutes is adequate to sufficiently free the coke drum of air. It is not necessary to check the contents of the coke drum for oxygen. To expedite this step, each coke drum should have its own atmospheric steam vent.

Insufficient steam purging has been found to result in rapid degradation of the refinery's circulating amine:
- Oxygen left in the coke drum mixes with coker wet gas.
- The oxygen reacts with mercaptans in the wet gas to form carbonylsulfides (COS).
- The COS reacts with amine to form heat stable salts which eventually decompose to carboxylic acids.
- The carboxylic acids (such as acetic acid) are corrosive and promote foaming in amine systems.

After 20 minutes, the vent is blocked in and pressure is allowed to build. It takes another 20 minutes until substantial pressure is reached. Both the top and bottom heads are now inspected for leaks. The decoking crew must accompany the operators during this inspection. If a head is found leaking, the decoking crew must quickly tighten the flange or replace the faulty gasket. Only when the steam test is passed should the decoking crew be dismissed. Note that this phase of the coke drum cycle requires two operators who are in close communication. Trying to do the job with one person will double the time required.

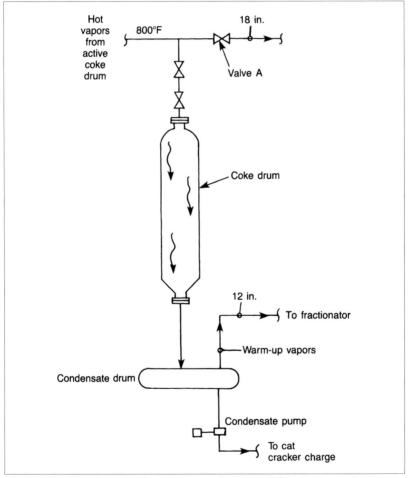

Fig. 2–13. Layout for coke drum warm-up (courtesy *Oil & Gas Journal*).

Warm-up

With the steam test completed the drum has been heated to approximately 200°F. Next, hot vapors from the drum being filled are cracked into the empty drum. For a 12-hour cycle the drum must be warmed to its preswitch temperature in one and a half hours (Figure 2–13).

Warming a drum to 700°F is certainly an aid to producing harder, better-quality coke. This temperature is an unreasonable target for a one-and-half-hour warm-up period. A temperature of 450°F–500°F can be achieved in this time frame if the warm-up facilities have been designed with this objective in mind.

The flow through the empty drum is typically 15% of the drum overhead vapor flow. When targeting for a 12-hour cycle (see table 2–2), approximately one-third of the hot vapor flow is diverted through the empty drum. The effect of this diversion of hot vapors to a cold drum is always noticeable on the coker fractionator. If one is heating a drum in one and a half hours and the fractionator is serving a single pair of coke drums, the effect on the fractionator operation will be striking. In particular, the pumparound duty

Table 2–2. Summary of 12-hour coking cycle.

	minutes
Switchover and start small steam flow	20
Small steam, line up quench water	30
Big steam	0
Bringing up little water rate	45
Full water rate	75
Coke cooling in place	60
Drain water and remove top head	90
Drop bottom, head and disconnect feed line	20
Raise skirt	10
Drill pilot hole	20
Retract stem and change head	10
Cut coke	75
Replace heads and reconnect food line	30
Steam purge	20
Pressure up and inspect flanges	20
Line up warm-up valving	15
Warm-up time	90
Slippage allowance	90
Total	12 hours

decreases by 35%–40%. Because many fractionator pumparounds are heat integrated with downstream equipment, upsets of such equipment can result. The temperature profile or top reflux rate of the coker fractionator drops, altering product fractionation. The overhead gas make briefly declines and may send the wet gas compressor into surge. Dynamic process control, such as a feed forward controller, and flexibility in pumparound heat removal can minimize these problems. If a fractionator serves several pairs of coke drums, its sensitivity to the warm-up cycle is markedly diminished.

To speed coke drum warm-up, the warm-up valve (shown in figure 2–14) must be partially closed to force the hot vapors through the empty drum. The vapor line connecting the condensate warm-up drum to the fractionation tower must be properly sized to handle the larger volume of vapor without a significant pressure drop. For example, if the main vapor line is 18 inches, the warm-up vapor line should be 12 inches.

Problems in Reducing Cycles

There are some very valid objections to reducing coking cycles in the manner described. These problems are related to coke drum life. There is no question that cutting coke drum cycles reduces drum life. A few historical observations on this subject are as follows:

- One unit operated 16-foot diameter coke drums on a 12-hour cycle for four years before drum replacement was deemed necessary. Previously, the drums had run on a 16-hour cycle for a decade. The drums are approximately 35 years old.

- A pair of 20-foot diameter drums were operated for one to two years on a 12-hour cycle without noticeable deterioration.

- Twenty-two-foot diameter drums experienced intermittent 15-hour cycles without signs of trouble.

- One unit operated routinely on an 18-hour cycle with 26-foot diameter drums.

All coke drums wear out in the sense that their diameters expand. This expansion, which is in part caused by the thermal cycling, eventually cracks the vessel wall. Such cracks, at least in carbon steel vessels, are a simple matter to repair and do not lead to a catastrophic failure.

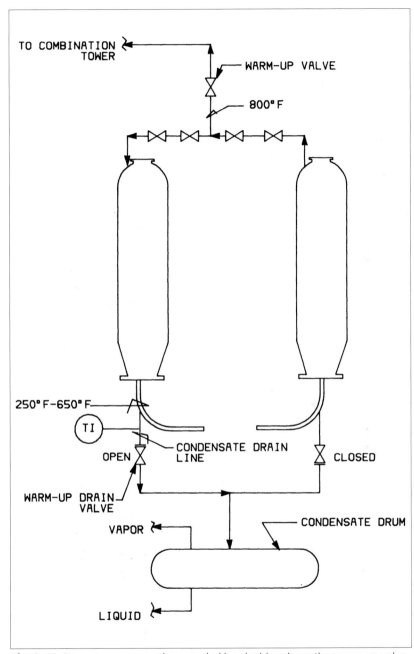

Fig. 2–14. Drum warm-up can be speeded by pinching down the warm-up valve.

Shortening the coke drum quenching period is the principal cause of reduced drum life. The combination of rapid temperature change and localized pressure surges due to the flashing of steam to water has been known to bulge a drum in just a few cycles. Cracking at the skirt attachment appears to be the most persistent problem. The usual solution has been to cut vertical slots at the top of the skirt to facilitate metal expansion and contraction. The drum support ring should sit on lubrite plates, and nuts on the anchor bolts should not be tightened. This allows for some movement of the support ring.[6]

The problem of drum cracking and bulging due to over-rapid quenching is exacerbated when making low (less than 10%) VCM coke. The observed deterioration in drum integrity, as correlated by growth in drum diameter, is ½–1 in/year for a carbon steel drum operating on a 12–16 hour cycle.

There does not appear to be any particular evidence that reduced warm-up cycle time increases the incidence of drum bulging or cracking.[7] The empty drum cannot really be warmed up by more than 200°F–250°F/hr. The rate of temperature change during quenching is much higher. The hundreds of tons of coke in a full drum interfere with the uniform cooling of the vessel with quench water. In contrast, the empty drum is heated relatively uniformly with hot hydrocarbon vapors. A word of caution: Rapid quenching during the production of needle coke is to be discouraged. The needle coke offers more resistance to the spread of quench water than does ordinary sponge coke. The dense structure of needle coke is the cause of such resistance.

Warm-up Condensate Recovery

A 1,000-T/D coker will generate roughly 500–600 B/D of warm-up coke drum condensate. In many refineries this relatively large volume is rerun as slop oil. This is a convenient but energy-wasteful operating practice.

The warm-up condensate is collected in the drum shown earlier in Figure 2–3. Initially, the condensate contains quite a bit of water and, hence, should be charged to FCCU feed tankage. The Conradson carbon and metals content of the condensate is in the range of heavy coker gas oil, and it will make acceptable FCCU feed.

When the condensate warms up to 350°F, it is safe to assume that it is water-free. It should then be charged to the combination tower, just above the heavy coker gas-oil trays, where it is fractionated into distillate and FCCU feed.

Green wax

The preceding process recommendation seems quite simple and logical. Unfortunately, the condensate produced toward the end of the warm-up cycle (i.e., when the drum is already heated to 650+°F) is a greasy, green-colored wax at ambient temperatures. If the process lines are not adequately heat traced, this wax sets up and plugs the lines.

When this happens, the shift operators may vent the condensate collection drum to a flare or even to the atmosphere because the normal vent to the combination tower is plugged. The condensate oil itself may be indirectly dropped into the sewer, where it mixes with coke fines to create an environmental mess.

Keeping track of orphan streams, such as coke drum warm-up condensate, is an unenviable but vital part of the operating engineer's job. The green wax when collected can, after all, be turned into gasoline.

In recent years, I have found an alarming tendency for major refinery operators to vent warm-up drum vapors to the atmosphere so as to speed coke drum cycles. The environmental implications of this are severe.

Troubleshooting Checklist for Delayed Coker Cycle Problems

Foamover Effects
High coke drum pressure
Plugging combination tower
  bottoms screen
Coked heater tubes
Coke lay-down on trays
Plugged blowdown system

Foamover Causes
Incompletely drained coke drum
Low overhead vapor temperature
Erroneous foam level indication
Failure of silicone antifoam system
Overfilling drums
Pressure surges
Foamover during steaming
Putting a drum on circulation
Caustic in coker charge

Coke Quality
High volatile combustible matter
Coke too hard
Drum overhead temperature
Time and pressure affect hardness

Shot coke
Needle coke

Long Coking Cycles
Watch full drum pressure
  during switch
Use little water instead of big steam
Add quench water to hold
  maximum drum pressure
Avoid plugged steam inlet nozzle
Steam purge through separate vent
Aggressive use of the
  warm-up valve
Plugged condensate drain line

Coke Drum Warm-up Condensate Recovery
Do not send to refinery slop
Wet condensate to FCCU feed
Hot condensate to
  combination tower
Green wax plugs system
Environmental effects
Venting to atmosphere

References

1. V. Mekler and M. E. Brooks, "New Developments and Techniques in Delayed Coking." Presented at API Division of Refining, New York, May 28, 1959.

2. NPRA Question-and-Answer Session on Refining Technology, Delayed Coker Operation, 1970, p. 51.

3. Ibid.

4. Ibid, p. 55 (Mr. Prosche).

5. R. W. Piazza et al., "Hydraulic Decoking-Design versus Cutting Time," ASME Publication 70-Pet-5, June 1970.

6. Heater Leigh, NPRA Question-and-Answer Session on Heavy Oil Processing, response to Question 48, 1982.

7. Ibid.

Delayed Coking Process

I t may be true that the wheels of progress turn slowly, at least insofar as refinery resid processing is concerned. The delayed coker—the historic refinery garbage can—is still the preeminent route for turning low-value resid into high-value distillates and gasoline.

Chapter 2 reviewed the problems and techniques associated with the coking cycle: coke drum filling, cooling, coke cutting, and drum warm-up. This chapter describes troubleshooting techniques relevant to the continuous aspect of the process:

- Combination tower fractionation
- Optimizing heat recovery
- Extending coking heater run lengths
- Wet gas compressor limitations

A flowsheet for a delayed coker is shown in figure 3–1. The process description corresponding to this flowsheet is reviewed in chapter 2.

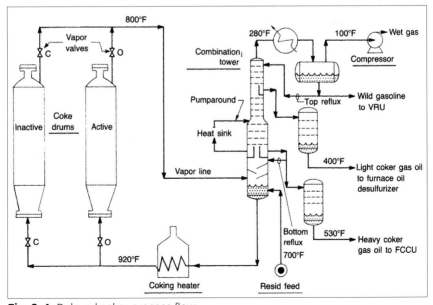

Fig. 3–1. Delayed coker process flow.

Coking Heater

The coking heater is the critical item of equipment to watch if long run lengths are going to be achieved. Since the heater outlet temperature is above 900°F, thermal degradation of resid to coke proceeds very rapidly. Lay-down of coke inside the heater tubes will cause local overheating of the tube walls. Eventually, these hot spots will bulge out, burst, and release oil inside the heater firebox. In the resulting fire, severe damage will be inflicted to the heater (see chapter 15).

A properly designed and operated coking heater will run for a year without any appreciable buildup of coke inside the tubes. Another heater can accumulate a thick coke deposit in 10 minutes. Consider the factors that prevent the coke that is formed in the tubes from laying down on the tube walls.

Most importantly, a continuously high mass velocity (pounds of resid per square foot of tube cross-sectional area per second) is the overriding factor in preventing the curtailment of the heater's operation because of coke lay-down. A well-designed heater will maximize this velocity consistent with the available pressure head. Unfortunately, excellence in design will not compensate for poor operations.

Mass velocity and heat flux

A well-designed coking heater will have a mass velocity ranging between 300 and 450 lb/ft²/sec. Higher mass velocities will create too high a front-end pressure at the heater inlet. A lower mass velocity will permit inadequate fluid shear at the tube wall to keep coke deposits from sticking to the tube ID.

The average radiant heat flux in coker service is calculated on the tube inside surface area for those tubes in the radiant firebox. A well-designed delayed coking heater will have a heat flux of 9,000 BTU/hr/ft². Over the years, I have learned through bitter experience that operating a coking heater with a 11,000 BTU/hr/ft² flux rate will produce heater run lengths from one to four months, depending on the thermal sensitivity of the charge stock, regardless of the mass velocity. Note that the flux rates mentioned above are based on the inside tube area.

Similarly, running at a mass velocity of less than 150 lb/ft²/sec will result in short heater run lengths even though the heat flux is only 7,500 BTU/hr/ft².

Feed interruptions

Operating experience indicates that rapid coking of the heater's tubes is not often caused by changes in resid feed quality, high outlet temperatures, or overfiring. Feed interruption incidents are the primary reasons for short heater run lengths. In particular, a stuttering feed interruption will quickly put a coking heater out of action.

A pump just starting to lose suction exhibits a stuttering discharge flow pattern. This on-again, off-again flow rapidly leads to tube coking. When the flow stops, the resid in the tubes thermally cracks and the resulting coke deposits in the tubes. Thirty seconds later, flow is briefly restored. The tubes refill with resid, the pump loses suction again, and then the cycle is repeated. The heat stored in the furnace's refractory walls can keep the tubes hot for quite a while, even with the firing rate cut back.

To prevent tube coking from a stuttering flow, the instrumentation shown in figure 3–2 is suggested. This arrangement automatically trips the coking heater charge pump and steam purges the tubes without waiting for operator intervention. Field observations have shown that a total and sudden loss of feed (such as are encountered during a power failure) rarely results in appreciable tube coking as long as steam is available to sweep out the heater tubes.

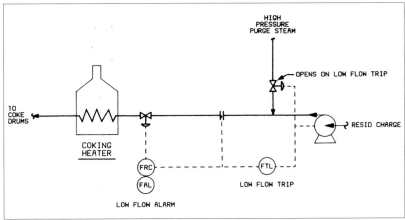

Fig. 3–2. Protecting the coking heater against feed interruptions.

Shift operators may repeatedly attempt to restart resid flow to a heater. After all, no one likes to have the unit shut down on one's shift. Operators need to be schooled not to do this. More often than not, these attempts result in coking up the heater. A good, steady, high-pressure supply of liquid flowing through the coking heater tubes is the only way to suppress coke lay-down inside the tubes.

Whenever feed is lost to a coking heater, the operators should be trained to make sure the heater outlet temperature is below 930°F (with steam flowing through the heater passes) before flow is restored. A short feed interruption will initially result in a heater outlet temperature of greater than 1,000°F due to radiation of energy from the refractory walls.

Also, try to avoid running a steam-turbine-driven pump as a coker heater charge pump. If the pump's steam supply comes from the same steam header, fluctuations in the steam header pressure (such as when steam is cut into a full coke drum) can cause a brief but deadly slippage in feed flow to the heater.

Velocity steam

Most coking heaters have several thousand pounds per hour of high-pressure steam injected into the heater tubes. The velocity steam should increase the linear velocity of resid in the tubes, shorten oil residence time, and thus reduce the thermal soaking time to which the resid is exposed. This ought to reduce the formation of coke in the heater tubes. In practice, there seems little correlation between the amount of velocity steam used and the rate at which coke builds up in the tubes.

Apparently, the steam flows at a high velocity through the center of the tubes, while the liquid resid creeps relatively slowly along the tube walls. This is called annular flow. The principal benefit from velocity steam may be to help sweep resid out of the tubes during a feed interruption. At mass velocities greater than 400 lb/ft²/sec, most refiners do not use velocity steam.

Sodium

Caustic is sometimes injected into a crude unit charge to control corrosion. The sodium salts formed remain in the crude unit bottoms and eventually are charged to the coker along with the resid.

Sodium speeds the thermal cracking of resid; sodium also deposits inside the coking heater tubes. Both factors tend to reduce heater run length. After burning out coke from inside the tubes (steam-air decoking), remember to water wash the tubes to remove residual sodium salts. You can see them as white deposits inside the outlet elbows.

Light resid

A barrel of 12°API resid takes perhaps 10% more furnace duty than a barrel of 8°API resid for identical heater inlet and outlet temperatures. Thus, a heater maintaining a constant outlet temperature will start firing much harder when it receives a slug of light feed. If overfiring results, the heater run length will be abbreviated.

Foamovers

Carryover from the coke drum will plug the heater charge pump suction screen. As the pump loses suction, flow will fall off and the stagnant resid in the tubes will coke. This difficulty is discussed in chapter 2 on delayed coking cycles. Trip off the heater charge pump on low flow (see figure 3–2).

Parallel passes

A coking heater will usually have two or four parallel tube passes. If an operator inadvertently pinches back the flow to one pass, this pass will be quite prone to coking up. Again, low mass flow through furnace tubes promotes coking and shortened heater run lengths.

In summary, once a coking heater starts plugging up with coke deposits, it is futile to try to keep going at reduced feed rates. Coking is accelerated because of the low mass velocities, and downtime is lengthened by the excessive coke accumulations.

Steam-air decoking

A typical coking heater run length is nine months to two years. At the end of the run the heater must be taken off-line for steam-air decoking. However, the very name "steam-air decoking" is misleading. Most of the coke (90+%) should not be burned off with air at all; it should be spalled by thermal shocking the tubes.

The idea behind steam spalling is simple. The coke deposits (typically ¼ inch thick) inside the coking coils have a small coefficient of thermal expansion compared to the tubes. The tubes are heated with a few burners to a high temperature (about 1,100°F). They are then rapidly cooled with steam. The tubes shrink faster than the coke and hence crack the coke off the walls. The high-velocity steam blows the coke out of the coil as it flakes off the tube walls.

The burn part of steam-air decoking is a final clean-up step to remove residual traces of carbon. If you can actually follow the progression of the burn by observing a red ring moving along the coil length, then this is an indication that sufficient time and care was not taken to spall the coke. Actual burn time should be six to eight hours. Burning is deemed to be complete when:

- The quenched effluent from the coil turns from black (indicating coke) to red (indicating iron oxide scale).
- The oxygen in the effluent lines out at 18%–20%.
- The CO_2 in the effluent lines out at 1%–2%.

Online spalling

The reader might very well wonder why not just spall a heater pass with steam without taking the furnace off-line? Why not simply blow the coke out of the coil with steam into the coke drum, while the other coil is charged with resid in the normal way? The answer is, this is done routinely in many refineries. It is called "on-line spalling." One typical procedure for a coker with two 4-inch heater passes is:

1. Start with steam and resid in the coils.
2. Five hours after switch time shut down half the burners.
3. Reduce the heater outlet to 760°F and pull the feed out of one pass but leave the steam going.
4. Raise the other pass outlet so as to maintain a combined outlet temperature of 900°F.
5. Use 5,000 lb/hr of steam in the coil to be spalled or keep the coil inlet pressure at 250 psig.
6. Slowly raise the coil outlet temperature to 1,100°F (about 200°F/hr).
7. Hold at 1,100°F for three hours. Raise the steam rate from 5,000 to 8,000 lb/hr to maintain a constant coil inlet pressure as the coke spalls.

8. Rapidly drop the coil outlet temperature to 850°F and keep increasing the steam flow to keep the coil inlet pressure constant.

9. After five minutes, rapidly return to 1,100°F–1,150°F coil outlet and hold at this temperature for two hours.

10. Do not switch coke drums during or just after steam spalling to avoid coke drum foamovers.

This procedure is based on ⅜-inch minimum wall thickness of 9% chrome tubes.

The one danger in frequent steam spalling is erosion to the last few tube return bends. These bends should be retrofitted with mule-ear return type plug headers which are highly resistant to erosion.

Wet Gas Compressor

Vapors from the coke drum must pressure their way through to the combination or fractionator tower reflux drum. Any restriction to their flow will increase the operating pressure of the coke drum. To avoid exceeding the coke drum relief valve pressure, some operators vent the reflux drum to a flare. This makes it appear as if the wet gas compressor is limiting the resid feed rate. Most often, though, the problem lies with upstream pressure drop.

A fouled overhead condenser

I was called to Axxoco Oil Co. to help increase the capacity of their delayed coker. The refinery was bottlenecked by resid processing capacity so that any incremental increase in coking capacity would permit the plant to increase crude runs substantially.

After explaining the need to increase coking capacity, the plant manager turned me over to the process manager. Apparently wishing to get an unbiased opinion on why coking capacity was limited, the process manager turned me over to the process engineer for the delayed coker.

The process engineer indicated that the capacity limit on the coker was coke drum pressure. He indicated that he was busy with an economic analysis of the coking unit and couldn't spend much time with me on the problem. He did say that incremental coking capacity at the time was worth $5.17/bbl. He suggested I work with one of the day-shift foremen.

Before contacting the day-shift foreman, I reviewed the unit log sheet (figure 3–3). It was apparent from the log sheet that the high coke drum pressure was caused by an excessively high pressure loss upstream of the wet gas compressor, combined with limited capacity of the compressor to handle the uncondensed coker vapors.

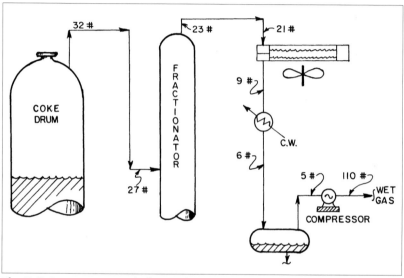

Fig. 3–3. Pressure survey indicates the need to clean fin fan air cooler.

I then met with the day-shift foreman on the unit, and we performed a pressure survey on the unit. The pressure survey revealed the location of the problem. The pressure drop across the tube side of the coker fin fan overhead condensers was excessive, causing low suction pressure at the compressor. The day-shift foreman was puzzled as to why the company had hired an outside consultant to run a simple pressure survey, but he also indicated that no one in the company had done one.

I then asked the day-shift foreman if he had any thoughts on the cause of the high pressure drop across the tube side of the condensers. He indicated he really didn't know, but he wondered if it could have something to do with coke fines that accumulated in the fin passages of the condenser tube bundle.

Further inspection revealed that the underside of the condenser tube bundle was encrusted with a thick layer of coke fines and dirt. I showed the foreman how to use strips of paper under the fin tube bundle to locate areas where the bundle was fouled enough to restrict air flow through the bundle.

I also demonstrated that temperature measurements across the top of the fin tube bundle would indicate hot areas where the bundle was fouled. These areas corresponded to the fouled areas indicated by the paper strips.

To inform the foreman further about the problem, I explained how fouling of the fin tube bundle decreased the heat transfer in the condenser, limiting condensing capacity. I explained how this would increase the volume of uncondensed wet gas flowing through the final tube pass, resulting in higher pressure drop and increased load on the compressor.

The high pressure drop had the effect of reducing compressor capacity by about 20%, resulting in a reduced coker charge rate of about 1,400 B/SD.

The solution to the problem of limited coker capacity was solved by jet blasting the fin tube bundle.

This case is a perfect example of why process engineers must go out to the unit frequently and discuss unit operations with the shift operators to keep units operating at peak efficiency. The problem could never have been solved solely by process engineers examining unit log sheets.

In this case, because it appeared that process engineers and operators rarely discussed unit operations and problems, the operators weren't informed enough to help solve the problem.

Vapor-line restrictions

Coke that builds up in the coke drum overhead vapor line is responsible for most back-pressure incidents. Operators find that tearing the insulation off these lines slows the rate at which coke deposits. A better method is to inject a heavy slop oil quench, as shown in Figure 3–4, into the vapor line to retard coke formation.

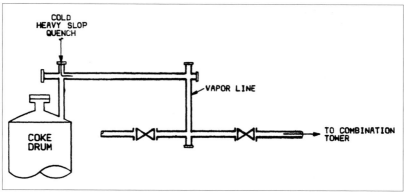

Fig. 3–4. Preventing coke buildup in the coke-drum vapor line.

The quench rate is about 3% on resid feed, or enough to lower the drum vapor temperature by around 15°F. In addition to reducing coke buildup, this is also an energy-free method to dispose of refinery slop oil. An added bonus is that shift operators report that slop injection makes the overhead vapor valves easier to turn.

Other causes of excessive coke drum back pressure are badly fouled combination tower overhead condensers, partially plugged trays, or insufficient tower pumparound heat removal. Fouled condensers and lack of pumparound heat removal overload the wet gas compressor. Plugged trays are best identified with a pressure drop survey.

Wet gas compressor rotor fouling

A variety of solids such as corrosion products, sublimed ammonia salts and coke fines can accumulate on a centrifugal wet gas compressor rotor. One indication of a problem is a high amp load on the motor drive when steam is first cut into a full coke drum.

The rotor fouling causes a loss in the polytropic compressor efficiency. To measure the *relative* polytropic efficiency at two points in time, calculate:

$$\text{Eff.} \cong \frac{(P_2/P_1-1)}{(T_2-T_1)}$$

where:

P_2 = Discharge pressure, psia

P_1 = Suction pressure, psia

T_2 = Discharge temperature, °F

T_1 = Suction temperature, °F

Eff. = A number proportional to compressor efficiency.

Frequently, the fouling deposits are found on the first stage rotor and the loss in polytropic efficiency due to fouling should be most notable on this first-stage rotor.

Combination Tower

A delayed coker combination tower is similar in design and function to an ordinary crude distillation tower; heat is removed and products are fractionated. Referring to figure 3–1, four products are made: wet gas, unstabilized naphtha or wild gasoline, furnace oil, and heavy gas oil.

The most common problem encountered in the combination or fractionation tower is tray damage. On start-up or during short unit outages, pockets of water can form. When this water contacts hot oil, it flashes, and the resulting pressure surge will upend trays. Corrosion damage to the furnace oil drawoff trays is also a possibility. The diagnosis of these problems as well as other fractionation difficulties is identical to those found on crude fractionators (see chapter 1).

On one coker, the combination tower trays were repeatedly upset due to water accumulating in the tower during short unit outages. After considerable investigation, the source of this water was found. Purge steam was being used under the vessel relief valves to prevent the relief valve inlet lines from coking up. While the unit was out of service for several hours, it would cool off. The steam condensed and formed potentially explosive pockets of water.

Explosion-proof trays

Coker fractionators are subject to damaging pressure surges due to accidental flashing of water. The pressure surges results in the tray decks "ripping" away from the tray support rings.

Actually, the tray decks are not ripped away from the support ring. Inspection of damaged trays indicates the decks have been bent at the ring. The trays are not bolted to the tray ring; they are clipped to the support ring (see figure 3–5). In the course of bending, the tray pulls the tray clips away from the support ring.

Ordinarily, 12-gauge tray decks have little structural strength. The real strength of a tray deck is located in the support lip shown in figure 3–5. Unfortunately, the support lip is not typically attached to the tray ring. Hence the force of a pressure surge must be borne by the 12-gauge tray deck where the tray is clipped to the ring. As 12-gauge steel does not have much strength to resist the bending moment developed against the tray by the pressure surge; the tray is bent at the tray ring.

To avoid this problem of intrinsic tray weakness, the support lips (also called "integral trusses") are attached to the vessel wall. As shown in figure 3–5, a bracket is welded onto the vessel wall below the ring. The support lip is attached to this bracket with a "shear clip." Now the trays' integral truss is attached to the vessel wall.

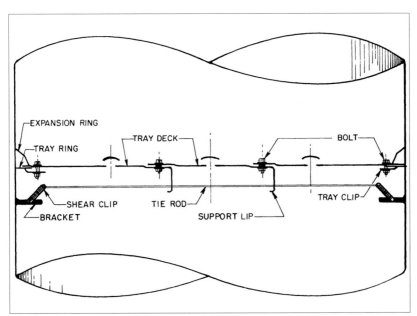

Fig. 3–5. Explosion-proof trays.

The shear clip must be designed to break away from the vessel wall at some predetermined stress. Otherwise, a powerful pressure surge could deform the wall of the vessel itself. However, if the shear clips can be designed to be strong enough to withstand a ΔP across the tray deck of about 0.6 psi (i.e., about 2 feet of hydrocarbon liquid) the downcomers would blow clear before the shear clips would fail.

Hence, the downcomers would act as a partial safety vapor bypass and protect the tray decks from damage. Back-to-back trays and bolted-in trays are other methods to protect tray decks.

Energy savings

The hot coke drum vapors are a valuable source of high-level heat. This energy, which is partially recovered in the combination tower heavy gas-oil pumparound section, can be used for generating steam, preheating coker feed, or reboiling a gas plant. The higher the temperature level at which this heat is made available for recovery, the greater is its potential value. This is the second law of thermodynamics in action.

Figure 3–6 illustrates how one combination tower was retrofitted to increase the pumparound draw temperature by 45°F. The heavy coker gas-oil feed to the steam stripper was modified to be withdrawn from the tower in two streams. The 3,100 B/D of relatively light, cool gas oil was trapped out and drawn off the tower before it could fall into the pumparound section. Computer calculations predicted that segregating this lighter gas-oil cut from the pumparound section would increase both the pumparound draw and return temperatures, with only a 3% reduction in pumparound heat duty.

After the retrofit, opening the drawoff line below tray 10 (refer to figure 3–6) increased the pumparound draw temperature from 575°F to 620°F without any noticeable effect on duty.

Coke Drum Cycles Affect Combination Tower Operation

At steady state, troubleshooting coker combination tower problems is no different than for a crude unit. Unfortunately, the combination tower heat and material balance are routinely upset due to the switching, steaming, and warm-up of the coke drum.

If a combination tower is serving only one pair of coke drums, the effect of the drum cycles will be quite noticeable on the tower operation. In particular, heat duties on the pumparound and top reflux will drop about 15% during coke drum warm-up. This can become a big problem if the operators are behind schedule and attempt to speed the warm-up time of the empty drum. By diverting more than the usual amount of hot vapors from the filling coke drum into the empty drum, the vapor feed to the tower can drop by 25%.

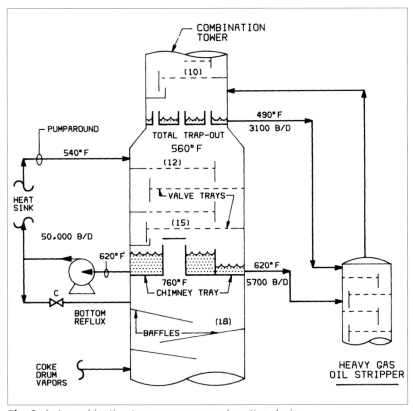

Fig. 3–6. A combination tower pumparound section design.

To avoid adversely affected naphtha and furnace oil end points, the following scheme is necessary:

1. Reduce the tower top reflux flow; maintain the tower top temperature.

2. Control the furnace oil drawoff rate to hold this drawoff temperature.

3. Reduce the pumparound circulation rate to control the pumparound drawoff temperature.

The only operating problem introduced by this control scheme is a rather drastic reduction in pumparound duty. If the pumparound heat is used to generate steam, this is not much of a debit. However, when adjacent distillation towers are reboiled with the hot pumparound oil, the reduced duty creates a problem.

On one gas plant, this difficulty was overcome by providing a supplementary trim steam reboiler. Referring to figure 3–7, the steam flow to this reboiler would increase when a coke drum was being warmed. Then when pumparound duty increased again, about an hour after switching drums, the steam reboiler would automatically cut back to keep the combination tower in heat balance.

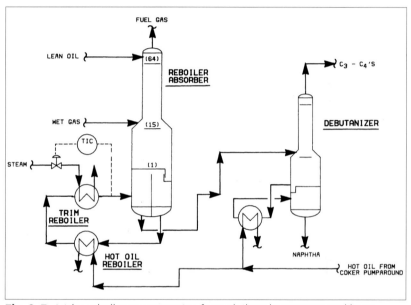

Fig. 3–7. A trim reboiler compensates for variations in pumparound heat.

For proper combination tower control, it would appear that a feed-forward computer-controlled system is required since conventional automatic control seems to lag behind the heat-loss effects caused by coke drum warm-up and switch-over. Many refiners have implemented such advanced control schemes with good success.

Minimizing coke yields

Table 3–1 summarizes rough rules of thumb for minimizing coke yields and maximizing liquid yields. There are no generally accepted values in the industry for these parameters.

Table 3–1. Coke drum yields.

Rules of Thumb for Delayed Cokers
1. Each 8 psi reduction in coke drum pressure reduces coke yield on feed* by 1.0 wt%.
2. Each 8 psi reduction in coke drum pressure increases liquid yields by 1.3 vol% on feed.
3. Each 10°F increase in coke drum vapor-line temperature decreases coke yield by 0.8% on feed.
4. Each 10°F increase in coke drum vapor-line temperature increases gas and distillate by 1.1 vol% on feed.
5. Each decrease of 1 wt% VCM requires an increase of 7°F to 9°F coke drum vapor-line temperature.
6. Reducing recycle by 10% on feed reduces coke yields by 1.2% on feed.
7. Reducing the virgin gas oil content of coker feed by 10% reduces coke yield by 1.5 wt% on feed.
8. Reducing coke yield by 1.0 wt% on feed raises liquid yields by 1.5 vol% on feed.
9. Decreasing cycle time by 6 hours increases coke VCM by 1.0 wt%.

*The term "feed" refers to fresh resid.

Troubleshooting Checklist
for Delayed Coking Process

Coking Heater Run Length
Low mass velocity
Feed interruptions
Stuttering resid flow
Loss of velocity steam
Sodium salts
Light resid
Foamovers

Wet Gas Compressor
Coke deposits in vapor line
Quenching vapor line
Plugged trays
Condensers fouled
Low pumparound duty
Rotor fouling
Air cooler plugging

Combination Tower
Fractionation problems same
  as crude unit
Water pockets upset trays
Check steam under relief valves
Maximizing pumparound
  drawoff temperature
Instability due to coke drum cycles
Explosion-proof trays
Low-temperature difference

Amine Regeneration and Scrubbing

O n July 23, 1984, 16 men and one woman were killed at Unocal's (now Citgo) Chicago refinery. An amine scrubber, being used to extract H_2S from a liquid hydrocarbon stream, experienced a weld failure. The vessel wall parted at a circumferential weld near the hydrocarbon feed inlet. The top portion of the vessel rocketed out of the refinery. A cloud of propane, propylene, and butylene vapors engulfed the unit and detonated. One cause that contributed to the vessel failure was circulating black, corrosive monoethanolamine (MEA), overloaded with hydrogen sulfide, through a vessel that had been fabricated without stress relieving the welds. We normally think of corrosion as a slow, if steady, process. Given the right environment, however, corrosion in the form of hydrogen-assisted stress corrosion cracking can cause a weld to crack in a matter of a few minutes.

Dirty amine is both a symptom and a cause of amine system failures. Circulating black amine will eventually curtail the refinery's ability to recover sulfur from process streams.

Figure 4–1 is a process flowsheet showing how amine solution is circulated to various refinery scrubbers to absorb H_2S. The lean amine chemically combines with H_2S (and unavoidably some CO_2) in the scrubbers. The resulting rich amine is stripped in the regenerator. Released acid gases (H_2S and CO_2) are charged to the sulfur recovery plant.

Amines are an organic base. When mixed with water, they turn pH paper blue. The two most common forms of amine used are monoethanolamine (MEA) and diethanolamine (DEA). MEA is the most powerful and reactive. DEA, MDEA, and a host of other less-reactive amines are also used in the industry. Unfortunately, MEA is the most corrosive of the amines. It is this corrosive aspect of amine solutions that makes the operation of amine systems a challenging job.

The objective in operating an amine system is to maximize H_2S recovery while keeping the solution inside the pipes and vessels and out of the sewer. Refinery effluent-treatment plants have a limited

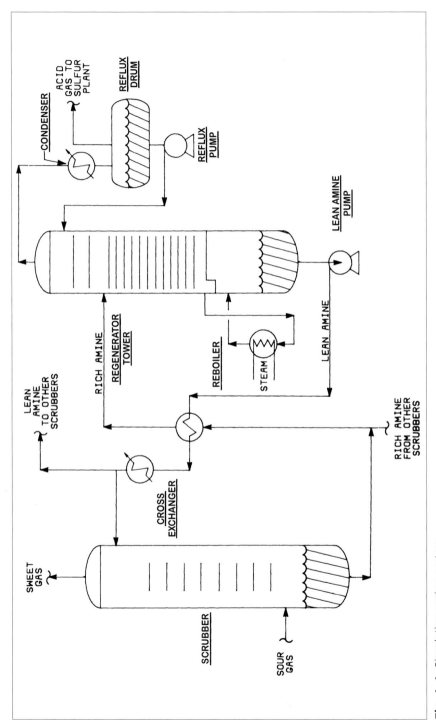

Fig. 4–1. Circulating amine system.

capacity to digest amine. Even small amounts in a refinery's outflow will violate environmental nitrogen limits.

Dirty Amine

Symptoms of a dirty, corrosive amine system are:

- Carry-over of amine from the scrubbers
- Dilution of the amine system with water due to reboiler leaks
- Plugging of instrument taps with particulates in the amine
- Loss of amine to the sewer because of leaks
- Rich amine leaking into lean amine in the cross exchanger

Problems related to improper operation of the regenerator are:

- Too much H_2S in sweet gas
- Excessive energy use in the regenerator reboiler
- Flooding in the regenerator tower

The operating engineer troubleshooting an amine system should first draw a sample of lean amine into a glass container. If the unit is in trouble, the amine will be dirty. Determining what has caused the buildup of dirt and what can be done to clean up the amine is the main subject of this chapter.

The seeds of destruction

Take a glass quart sample bottle and fill it with unfiltered lean amine. You can rate your amine quality as follows:

- **Bright and clear.** Amine is in excellent shape. A yellow tinge indicates the presence of iron, but this is of no great consequence.
- **Grayish cast.** Solution is a pale, dull gray. Objects can be seen through the bottle without difficulty. This is okay; however, do not let the amine solution get dirtier.
- **Translucent black.** Objects can barely be seen through the bottle. Upon standing 10 minutes, a small amount of sediment is visible. You are now in trouble. Erosion-corrosion is generating particulates faster than they can be removed.
- **Opaque black.** Give the bottle a good shake. If you can't see through it, start polishing up your résumé. You will notice a lot of particulates settling in the bottle. A similar fouling process is proceeding in refinery vessels.

If the amine looks brownish, air is getting into the system. Oxidized amine is corrosive. A blue tinge in the lean amine indicates the presence of cyanides. The cyanides are coming from the FCCU wet gas. They should have been scrubbed out of the wet gas with an aqueous-phase polysulfide wash. These cyanides will promote corrosive failures in amine regenerator overhead system and hydrogen blistering or cracking in piping and vessels that contact the lean amine.

Submit a sample of unfiltered lean amine to the lab to determine quantitatively the wt% particulates. A good solution should be less than 0.01 wt% (i.e., 100 ppm).[1]

Dirty Amine Ruins Operation

Running a sulfur recovery operation with dirty amine is analogous to deficit spending. You are borrowing against the future, but the day of reckoning will surely come.

The insidious aspect of circulating dirty amine is its erosive nature. Carbon steel is corroded by clean amine. However, the sulfide products of corrosion stick to the metal surfaces and inhibit further attack. Particulates in the circulating amine erode this protective layer. New metal is exposed to corrosion and then more particulates are generated as the corrosion-erosion cycle perpetuates itself. This environment is manifested by several signs.

- **Foaming.** Dirt reduces the surface tension of liquids. Particulates will cause amine to foam. Foaming in regenerators results in high amine concentrations in the regenerator reflux water. Foaming in the scrubbers causes amine to be carried overhead with the hydrocarbon being scrubbed.

- **Plugged instrument taps.** Flow rates in dirty amine systems tend to be erratic. Orifice taps on flow meters and level taps on float chambers often plug. Level control in the bottom of the scrubbers becomes unreliable and massive carry-overs of amine are frequent.

- **Condenser fouling.** Rich amine regenerator feed splashes overhead. Particulates accumulate in the regenerator condensers, heat transfer is impaired to a certain extent, and the reflux temperature rises.

- **Reboiler tube failures.** Enhanced corrosion rates are most evident in the regenerator reboilers. Dirty amine has caused tube failures after six months of service.
- **Filter plugging.** The dirtier the amine, the shorter the filter life. The shorter the filter life, the dirtier the amine. For really bad amine, filter pressure drop can increase by 1 psi per hour.
- **Regenerator flooding.** Eventually, dirty amine plugs the regenerator trays, and the massive carryover of liquid which follows shuts down the sulfur plant (chapter 14).

Cleaning Up Amine

The iron sulfide particulates circulating in a dirty amine system have built up from a combination of factors. Foremost among these is inadequate filtering.

There are three common types of filters: rotary precoat filters, cartridge filters, and stacked paper plates. In practice, paper plates are best. In particular, Sparkler stacked filters are easy to maintain. Operation of rotary precoat filters is too complex for many refinery applications. Cartridge filters are good, except that the cartridge cost can be high when frequent changeouts are necessary. Stacked paper plate filters are quite simple: the paper is discarded after each use. On one unit it was estimated that 1,000 pounds of particulates had accumulated in one 50-plate Sparkler filter. In 2009, probably 90% of filters are cartridge types.

The trick to successful filter operation is always to filter 10% of the circulating amine. In practice, two filters are piped up in parallel. When one filter cannot pass 10% of the flow without exceeding 50-psi pressure drop, switch filters. Then immediately change the cartridges in the spent filter.

For a clean, well-designed unit, switching on a one-month cycle is typical. When amine is black, one may be switching filters every day for a month before solution cleanliness is improved. It will take two people three hours to renew a large paper-plate unit and one person two hours to change cartridges. The operating engineer needs to convince the unit superintendent of the necessity to invest in this maintenance manpower to change filters regularly on a priority basis.

The purpose of the charcoal filter is not to remove particulates. The charcoal is there to absorb heavy hydrocarbons or soluble materials that have contaminated the amine. Only filtered amine should be charged to the charcoal filter. The effluent from the charcoal filter should be refiltered in a sock-type filter to remove charcoal particulates or be recirculated back to the primary lean amine filter.

Corrosion inhibitors

To clean up amine, the rate of particulate generation must be slowed and the iron sulfide already in the amine filtered out. Reducing corrosion rates is the way to do this.

In one Midwestern refinery, the amine solution was thick with iron sulfide. Within two weeks, the solution was restored to a light gray by constant filtering, maximum reclaimer operation (see following section), and use of a corrosion inhibitor. Twenty-five percent MEA was being circulated. A high molecular weight film-forming amine corrosion inhibitor was slugged into the regenerator reboiler inlet. Corrosion rates, as monitored by a probe on the reboiler outlet, fell from 50 mil/yr to 3 mil/yr.

Reboiler corrosion

Hot rich amine eats carbon steel tubes. To minimize corrosion rates, the operating engineer must be sure that incompletely stripped rich amine is not reaching the reboiler.

If substantial reflux is provided, you can be sure that the amine is well stripped before it is drawn off the reboiler trapout tray. The regenerator reflux rate (lb/hr water) should be 10%–30% of the reboiler steam rate. To double-check stripping efficiency, pull samples of lean amine and reboiler feed. Both should have the same H_2S concentration. Remember, H_2S must be stripped out of the regenerator trays, not in the regenerator reboiler. This is important enough to repeat: keep a decent reflux rate in the regenerator to prevent reboiler tube corrosion.

Make sure that the steam condensate level in the reboiler channel head is below the bottom row of tubes. Also check that the reboiler steam supply has been completely desuperheated with clean steam condensate. Preventing water submergence of tubes and using desuperheated steam results in cooler reboiler tubes. This slows down corrosion and particulate generation.

Regenerator feed temperature

Rich amine is heated in the cross exchanger (see figure 4–1). Overheating causes acidic vapors to flash out of the rich amine. These wet acidic vapors will corrode the cross exchanger. Keep the rich amine below 190°F by diverting the lean amine around the exchanger.

Reclaimer Operation

Only MEA solutions can be reclaimed at ordinary regenerator pressures. Twenty-five percent MEA has been successfully used in refinery applications when such use has been accompanied by a consistent reclaimer operating program.

MEA is the most reactive of all amines, and hence most subject to degradation. For MEA service, the reclaimer can be the most important piece of equipment on the unit. Figure 4–2 shows a properly operating reclaimer. MEA exposed to oxidizing agents (COS, SO_2, O_2) reacts to form soluble products of degradation. These products are corrosive and must be removed from the circulating solution.

The reclaimer is simply a pot boiled with a steam coil. Vaporized MEA solution is vented back to the regenerator. The products of degradation remain behind. When soda ash is added to the reclaimer, the MEA tied up in the products of degradation are released. Reclaimer duty should be maximized consistent with available heat-transfer capacity. However, don't overheat the amine—300°F is the maximum temperature limit.

Reclaimer duty may be defined as pounds per hour of solution vaporized. This is equal to the steam flow to the reclaimer. If the steam meter isn't working, the operating engineer can measure the duty as follows:

1. Mark the amine liquid level on the reclaimer gauge glass.
2. Block in the amine feed to the reclaimer; keep the steam flow constant.
3. Measure the time it takes for the gauge glass level to fall 4 inches.
4. From the reclaimer geometry, calculate the volume of liquid boiled off in lb/hr. This is the reclaimer duty.

A reclaimer duty equal to 1% of the regenerator feed is acceptable. A more rigorous method is to analyze the lean amine

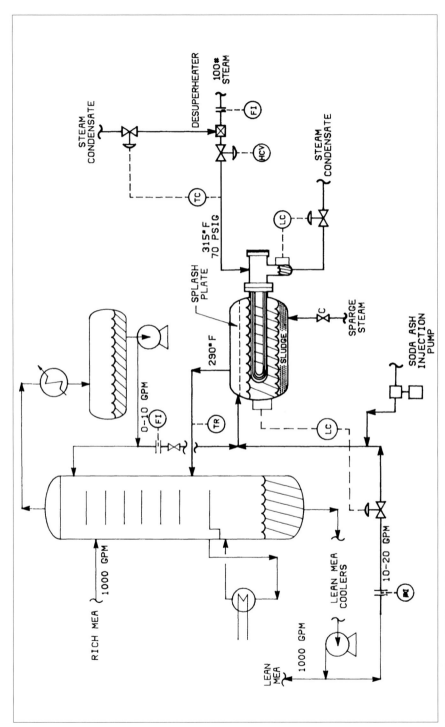

Fig. 4–2. Proper MEA reclaimer operation.

for thiosulfate. The thiosulfate level is a measure of the products of degradation in amine. Keep its level below 0.05 wt%.

As the amine reacts with various oxidizing agents, it forms heat-stable salts. This degraded amine cannot absorb H_2S. In any amine system, the concentration of heat-stable salts should be kept well below 10% of the circulating amine strength. Not only will these salts promote foaming, they will also degrade to carboxylic acids such as acetic acid. These acids promote hydrogen-assisted stress corrosion cracking in high-pH environments. Adding caustic directly to the circulating amine will not remove heat stable salts. More than a few percent of caustic in the amine, however, will promote fouling, plugging, and foaming.

Washing the reclaimer

A reclaimer first put online will have a vapor outlet temperature of about 270°F. When this temperature reaches 295°F, the reclaimer should be taken off-line and cleaned. You can delay removing the reclaimer from service by adding steam or water to the shell side. Regenerator reflux water is the usual source of water.

The water dilutes the amine salts in the reclaimer, thereby reducing the boiling point. If regenerator reflux water has been used, the amine system water balance remains unchanged. Unfortunately, the pounds per hour of dilution water added must be subtracted from the reclaimer duty. When the reclaimer duty is 40%–50% of its clean duty, it should be washed. The heavy ammonia odor emitted from the reclaimer as it is steamed and washed is normal.

How much soda ash to use

Soda ash or caustic may be used interchangeably to neutralize the acid products of degradation in a reclaimer. To calculate the moles per day of Na_2CO_3 or NaOH required, measure the reclaimer feed rate (assume it is equal to reclaimer duty in lb/hr). Multiply this by 24 hours. Next, submit a sample of lean amine to the laboratory for thiosulfate concentration analysis. From the volume and normality of KOH used in the lab to titrate for thiosulfate, calculate the equivalent pounds of NaOH or Na_2CO_3 needed to titrate one pint of lean amine. Soda ash will neutralize twice as much thiosulfate per mole as does caustic. Finally, multiply the reclaimer duty (lb/day) by the previously calculated value.

Add the required amount of soda ash or NaOH on a batch basis once per day. A typical value for a refinery unit circulating

1,500 gpm is 10–40 lb/day of soda ash. Precise regulation of soda ash addition is not necessary.

Extending reclaimer tube life

Reclaiming MEA is a tough service for carbon steel tubes. Certain stainless steels (S.S.) cannot be used in most refinery applications because chlorides concentrate in the boiling liquid and cause stress corrosion cracking (see chapter 22). (316 S.S. is one of the better materials used for bundle retubing in this service.)

Retubing bundles every two years is not unusual. To prevent excessive tube failures, keep the tube bundle submerged in liquid. Place a prominent red line on the level gauge glass to show the operators the height of the tube bundle. A pattern of tube failures near the top of the bundle indicates the operators are often running at too low a liquid level.

Use desuperheated 60–90 psig steam to reboil the reclaimer. Check this with a surface thermometer and pressure gauge. If the steam is not saturated, your condensate supply to the desuperheater is not working. The chemistry of MEA degradation and reclaiming is discussed in detail by Polderman et al.[2]

Using a reclaimer instead of a filter

In one large Gulf Coast refinery, filtering the MEA solution had been temporarily discontinued. The amine had turned black. Consistent use of the reclaimer alone effectively cleaned up the solution. Twenty gpm of the 1,400 gpm lean-amine circulation was charged to the reclaimer.

To extend reclaimer washing cycles, steam was injected directly into the shell side of the reclaimer. This diluted the amine and reduced the reclaimer temperature. To maintain the system water balance, only 100% MEA was used as make-up to the lean amine storage tank. When the reclaimer contents had noticeably thickened (usually after two weeks), it was washed out. After three such cycles, the particulate content of the lean amine had been reduced by 80%.

Foaming

Most large amine spills are a consequence of scrubber foaming. The amine solution froths up on the trays and is carried overhead with the sweet gas. This leads to the following scenario, beginning at the refinery's fuel gas scrubber:

1. The sweet fuel gas is distributed to many process furnaces.
2. At each furnace, the fuel gas knock-out drum fills with amine.
3. The furnace operators drain their knock-out drums to the refinery slop system. The amine mixes with large quantities of water in the slop system.
4. The water in the slop system is drawn off to the refinery effluent treating plant where the operators note a tremendous increase in the nitrogen level in their effluent.

Typically, the chief environmental engineer then notified the operating engineer that he had better find the trouble quickly before they were both terminated.

Early warning system

Foaming in a scrubber can be stopped by reducing the hydrocarbon flow to the scrubber. Install a ΔP cell across the scrubber trays to observe when the scrubber is foaming over. If the indicated pressure drop (in feet of water) equals 40%–50% of the height between the top and bottom trays, the scrubber is foaming over. Use a strain gauge to connect the ΔP cell to the scrubber, which eliminates the problem of plugged pressure taps. Have the pressure drop readout activate an alarm in the control room.

Causes and cures of foaming

A wide variety of factors cause foaming inside an amine scrubber:

- Corrosion products
- Carbon particulates from the carbon filter
- High amine concentrations
- Overloaded amine
- Excessive concentrations of heat-stable salts
- Caustic in circulating solution
- Heavy aromatic hydrocarbons

These components all act to reduce the surface tension of the circulating amine. Consistent filtering and reclaiming can control most of these problems.

Corrosion inhibitors cause foaming. Pour some freshly made amine solution into a clean bottle. Add a typical concentration of inhibitor and shake vigorously and then see if any unusual foaming appears. Sometimes, the paper used in filters is coated with a chemical that can cause foaming. Tear off a piece of this paper and shake it with amine to see if the solution foams up.

Silicone antifoam agents, when used in excess, cause foaming. A squirt of silicone defoamer will temporarily stop foaming; too much will make the problem worse.

Charcoal filters need to be changed. You should be passing several percent of the circulating solution through a charcoal bed. The charcoal removes surfactants. Change the charcoal at the first indication of foaming, but be sure not to let charcoal fines enter the regenerator and scrubbers, as these fines also cause foaming.

Hydrocarbons condense in cold amine. The lean-amine temperature must be 10°F above the dew point of the sweet gas. Cooler amine will condense liquid hydrocarbons and initiate foaming. Cold amine (70°F) will in itself have a tendency to foam.

Detergent cleaners may have been used during a unit turnaround. When the unit is put back online, the detergent finds its way into the amine system.[3] The charcoal filter should absorb this detergent.

Possibly the most frequent cause of flooding or foaming in refinery fuel gas scrubbers is high liquid levels in the bottom of the tower. Referring to figure 4–3, when the liquid level in the bottom of the scrubber rises to the level of the sour-gas inlet, the tower will flood. This occurs because the incoming gas will entrain the bottoms liquid onto the bottom tray. This entrained liquid will cause the bottom tray to flood, and the flooding will back up the column until liquid amine is carried overhead with the sweet fuel gas.

Actually, the liquid in the bottom of the tower always consists of two phases: amine (1.0 sp gr) and hydrocarbon (0.6 sp gr). The hydrocarbon accumulates in the bottom of the tower due to a small amount of liquid entrainment in the inlet gas. Because the hydrocarbon is lighter than the amine, it simply floats on top of the amine.

If only a single overall gauge glass (as shown in figure 4–3) has been provided, only amine (1.0 sp gr) will be present in the gauge glass. As the average specific gravity of the liquid in the tower is lower than the liquid in the gauge glass, the gauge glass will record a lower level than the real liquid level in the tower.

To find the true liquid level in the tower, you can determine the vapor-liquid interface by touch. The vapor inlet will be 20°F to 40°F cooler than the bottoms liquid. This temperature gradient level will correspond to the true liquid level in the tower. A properly designed external liquid level indicator is shown in figure 4–4.

Liquid–liquid amine scrubbers

Amine carryover from such columns is common. Field observations have shown that the best way to stop the carryover is to increase the rate of amine circulation.

Reducing the amine acid gas loading by increased circulation reduced the tendency of the amine to emulsify in the hydrocarbon phase and also minimized the hydrocarbon-amine interface level. Strangely, increasing the amine circulation on one such contactor dramatically decreased the top liquid-liquid interface level by promoting settling in the column's packed bed.

Declining amine strength

The 100,000-gallon lean amine surge tank was filling fast. After a few days, it reached its maximum capacity. Finally, the operators decided emergency action was required, so they called the operating engineer for help. He got there just in time to watch the tank overflow to the sewer.

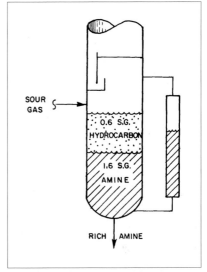

Fig. 4–3. A false external-level indication causing flooding.

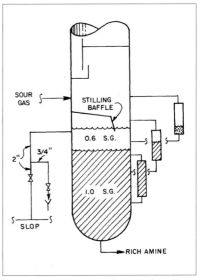

Fig. 4–4. A properly designed amine fuel-gas scrubber.

When an amine surge tank level is going up and amine strength is dropping, water or steam is leaking into the system. In a complex refinery, amine may be circulated to a dozen units scattered over several square miles. Finding a large water in-leakage (5 gpm) is a tough job.

It is not likely that you can find this type of leak by sampling for amine strength at different points in the refinery. The entire amine system volume may circulate through the scrubber and regenerator 20 times a day. A small loss in amine strength per pass cannot be detected by the lab. This small loss in strength is, however, multiplied by 20 each day.

To determine why amine strength is weakening, send a sample of lean amine to the lab for total dissolved solids (TDS). Also, submit the water used for the amine system to check for TDS. A high TDS for the amine sample indicates a leaking amine water cooler. For each cooler, check the inlet water pressure and the outlet amine pressure. If the water pressure is higher, proceed as follows:

1. Block in the cooling water inlet and outlet valves (warm amine to a scrubber won't hurt for a few minutes).

2. Open the water drain valve on the exchanger.

3. When the exchanger is almost drained down, check the drain with pH paper. If it turns blue, the exchanger is leaking.

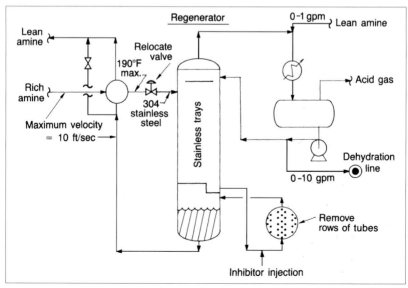

Fig. 4–5. Small changes reduce amine system corrosion.

Alternately, a low TDS result for lean amine shows that steam or condensate is entering the amine system. Verify that the condensate make-up valve is closed. Then block in any cross connections between the amine system and sour water pump-out lines, and so on. Next, block in the reclaimer vapor outlet and amine inlet. A leak in this exchanger will then manifest itself by the presence of amine in the channel head drain.

If none of these steps has identified the problem, it is likely there is a regenerator reboiler tube leak (see figure 4–5). When the regenerator has one reboiler, it cannot be taken off-line for leak testing. In this case, inject a small quantity of a tracer chemical into the steam supply. Then check the regenerator bottoms by chemical analysis for the tracer. Sulfur hexafluoride or lithium bromide can be used.[4]

Retrofitting Tips

A few changes can reduce the amine unit's potential to corrode:

- **Velocities.** Revamp process piping so that velocities are less than 10 ft/sec. For exchangers, hold velocities below 3 ft/sec.[5] Remember, rapidly flowing amine is erosive.

- **Rich amine flow control.** Locate the regenerator feed control valve downstream of the cross exchanger. This will minimize vaporization of corrosive gases in this exchanger. The control valve internals and piping downstream of the control valve should be 300-series stainless steel.

- **Reflux water pH.** To protect the regenerator condensers from the corrosive effects of low pH water, a small amount of amine entrainment is needed. The reflux water must have at least one-half wt% amine. To ensure this concentration, tie in a 1-inch line from the lean amine pump to the inlet of the condenser. Incidentally, ammonia often accumulates in the reflux water and can be accidentally titrated as amine by the lab. This will not hurt anything.

- **Reboilers.** Remove tubes to form vertical or diagonal vapor passageways through the bundle. This reduces turbulent boiling in the shell. Retubing with 304 stainless tubes will reduce corrosion if the amine is low in chlorides.[6]

- **Regenerator internals.** Replace all carbon steel trays and tray parts with 304 stainless steel. Use Teflon gaskets. Remove all copper and copper alloy equipment from rich amine service.

- **Amine dehydration.** A steam or water leak into the system will dilute the amine. Weak, rich amine becomes acidic and corrosive. To increase amine concentration, reflux water can be diverted to a closed drain. This will require piping the reflux water to a place where H_2S can be safely flashed off. Do not drain this water to an open sewer; it is supersaturated with deadly H_2S. This is one mistake nobody makes more than once.

Cut Reboiler Steam Usage

The principal energy requirement in a sulfur recovery complex is the amine regenerator reboiler steam. The operating engineer can significantly reduce energy usage through some simple field work. All he needs is a test kit to measure H_2S and CO_2 concentrations from 0.01% to 10% (Dräger or equivalent):

1. Make a list of the scrubbers in the refinery. Determine the permissible H_2S concentration for sweet gas from each scrubber. For fuel gas scrubbers, the specification is usually 150 parts per million (ppm).

Safety Note

All piping and vessels used in MEA service must be stress relieved to guard against hydrogen-induced stress corrosion cracking. The inevitable corrosion that takes place in all amine systems produces ionic hydrogen. This hydrogen migrates through the vessel wall and recombines at imperfections in the lattice structure of the steel. Usually, most of the ionic hydrogen recombines to innocuous molecular hydrogen at the surface of the vessel. However, cyanides retard this recombination. Then the ionic hydrogen recombines inside the vessel wall to form stress-producing molecular hydrogen. Typically, this will cause the vessel wall to blister and shorten the usable life of the vessel. However, if the hydrogen accumulates in the hardened, heat-affected zone of a weld, which has not been stress relieved, rapid weld cracking can result.

To a lesser extent, the above warning applies to all amine systems. The presence of carboxylic acids produced from heat-stable salts, overloaded amine, and cyanides accelerates the process of hydrogen-assisted stress-corrosion cracking.

2. Multiply each scrubber's top pressure (psia) by its sweet gas H$_2$S spec (ppm). Use the same calculation for liquid propane treaters.

3. Divide the value obtained above by a temperature correction factor (use only between 75°F to 130°F):

Correction factor = 1.5 × scrubber top temp. (°F) ± 90

For example, for a fuel gas scrubber with a tower top pressure and temperature of 70 psig and 110°F:

$$SR = \frac{150\,ppm \times (70\,psig + 14.7)}{(1.5 \times 110°F) - 90} = 169$$

where:

SR = Scrubber rating factor

4. The scrubber with the smallest SR factor is your critical operation. Let's call this the critical scrubber.

5. With your H$_2$S test kit, measure the H$_2$S concentration in the sweet gas from the critical scrubber.

6. Your regenerator reboiler duty should be controlled on the basis of pounds of steam per gallon of lean amine circulated (lb/gal); proceed by cutting reboiler steam 0.1 lb/gal. *Note:* Scrubber sweet gas purity is controlled by the regenerator stripping steam rate, not by the amine circulation rate to the scrubber.

7. Repeat steps 5 and 6 until the H$_2$S in the sweet gas from the critical scrubber reaches the maximum specification.

8. Using the test kit, measure the H$_2$S and CO$_2$ concentrations in the sour and sweet gas at each scrubber. Calculate the moles per hour of H$_2$S and CO$_2$ picked up by the amine.

9. Calculate the amine circulation required at each scrubber to obtain 0.4 to 0.5 moles of acid gas (H$_2$S + CO$_2$) per mole of amine circulated (see table 4–1 for conversion factors).

10. Reduce the actual amine flow to each scrubber to the calculated value. The 0.4 to 0.5 molal ratio includes a reasonable safety factor.

Don't forget to reduce the steam to the regenerator reboiler again, as per step 6. However, never reduce reboiler duty below the amount needed to maintain a minimum regenerator reflux rate. This is 10% of the reboiler (plus reclaimer) steam usage, in lb/hr.

Table 4–1. Conversion factors.

$$\frac{\text{grains of } H_2S}{\text{gal of solution}} \times \frac{0.00323}{\text{wt\% MEA}} = \text{moles of } H_2S/\text{mole of MEA}$$

$$\frac{\text{grains of } H_2S}{\text{gal of solution}} \times \frac{0.00557}{\text{wt\% MEA}} = \text{moles of } H_2S/\text{mole of DEA}$$

$$\frac{\text{cu ft of } CO_2}{\text{gal of solution}} \times \frac{1.91}{\text{wt\% MEA}} = \text{moles of } CO_2/\text{mole of DEA}$$

$$\frac{\text{cu ft of } CO_2}{\text{gal of solution}} \times \frac{3.28}{\text{wt\% MEA}} = \text{moles of } CO_2/\text{mole of MEA}$$

$$\text{Percent } H_2S \text{ (by weight)} \times \frac{1.8}{\text{wt\% MEA}} = \text{moles of } H_2S/\text{mole of MEA}$$

$$\text{Percent } H_2S \text{ (by weight)} \times \frac{3.1}{\text{wt\% MEA}} = \text{moles of } H_2S/\text{mole of DEA}$$

$$\text{Percent } CO_2 \text{ (by weight)} \times 0.72 = \text{cu ft of } CO_2/\text{gal of solution}$$

$$\text{Percent } H_2S \text{ (by weight)} \times 558 = \text{grains of } H_2S/\text{gal of solution}$$

Minimizing CO_2 recovery

FCCUs produce a sour fuel gas stream rich in CO_2. When scrubbed for H_2S removal, most of the CO_2 is incidentally picked up by the amine.[7] Absorbing CO_2 from fuel gas in a refinery wastes energy and sulfur recovery capacity.

To minimize CO_2 recovery, reduce the number of trays used to absorb H_2S. Install several new feed nozzles at various locations and experiment.[8] Figure 4–6 summarizes the results of one such experiment on a refinery fuel gas scrubber.

MDEA and DIPA

For enhanced rejection of CO_2 use of a 45 wt.% MDEA is suggested. However, do not switch from using ordinary DEA to the proprietary MDEA or DIPA amines. Not only are these solvents more expensive but they tend to foam worse than DEA at a similar molar concentration. My experience teaches that a good design basis for a refinery amine system with an FCU is:

- 35 wt.% DEA
- 0.45 moles of total acid gas $(CO_2 + H_2S)$ per mole of active amine
- Heat stable salts not to exceed 2 wt.% even for short periods

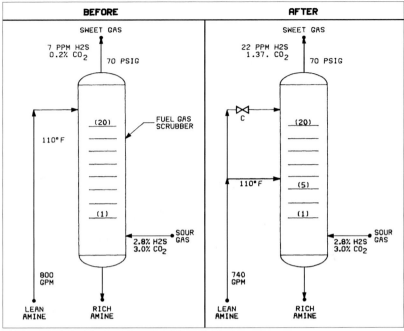

Fig. 4–6. Minimizing CO_2 removal from fuel gas.

Higher acid gas loadings are okay if particulates, cyanides and heat stable salts are very low. But, in practice, this is often not the case. Some vendors claim that proprietary amines will save energy and enhance capacity; sadly, these claims are false.

Other Problems

- **Poor sweetening.** If scrubbed fuel gas is off spec (high H_2S), raise the regenerator reboiler steam. Check for excessive acid gas loading by calculation (as discussed earlier). If this doesn't help, check the cross exchanger for leaks.
- **Brown amine.** Exposure of amine (especially MEA) to oxygen will produce high concentrations of degraded amine. A strong ammonia odor and a brown tinge is the tip-off that the amine is exposed to an oxidizing agent.[9] Check the inert gas or kerosene blanket on the lean amine storage tank. Ask the FCCU operators to review the operation of their regenerated catalyst steam stripper. Poor deaeration of catalyst results in an increase of oxygen compounds in the FCCU fuel gas product.

- **COS removal.** Propylene produced from cracking plants will contain carbonyl sulfide (COS). When treating propylene to produce a polymer-grade product, one has to meet a stringent sulfur specification. If COS is reported in propylene product, raise the temperature of lean amine to the scrubber 10°F. This usually eliminates the problem.

- **Hydrocarbon skimming.** The single biggest problem in operating a refinery Claus sulfur recovery plant is hydrocarbons in acid gas feed (see chapter 5). When the rich amine charged to the regenerator contains light liquid hydrocarbons, coking of the sulfur plant catalyst is almost certain. The hydrocarbons will distill overhead in the regenerator and combine with the acid gas product.

If naphtha is accumulating in the regenerator reflux drum, excessive concentrations of propane and butane will occur in sulfur plant feed. A commercially proven method to eliminate this problem is detailed in figure 4–7. The rich amine surge drum is retrofitted with baffles. Figure 4–7 is roughly drawn to scale. This baffle arrangement will automatically skim off the hydrocarbons.

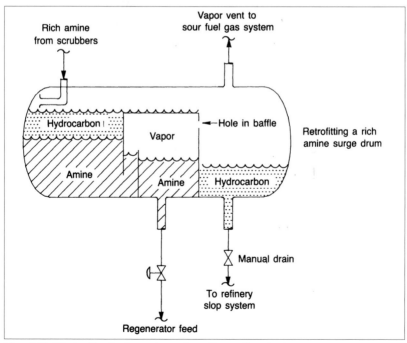

Fig. 4–7. Eliminating hydrocarbons from the sulfur plant.

The first compartment separates amine from hydrocarbons. A residence time of 10 minutes is about right. A second compartment takes the swings in rich amine flow rates. The last compartment stores recovered liquid hydrocarbons until they can be manually drained.

Troubleshooting Checklist for Amine Systems

Dirty Amine
Cyanides
Amine color and
  solids concentration
Inadequate filtering
Corrosive amine
Regenerator feed temperature
  too high
High velocities
Stress cracking

Reboiler Corrosion
Corrosion inhibitor
Condensate backup in channel head
Regenerator reflux rate too low
Superheated reboiler steam

Reclaimer Operation
High S_2O_3 levels
Soda ash addition rate
Diluent water rate
Tubes not submerged in liquid

Energy Reduction
Minimize CO_2 recovery
Cut amine circulation
Reduce reboiler steam

Foaming in Scrubbers
High N_2 in refinery effluent
Scrubber pressure drop
Dirty amine
Excessive concentrations
  of silicone or inhibitor
Charcoal filter
Condensing hydrocarbons
Extraneous surfactants
Plugged level control taps
Liquid-liquid scrubbers

Loss in Amine Strength
Amine cooler leaking
Reboiler or reclaimer leaking
Condensate make-up
  valve open
Test with tracer chemical

Poor Sweetening
Leaking cross exchanger
Regenerator reflux rate too low
Amine degraded
COS in treated propylene

References

1. D. Bailard, "How to Operate an Amine Plant," *Hydrocarbon Processing*, June 1967, p. 171.

2. L. D. Polderman et al., "Degradation of MEA in Natural Gas Treating Service," *Oil & Gas Journal*, May 10, 1955.

3. Melvin White, "Corrosion Experience at Arkansas Fuel Oil Columbia Plant," Carbide Company Information Bulletin, 1949.

4. B. Fries, "Nonradioactive System Measures Heat Exchanger Leaks," *Oil & Gas Journal*, Dec. 3, 1979.

5. Bailard, 1967.

6. L. D. Polderman, "Stainless Steels for Exchangers in MEA Service," Carbide Company Information Bulletin, November 1954.

7. A. L. Kohl, "Plate Efficiency for CO_2 Absorption in MEA," *AIChE Journal*, 1956.

8. Ibid.

9. Bailard, 1967.

Sulfur Recovery

Not too long ago, the function of a refinery sulfur recovery unit was to produce a low-value byproduct. Today, sulfur dioxide (SO_2) emissions must be strictly controlled. Pollution abatement has become as important as profitability, and effective sulfur plant troubleshooting is a vital component in keeping a refinery onstream.

The process operating engineer can expect to find himself in the refinery manager's office if the sulfur plant falters. Environmental authorities have shut down entire refineries because of sulfur plant outages. Even if the only consequence is that low-sulfur crude is substituted for a less-expensive high-sulfur crude, the economic effects can be staggering.

Modern sulfur recovery technology was developed by natural gas producers. Recovering sulfur in a petroleum refinery is a tougher job because of the variability in quantity and composition of the sulfur plant feed, which is called *acid gas.*

Acid gas consists primarily of H_2S and CO_2 stripped out of the refinery's circulating amine. The flow of acid gas will bounce with every change in refinery operations. Also, the acid gas will contain variable quantities of hydrocarbons, COS, and CS_2. The ratio of H_2S to CO_2 can also shift with unexpected suddenness.

Sulfur Recovery Chemistry

The conversion of H_2S and the recovery of sulfur is one of the more simple chemical reactions that take place in a refinery. First, an aqueous solution of amine, which is an organic base, scrubs H_2S and CO_2 out of a hydrocarbon stream. Then the acid gas (H_2S plus CO_2) is stripped out of the amine with steam. The acid gas is charged to the sulfur plant where the H_2S is partially oxidized to water and sulfur according to the primary equilibrium reaction:

$$H_2S + \tfrac{1}{2} O_2 + 2N_2 = H_2O + S_{(L)} + 2N_2$$

Too much air leaves excess SO_2 in the tail gas; too little leaves some H_2S unconverted.

Finally, the liquid sulfur is recovered. Tail gas exiting the last reactor contains unconverted H_2S and SO_2, as well as CO_2, N_2, and H_2O. Varying amounts of sulfur vapor, COS (carbonyl sulfide), and CS_2 (carbon disulfide) are also present. Environmental restrictions for new sulfur plants require further processing of this tail gas.

Process flow

Figure 5–1 shows a single-stage sulfur plant (or Claus plant, named after the inventor of the process, Carl Friedrich Claus). Two-and-one-half moles of air are used to burn 1 mole of H_2S in the reaction furnace. About 65% of the H_2S is converted to sulfur in this furnace. The hot gases are cooled, allowing the sulfur vapors to be condensed. The sulfur drains out of the condenser through pipes submerged in liquid sulfur (seal legs).

The cooled gases still contain large concentrations of H_2S and SO_2 and smaller amounts of COS and CS_2. To reduce these concentrations, the gas stream is reheated and passed through a catalyst bed. Figure 5–1 shows a single catalyst bed. This is designated a single-stage Claus plant.

What Can Go Wrong

Most of the troubleshooting assignments encountered in sulfur plants are related to pressure drop. Plugged seal legs, disintegrated catalyst, carbon deposits, and boiler leaks are the usual causes. Excessive pressure drop will result in an air deficiency or even blown seal legs.

Inadequate conversion of H_2S to liquid sulfur is the other problem. Improper air-acid gas ratio and loss in catalyst activity are the most common causes. Increased SO_2 in the incinerator stack will be the end result of reduced conversion.

Finding lost conversion

H_2S is an unstable molecule. It readily converts to elemental sulfur and water when exposed to O_2 or SO_2. The basic Claus reaction:

$$2H_2S + SO_2 \longrightarrow 2H_2O + 3S$$

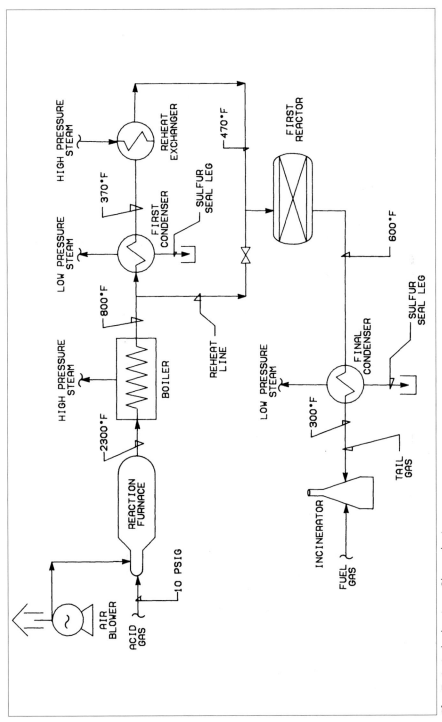

Fig. 5–1. A single-stage Claus plant.

can be driven to high conversions. With the one reactor shown in figure 5–1, a conversion level of H_2S to sulfur of 85% is expected. A two-stage unit, with two reactors, produces a conversion of 95%. Up to 98% is possible with three reactors.

Measuring sulfur losses

To measure sulfur losses, check for H_2S and SO_2 in the final condenser effluent. Dräger tubes are a simple and reasonably accurate way to get this data. Don't submit sulfur plant tail gas samples to the lab for analysis. The H_2S and SO_2 will react in the sample container to form solid sulfur and water. To calculate conversion from these measurements:

1. Add the ppm of H_2S plus SO_2.
2. Add 2,000 ppm to the preceding (this allows for COS, CS_2, sulfur vapors, and entrained sulfur droplets).
3. Divide the total ppm of sulfur as obtained above by 300,000.
4. Express this result as percent.
5. Subtract this percent from 100%.

Poor conversion is 95% for a three-stage unit, 90% for a two-stage unit, and 80% for a single-stage plant.

Wrong air ratio

Using too much air is the easiest way to lose conversion. For best conversion, the ratio H_2S/SO_2 is 2:1. This ratio is measured in the tail gas from the final condenser. Sulfur plants are best run on closed-loop tail gas analyzer control. In practice, other methods are often needed to adjust air to the reaction furnace.

If air flow to the reaction furnace is much too high, SO_3 will be formed in the incinerator and a white plume will result. A yellowish plume indicates insufficient air to the front end of the sulfur plant.

Incinerator temperature can also be used to indicate air required in the reaction furnace. A high incinerator temperature, coupled with low incinerator fuel use, is a sure sign of insufficient air to the reaction furnace. Alternatively, if large amounts of fuel are required to maintain incinerator temperature, air to the reaction furnace is in excess.

If the tail gas analysis showed SO_2 equal to H_2S, one could cut sulfur losses by 5% just by reducing air. A complete treatment on the effect of air ratios on conversion has been reported by Kerr.[1]

Plenty of catalyst

Sulfur plants usually have excess catalyst. Typically, the catalyst bed will be 3 feet deep. For catalyst in good condition, equilibrium conversion is reached in the top 6 inches.

The operating engineer, when troubleshooting a conversion problem, can depend on the following: overall conversion does not decrease noticeably with higher throughput. Plant tests on one unit, conducted over a range of 30% to 120% of design capacity, actually showed better conversions at the 120% rate. Therefore, start a troubleshooting program by assuming that cutting throughput will not help.

Reactor problems

It's hard to do much harm to sulfur plant catalyst without first causing excessive pressure drop. If you suspect reduced recovery because of lost catalyst activity, check the temperature rise across each reactor. For example, for a three-stage Claus unit:

Reactor	T° rise (outlet-inlet)
1st	125°F
2nd	40°F
3rd	5°F

This is a good profile. If one day you find the heat of reaction has shifted downstream, you may see the following:

Reactor	T° rise (outlet-inlet)
1st	90°F
2nd	55°F
3rd	20°F

This temperature shift means the effluent from the first stage is not reaching equilibrium. Sulfur formation in the first reactor has decreased 30%. In this reactor, overall catalyst effectiveness has declined.

The problem may be due to catalyst deactivation caused by sulfur precipitation. This is a result of low reactor feed temperatures. Check the operation of the reheat exchanger upstream of the reactor with the reduced temperature rise. Raise the reactor inlet temperature 30°F. After a few days, this will dissipate the offending sulfur deposits.

If catalyst activity has been irreversibly lost, you may want to change the catalyst. A following section describes a procedure to help make this decision.

COS and CS$_2$

The presence of hydrocarbons and CO$_2$ in the acid gas promotes the formation of COS and CS$_2$ in the reaction furnace. These compounds can significantly contribute to SO$_2$ in the incinerator.

One can essentially destroy both COS and CS$_2$ by operating the first reactor at an outlet temperature of 650°F. At this temperature the compounds are hydrolized to H$_2$S and CO$_2$. An increase in SO$_2$ emissions, accompanied by a lower than normal first reactor inlet temperature, is likely due to COS and CS$_2$.

Leaking reheat exchanger

Some multistage sulfur plants reheat third-stage reactor feed with first-stage reactor effluent. If your plant has one of these heat exchangers, check it for leaks, which contribute to lost conversion. A large increase in the third-stage inlet temperature indicates this type of reheat exchanger is leaking.

Sulfur fog

Sulfur plants have the peculiarity of converting a lower percentage of hydrogen sulfide to sulfur as the unit charge drops. One of the reasons for this is sulfur fog. Sulfur should condense on the walls of the tubes. However, at low tube-side gas velocities, the sulfur precipitates in the gas stream itself. A sulfur fog is formed.

This fog does not drop out of the end of the condensers. Eventually, much of it appears as SO$_2$ in the incinerator. Damage to the final condenser demister may also allow entrained sulfur to escape to the incinerator. This demister can be extensively damaged from sulfur fires during start-up.

Cold reheat gas

Hot gas from the high-pressure boiler (figure 5–1) is often used to supplement the reheat exchangers. This hot gas contains sulfur vapors. About 65% of the H$_2$S in acid gas is converted to sulfur vapors in the reaction furnace. Therefore, this hot reheat gas increases the partial pressure of sulfur in the reactor.

When reheat gas is used, equilibrium in the reactor is adversely affected. At reduced plant charge, the gas outlet temperature from the high-pressure boiler drops. This means more reheat gas is needed to compensate for its lower temperature.

In practice, a change in reheat gas temperature has a noticeable effect on conversion only when hot reheat gas is used in the last reaction stage.

When to change catalyst

A favorite question put to an operating engineer by the sulfur plant supervisor is, "Do we need to change catalyst during the unit turnaround?" With a little luck, he may remember to ask you before the plant is shut down.

If pressure drop is normal through the catalyst beds, this will be a tough question. With adequate instrumentation, you can obtain a vertical temperature profile through the first catalyst bed and then develop data to make a firm decision. Figure 5–2 illustrates the method. For catalyst that is in good condition, 90+% of the heat of reaction is released in the top 6 inches. If catalyst activity is impaired, the reaction is shifted down the bed.

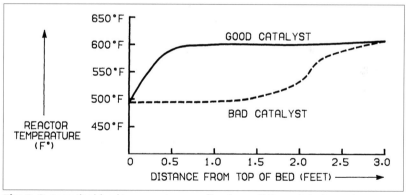

Fig. 5–2. A vertical-bed temperature profile shows if the catalyst needs to be changed.

Damage to catalyst and reduced conversion can be a consequence of many other factors besides lost activity: carbon deposits, leaking condenser tubes, damaged support screens, sulfuric acid formation, or operation at the sulfur dew point. These problems are, however, invariably associated with increasing pressure drop.

Pressure Drop

It is of utmost importance to watch for high-sulfur plant pressure drop. Sulfur plants don't suddenly plug without a prior pressure drop increase. Troubleshooting a sulfur plant requires foresight. The operating engineer will want to have the data plotted, as in figure 5–3, for his unit. This figure illustrates the use of the capacity ratio parameter, calculated as follows:

$$\frac{\Delta P_C}{(F_{C^-})^2} = X_C$$

$$\frac{\Delta P_D}{(F_D)^2} = X_D$$

$$X_C/X_D = \text{capacity ratio}$$

where:

ΔP_c = Current pressure at the reaction furnace inlet, psig

F_c = Current air flow to reaction furnace, standard cubic foot per hour (SCFH)

ΔP_D = Design pressure at the reaction furnace inlet, psig

F_D = Design air flow to reaction furnace, SCFH

Pressure drop in a sulfur plant is proportional to the square of the throughput. When you find your plant not adhering to this rule, there is something gone awry with your unit.

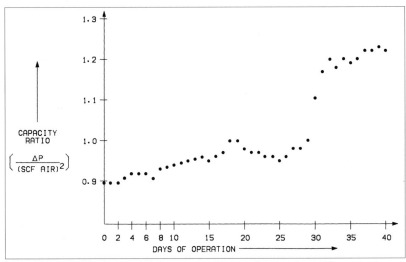

Fig. 5–3. A sudden increase in pressure drop spells trouble on Claus plants.

Carbon deposits

The data plotted in figure 5–3 were actually not assembled until after the catastrophic pressure rise. The plant operators had not noticed the increase in the reaction furnace pressure. Only when they tried to increase acid gas charge and ran short of air blower capacity did they realize something was amiss.

An abnormality had been reported on the 30th day. A quantity of light hydrocarbon was skimmed off the amine regenerator reflux drum. When a sample of this hydrocarbon was drawn, it bubbled in the sample container.

Light hydrocarbons had accidentally entered the amine regenerator, along with the rich amine. The hydrocarbon was stripped overhead. Some was condensed in the reflux drum; the rest remained as a vapor and was charged, along with H_2S, to the sulfur plant.

Ten times more air is needed to oxidize a mole of propane than a mole of H_2S. When the light hydrocarbon vapors reached the sulfur plant, carbon black was made in the reaction furnace:

$$C_3H_8 + 2O_2 \longrightarrow 3C_{(s)} + 4H_2O$$

The carbon black deposited on the top of the first catalyst bed. Gas flow was restricted, and high pressure drop resulted. Providing sufficient combustion air to the reaction furnace could have prevented this incident.

The operating engineer can determine if the increasing pressure drop is due to carbon accumulation on catalyst by making the following observations:

- Is the SO_2 concentration in the sulfur plant tail gas very low (less than 1,000 ppm)? Low SO_2 is a sign of insufficient air to the reaction furnace.
- Are light hydrocarbons accumulating in the amine regenerator reflux drum?

Having determined that the catalyst bed is plugged with carbon, the engineer will want to correct the situation. Over a period of time, the carbon will react with SO_2. Unfortunately, this reaction proceeds slowly at low temperatures. Maximizing reactor inlet temperature and SO_2 levels will help. Significant (10%) reductions in pressure drop can take weeks. Shutting down and changing out the catalyst may be more practical.

The best solution to this incident would have been to keep carbon black from forming in the first place. This could have been done by alert operators raising air to the reaction furnace. One might even have hoped that the tail gas H_2S/SO_2 ratio analyzer would have automatically increased air flow.

In the real world, there is only one reliable way to prevent such situations. Liquid hydrocarbons must be separated from the rich amine upstream of the amine regenerator (see chapter 4).

Leaks cause pressure drop

A tube leak in the high-pressure steam boiler can lead to disaster. The high-pressure water will erode the metal, and the flow of water into the hot gas stream will rapidly increase. Water quenches the sulfur-bearing gases. If the direct reheat line is open, sulfur precipitates on the catalyst. This stops gas flow through the plant.

The worst thing that can happen to a sulfur plant is a crash shutdown. Sulfur plants should be cleared of sulfur by burning natural gas instead of H_2S before a shutdown. Continue natural gas firing until the amount of liquid sulfur overflowing from the seal legs is reduced to a trickle. When the plant suddenly shuts down, precipitating sulfur solidifies in the catalyst beds. Then flow through the unit cannot be reestablished.

An alert operating engineer must identify boiler tube leaks before it is too late. The capacity ratio plot (figure 5–3) is the key. A gradual increase in pressure drop is an early warning sign. When this happens, check for low steam production rates from the high-pressure boiler. Another tip-off is a low gas outlet temperature from this boiler. If both steam production and outlet temperature are low and pressure drop is high, shut down the plant. There is a tube leak.

On one unit, a high pressure drop was observed. The operators suspected a plugged condenser sulfur seal leg. They opened a drain on the condenser with the intent of drawing off excess sulfur. Steam, not sulfur, discharged from the drain. Six days later, the plant shut down with a giant leak in the high-pressure boiler tube sheet.

Water (steam) leaks also reduce conversion of H_2S to sulfur. The Claus reaction shows that equilibrium is shifted to the left as the water partial pressure increases.

Preventing boiler leaks

There are two simple rules to minimize boiler leaks: 1) keep total dissolved solids (TDS) in the boiler blowdown below 2,500 ppm; 2) keep all boiler tubes submerged in water. Watch the steam drum liquid level closely. For forced circulation boilers, circulate 10–15 pounds of water for each pound of steam generated. Damage to the ceramic ferrules in the front-end tube sheet is another common cause of boiler leaks.

Condenser leaks

Tube leaks may also occur in the low-pressure steam condensers. The leaking condenser is identified through a pressure drop survey. Measure the pressure drop across each condenser. The first condenser in the train exhibiting a disproportionately high pressure drop is the leaker. If the leaking tubes are found in the bottom of the condenser, formation of sulfuric acid may be the cause (see the section "Start-Up Tips").

Routine pressure surveys

The preceding experiences illustrate the need for routine pressure surveys on sulfur plants. A single 0–15-psig gauge is used. Sulfur will plug pressure taps very quickly. They can be drilled out or melted with a propane torch.

Take a complete pressure survey just after the unit comes onstream after a turnaround. This will give you a base point from which to judge future problems. When comparing pressure drops at different throughputs, normalize the data by:

$$\Delta P \text{ is proportional to (air flow)}^2$$

Plugged seal legs

Liquid sulfur is drained from the condensers through seal legs submerged in sulfur to prevent gas in the condenser from blowing them. Required seal depth is:

$$\text{Seal depth (ft)} = \frac{\text{condenser pressure (psig)} \times 2.31}{\text{sp gr of sulfur at condenser temperature}}$$

[The sp gr of sulfur between 250°F to 350°F is normally about 1.79.]

When a seal leg plugs, liquid sulfur backs up in the condenser. This restricts gas flow and results in high pressure drop. Again, the best indication of this problem is a routine pressure survey. Having determined that a condenser has excessive pressure drop:

1. Locate the condenser tube-side drain connection.
2. With suitable breathing protection from H_2S (i.e., Scott Air Pac), unplug and open the drain.
3. If a steady flow of liquid sulfur is observed, the seal leg is plugged.

Do not try to keep the condenser drained down in this manner. Sooner or later, H_2S-rich gas will blow through and create a potentially fatal hazard. Plugged seal legs must be unplugged. Packing glands, used with valves that bolt onto the seal leg flanges, can be used to drill out seal legs onstream.

Plugged seal legs are often a problem just after start-up. Minerals in the refractory are leached out by the moisture and acid produced during heat-up. These minerals, as well as corrosion products, accumulate on the liquid sulfur surface in the seal legs. A steady increase in pressure drop, shortly after start-up, is often caused when these deposits solidify in the seal legs.

Shortened seal legs

Trash sometimes accumulates in the sulfur seal pit. When the seal legs are pulled for replacement or repair, the maintenance forces find the legs are too long to fit back into the seal pit, so they shorten the legs.

Do not shorten the seal legs! This can permit blowing the seal legs at an unexpectedly low plant pressure. One Gulf Coast refiner (without the knowledge of the operating manager) actually shortened the sulfur seal legs by 18 inches. When the seal legs blow, they vent H_2S directly in the process area.

Catalyst support screens

Sulfur plant catalyst is supported by thin flexible screens. These screens are lapped and folded over to keep catalyst from leaking through the support grating. Improper installation of screens occurs frequently when catalyst is changed. The catalyst may wash down into a seal leg and plug it. When you find normal sized catalyst balls in the condenser drains, you can count on a shutdown to repair the support screens. Don't forget to seal the screens to the walls of the reactor. On one unit, small but intact catalyst balls were found in the seal legs. This indicates poor quality control in the manufacture of the catalyst.

Start-Up Tips

Most damage to sulfur plants occurs during start-up. You can reduce troubleshooting activity later by closely monitoring the fuel gas firing phase during heat-up.

Keep O_2 levels in flue gas at about 1%. Excess O_2 will form sulfuric acid when mixed with sulfur and moisture in catalyst beds. The H_2SO_4 disintegrates the catalyst. Low conversion and increased pressure drop result. The acid may also attack the catalyst support screens.

This leads to plugged seal legs. Even worse, the lower tubes in the condensers will get a diluted sulfuric acid bath, and consequently corrode.

Avoid deficient oxygen

Burning a hydrocarbon with insufficient air produces sooty smoke. The soot deposits on the first catalyst bed. To do a thorough job of plugging a bed with this technique takes about eight hours.

A foolproof way to make sure you are not badly oxygen deficient is to connect a piece of tubing to the back end of the high-pressure boiler. Then attach the other end of the tubing to a bottle filled with clean, damp cotton. Observe the cotton. If it starts turning black after a few minutes, you are running oxygen deficient. Personnel who have been unable to master other analytical techniques find this method useful.

Start-up atmospheric vent

Putting a cold sulfur plant online, when done properly, can take almost two days. During the initial portion of the start-up, firing must be carefully controlled to avoid damaging the refractory in the decomposition furnace (i.e., the thermal reactor) due to a too-rapid heat-up. Therefore, only a small volume of flue gas exits from the decomposition furnace and the high-pressure steam generator, during the first 12 hours of the start-up. This amount of flue gas is really not sufficient to warm the large weight of catalyst appreciably in the fixed-bed reactors. Hence, the water in the flue gas, produced by the combustion of hydrogen in the decomposition furnace, may condense on the catalyst beds.

In the presence of sulfur (which is always present in the reactor beds after the unit is commissioned) and oxygen, sulfuric acid is formed. This acid leaches out minerals from the reactors' refractory walls, corrodes the condenser tubes, forms a pressure drop-producing crust on the top of the catalyst beds, and may result in seal leg pluggage.

Excluding excess oxygen from the combustion gases can reduce the formation of sulfuric acid. Of course, one then encounters the danger of converting the front end of the Claus train into a carbon-black plant and plugging up the catalyst bed in the first fixed-bed reactor with coke.

All of these invidious possibilities may be easily avoided by installing a removable start-up stack. For one unit, a 12-inch diameter, 20-foot length of pipe was bolted onto the manway entrance to the backend of the fired tube boiler. The large butterfly valve at the outlet of the Claus train was closed and the decomposition furnace was heated in a normal manner, but without taking care to minimize excess oxygen, as the entire flue gas stream exited from the new temporary stack. Once the decomposition furnace had been heated to 1,400°F, the temporary stack was removed and normal Claus plant warm-up procedures were followed. By now, however, the rate of natural gas firing did not have to be controlled carefully to avoid over-rapid heat-up of the bricks in the decomposition furnace.

Thus, a large volume of flue gas could be generated, which rapidly heated the catalyst beds past the water dew point temperature. Also, excess oxygen of several percent was tolerable, until sulfur fires were ignited in the catalyst beds. However, this condition was readily apparent from the reactor's temperature profile and corrected by pinching back on the combustion air to the decomposition furnace.

Maximizing Plant Capacity

Innovative changes in the design and operation of some existing sulfur recovery plants can produce large increases in capacity. The capacity of the majority of sulfur plants is limited by front-end pressure (the acid gas pressure at the reaction furnace). Although in theory, conversion of hydrogen sulfide to sulfur liquid is slightly reduced at higher gas throughputs due to reactor and condenser limits, in practice, such effects are quite small. For example, the temperature profile of one lead fixed-bed reactor on a Claus plant (figure 5–4) shows almost all the reaction taking place in the top 30% of the bed.

The parameters which limit maximum front-end pressure are:

- Seal leg depth
- Air blower maximum head
- Reaction furnace design pressure
- Acid gas supply pressure

Front-end pressure will vary with the square of the moles of air plus acid gas that enters the reaction furnace.

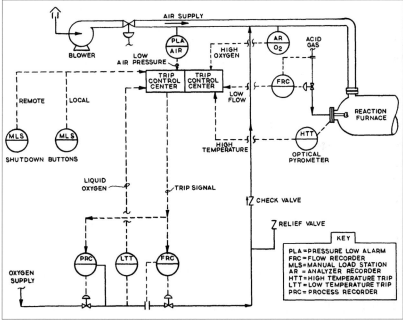

Fig. 5–4. Safety controls for O_2 enriched air in Claus unit.

The principle constituents of these streams are:

- H_2S from natural gas
- CO_2 absorbed from natural gas

- O_2 from combustion air
- N_2 from combustion air
- H_2O (vapor)-contained in the acid gas from the amine regenerator reflux drum
- Miscellaneous hydrocarbons and mercaptans

Oxygen Enrichment

The use of oxygen to enrich process air for combustion purposes is a common practice. For instance, oxygen has been added to the air blower discharge of FCCUs in many refineries.[2] Enrichment concentrations of 30% to 40% oxygen are typical.

Oxygen enrichment of the air supply to one 50-T/D Claus plant was initiated by the author in 1981 to reduce the amount of nitrogen flowing through the reaction train. The overall Claus reaction for a typical acid gas stream having the composition shown in table 5–1 will yield an effluent with the following composition (in mole%): N_2, 60; H_2O, 30; CO_2, 5; H_2 and other, 5.

Table 5–1. Typical acid gas composition.

	Mole%
CO_2	12
Hydrocarbons*	1
H_2O	8
H_2S	78
Other	2
Total	100

*Average molecular weight of hydrocarbon is typically equal to propane.

Because the capacity of a Claus plant is essentially proportional to the volume of the effluent gas, substituting oxygen for air will result in a large capacity increase. For the 50-T/D Claus plant discussed in this chapter, an oxygen enrichment of up to 31% was demonstrated.

Naturally, the use of enriched air resulted in an increase in reaction furnace temperature. Both the calculated increased capacity and the theoretical temperature rise in the reaction furnace have previously been published.[3] The observed changes in capacity and temperature for the 50-T/D Claus plants are summarized in table 5–2. Further information on oxygen enrichment in Claus plant operations can be found in other publications.[4,6,7]

Table 5–2. Observed effect of oxygen enrichment on a 50-T/D Claus unit.

Oxygen concentration	21%	29.5%
*Thermal reactor temperature	2,050°F	2,170°F
*Front-end pressure	14 psig	10 psig
*Temperature rise across three fixed bed reactors	151°F	169°F

*Equates to an increase in capacity of 18%.

Fail-Safe with O_2

The hazards of oxygen or enriched air are well known in the industry. The unique safety problems associated with use of enriched combustion air on a Claus unit were evaluated. The results of this study are summarized in figure 5–4. The safety system shown was installed and functioned satisfactorily on the 50-T/D Claus plants at oxygen concentrations of up to 31%. The principle features of this system were:

- Reaction furnace temperature monitored by an optical pyrometer
- High-reaction furnace temperature trips off oxygen flow
- High oxygen concentrations trip off oxygen flow
- Low flow of acid gas trips off oxygen flow
- Low oxygen supply temperature, indicating possible liquid oxygen in the supply gas, trips off oxygen flow
- Low air supply pressure trips off oxygen flow
- Oxygen flow control on flow recorder reset manually based on concentration of oxygen in air supply to the reaction furnace
- Oxygen pressure to preceding Flow Recorder Control (FRC) set by a pressure recorder
- Oxygen flow could be shut-down from either the control room or the field

Bypass Reheat Exchanger

The feed to the first fixed bed reactor must be reheated from 370°F to 440°F. On many sulfur units, this is accomplished by a heat exchanger utilizing high-pressure steam (refer to figure 5–1).

While this type of "indirect reheat" exchanger is a fine way to expedite sulfur plant start-ups, it does very little to improve the capacity of conversion of H_2S/SO_2 to liquid sulfur. For the 50-T/D Claus unit, a bypass to direct reheat gas to the first fixed bed reactor was installed. Figure 5–5 illustrates the location of the bypass.

The partial bypassing of both the reheat exchanger and the first stage condenser reduced the Claus train front-end pressure from 10.2 psi to 9.2 psi. This reduction in front-end pressure expanded capacity by 5%.

Theoretically, the increase of sulfur vapors to the first reactor would reduce conversion. In practice, a small shift in the reactor temperature rise from the lead reactor to the second and third reactors was noted. The theoretical reduction in conversion was too small to observe with a Dräger tube analysis of the sulfur plant tail gas.

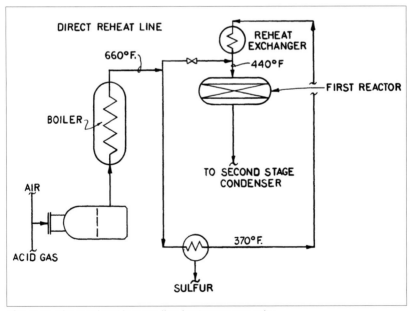

Fig. 5–5. Direct reheat bypass line increases capacity.

The use of oxygen-enriched air and the partial bypassing of the first stage condenser and reheat exchanger resulted in a combined capacity increase of 24%. However, these two operating parameters were only used during periods of limited sulfur recovery capacity.

Enriched air was limited because of oxygen cost. Use of the reheat bypass line required frequent and inconvenient adjustments to the reheat valve, shown in figure 5–5, in order to control the first fixed bed reactor outlet temperature.

Increased Front-End Pressure

The 50-T/D Claus unit was designed for a maximum operating pressure of 15 psig for vessel mechanical integrity and air blower discharge pressure. Unfortunately, the front-end pressure was limited to 10 psig by the depth of the sulfur seal leg drains. Above 10 psig, the process gas in the reaction furnace would blow out through the sulfur drain leg, and toxic vapors would be emitted to the atmosphere.

To permit a 14 psig front-end operating pressure, the seal legs were "cascaded" as shown in figure 5–6. Sulfur drains from the boiler and the first stage condenser were looped into the second stage condenser. This prevented a seal leg blowout from occurring until the second stage condenser reached a pressure of 10 psig. The effect of this change, which allowed an operating front-end pressure of 14 psig, was to up plant capacity by 23%.

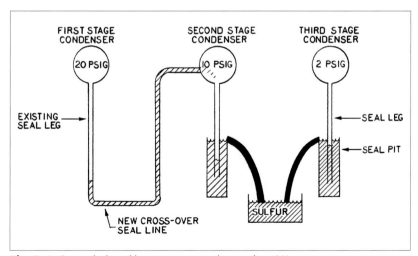

Fig. 5–6. Cascaded seal legs may expand capacity 40%.

The 4-psi increase in the reaction furnace pressure increased the acid gas pressure by an equivalent amount. This pressure rise backed up through the amine regenerator and raised the amine reboiler temperature by 6°F. Because the reboiler tube metallurgy was carbon steel, this increased temperature was of concern in regard to decreased reboiler tube life. To avoid accelerated corrosion, a corrosion inhibitor was injected into the reboiler inlet line. As indicated by a corrosion probe, the inhibitor effectively controlled the rates of corrosion at the higher reboiler temperatures.

Hydrocarbon in Acid Gas

A single mole of propane will consume as much sulfur plant capacity as 10 mol of H_2S.

Propane dissolved or entrained in the amine regenerator feed will appear in the acid gas feed to the Claus train.

To minimize the hydrocarbon content of the acid gas, the rich amine flash drum (i.e., the amine regenerator feed drum) was modified to operate at 10 psig instead of 50 psig. The resulting reduction in the hydrocarbon content of the acid gas was difficult to gauge because of its normal variability.

The average reduction approximated 0.5 mol% which represented a 5% gain of sulfur recovery capacity.

Water Vapor and Carbon Dioxide

Reducing the water vapor content of acid gas to a minimum also increased Claus capacity. During periods when one of the two sulfur trains was out of service, the amine regenerator reflux drum temperature was reduced from 135°F to 110°F. This was achieved by spraying treated water on the exterior of the amine regenerators' overhead fin fan tube bundles. This reduced the water content of the acid gas from 10% to 5% and thus increased sulfur recovery capacity by 2%.

By changing the feed location of the lean amine (MEA) to the gas scrubber, the rejection of CO_2 to sweet fuel gas was increased from 5% to 60%.

The CO_2 content of the acid gas dropped from 12% to 6%. This resulted in an increase of 2% in overall sulfur plant capacity.

The feed point change on the natural gas scrubber consisted of dropping the MEA inlet nozzle down from tray 20 to tray 5. Note that an even greater reduction of CO_2 to sweet fuel gas may be achieved by substituting MDEA for MEA.[5]

Reactor Inlet Baskets

Anyone who has ever inspected a Claus reactor after several years of use will note the crust deposited on the top of the reactor bed. This crust may account for a large portion of the pressure drop buildup seen as a run progresses.

Figure 5–7 illustrates one method to mitigate this problem. Baskets, partially filled with catalyst support balls, are inserted in the Claus plant catalyst bed. The depths of the baskets are sufficient to double the exposed surface area at the top of the bed. While the effect on the initial reactor pressure drop is small, during the course of a one-year run, the average reduction in pressure drop was estimated to be 30%. The baskets shown in figure 5–7 were only installed in the first reactor, as encrustation at the top of the second and third reactors is less of a problem.

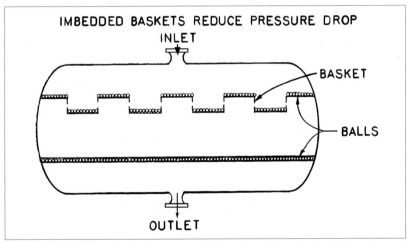

Fig. 5–7. Imbedded baskets reduce pressure drop.

The loss in overall conversion of H_2S to liquid sulfur due to the shorter average catalyst bed depth of the first reactor was too small to observe. Also, no shift in reactor temperature rise from the first to second reactors was observed. This too indicated that the shorter average reactor bed did not adversely affect conversion. The average 30% reduction in the first reactor pressure drop resulted in an approximate increase in Claus capacity of 2%.

Other Modifications

A few other changes which supported the increases achieved above by increasing the process air pressure to the reaction furnace were:

- The air blower check valve internals were modified to eliminate a substantial pressure drop.
- The air blower suction filters were altered to reduce pressure drop.
- The automatic vent on the blower discharge was overhauled to prevent unintended leakage of process air.

Table 5–3 summarizes the net capacity gains achieved by the process modification detailed previously.[4]

Table 5–3. Cumulative effect of modifications to increase Claus plant capacity[4].

1. Oxygen enrichment	18%
2. Direct first-stage reheat	5%
3. Cascaded seal legs	23%
4. Minimize hydrocarbons in acid gas	5%
5. Minimize water vapor content of acid gas	2%
6. Minimize CO_2 conetent of acid gas	2%
7. Reactor baskets	2%
	171%*

*Items 1 through 7 were multiplied together to arrive at the cumulative effect.

Pyrophoric Iron

During a sulfur plant outage, quantities of SO_2 may be observed emanating from open manways. Also, a blue fire can sometimes be seen in cold process vessels and lines. Pyrophoric iron is the cause of these phenomena. When dry, it spontaneously ignites after exposure to air. The heat evolved from pyrophoric iron combustion in catalyst beds can lead to high reactor temperatures. Also, irritating SO_2 vapors interfere with maintenance personnel working to repair the unit. (At 1,000 ppm, SO_2 is fatal to breathe.)

A small nitrogen purge, connected to the top of each reactor, will effectively suppress pyrophoric iron fires. Remember, however, that if you put out such a fire with water or nitrogen, the pyrophoric iron will reignite when it dries out and is again exposed to air. The nitrogen purge on the reactors also prevents the infiltration of air and moisture into the catalyst during a unit outage. Air, moisture, and sulfur make sulfuric acid, which causes catalyst deterioration and eats condenser tubes.

A typical three-stage Claus train, showing a normal temperature and pressure profile, is illustrated in figure 5–8. Pressure drops or temperature differences, which vary widely from the parameters shown, are likely indicative of a process problem.

Safety Note

Inhalation of H_2S for a few minutes is fatal. H_2S is insidious. Ambient concentrations of H_2S high enough to be dangerous will desensitize your olfactory nerves. A deadly concentration of H_2S has no odor.

Tail Gas Clean-Up

- **Beavon-Stretford.** Residual sulfur compounds (SO_2, COS, CS_2, S) are all reduced to H_2S in a catalytic hydrogenation reactor. The H_2S is oxidized in an aqueous-phase scrubber to sulfur and H_2O. The sulfur is then removed from the circulating aqueous phase. The oxygen is carried by an iron salt dissolved in the water and is re-oxidized with air.

- **Shell Claus Offgas Treating (SCOT).** The sulfur compounds are reduced to H_2S as previously described. However, the H_2S is selectively absorbed (i.e., CO_2 absorption is minimized) by a proprietary amine (ADIP or MDEA). The H_2S is then stripped in an ordinary amine regenerator and recycled back to the front-end of the sulfur plant.

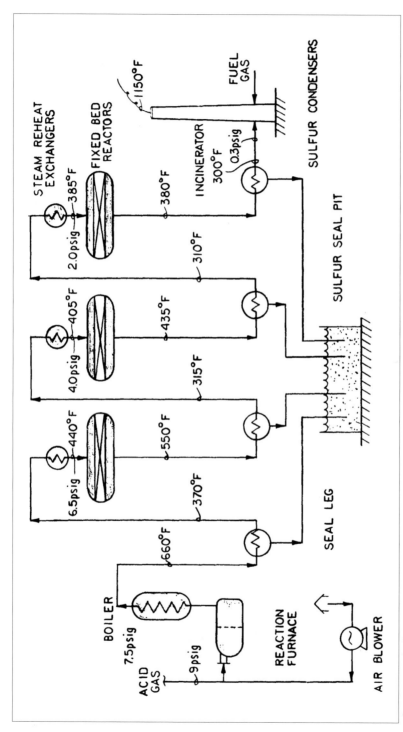

Fig. 5–8. Typical temperature and pressure profile for a gas-field sulfur-recovery train processing acid gas with 85% of H_2S through a three-stage Claus plant.

Troubleshooting aspects of the SCOT process are essentially the same as described in chapter 4 on amine treating. The operating characteristics of the Beavon-Stretford process have not been very successful. A number of U.S. refiners have totally given up on the process and converted their existing facilities to the SCOT process, which is much more common.

A somewhat similar but apparently more successful Beavon-Stretford type process is the LOW-CAT process promoted by ARI Technology. The French tail-gas treating process marketed by Institut Francaise du Petrole (IFP) also cannot be made to work in a predictable fashion.

Troubleshooting Checklist for Sulfur Plants

Measuring Conversion
Analysis of tail gas
Calculate conversion

Finding Lost Conversion
H_2S to SO_2 ratio at 2:1
Incinerator plume color
Incinerator temperature
Total temperature rise
 across reactors
Shift in reactor temperature profile
Sulfur precipitation on catalyst
Leaking reactor feed—
 effluent exchanger
Low velocity (sulfur fog)
Cold reheat gas
Catalyst condition
High COS plus CS_2

Start-up Tips
Heat up reactors quickly
Avoid H_2SO_4 formation
Avoid making carbon black

Increased Pressure Drop
Measure amount of lost capacity
Coked catalyst
High-pressure boiler leaks
Condenser tube leaks
Plugged seal legs
Improperly installed catalyst
 support screens

Maximizing Plant Capacity
Cascade seal legs
Maximize use of hot reheat gas
Reduce air blower check valve
 pressure drop
Change air blower suction filters
Minimize the H_2O in acid gas
Increase air blower
 driver horsepower
Reduce CO_2 in acid gas
Tail-gas treating
SCOT
Beavon-Stretford

References

1. E. K. Kerr, "The Claus Process." *Energy Processing/Canada,* July-August 1976.

2. T. S. Hansen, "Oxygen/Resid Relationships in FCC Operations," *Oil & Gas Journal,* Aug. 15, 1983.

3. M. R. Gray, "The Profitability of Oxygen Combustion Air Use in Claus Sulfur Plants," M. Eng. Thesis, University of Calgary, 1980.

4. Linde, Union Carbide, "Claus Plant Oxygen Enrichment," Union Carbide technical bulletin, 1983.

5. G. R. Daviet, et al., "Switch to MDEA Raises Capacity," *Hydrocarbon Processing,* March 1979.

6. H. Fischer, "Sulfur Costs Vary With Process Selection," *Hydrocarbon Processing,* March 1979.

7. H. Fischer, "Here's How the Modified Claus Process Treats Low Sulfur Gas," *Oil & Gas International,* July 1971.

8. B. G. Goar, et al., "Claus Plant Capacity Boosted By Oxygen-Enrichment Process, *Oil & Gas Journal,* Sept. 30, 1985.

Alkylation

An acid runaway on a sulfuric acid alkylation unit is an unforgettable experience. The alkylate turns purple, numerous hydrocarbon leaks appear in the reactor effluent piping, and acid strength declines at an uncontrollable rate. Most disturbing of all are the irritating fumes emitted from the spent acid tanks as the low-strength runaway acid is pumped out of the alkylation unit.

Few processes are as unforgiving of errors or lack of attention as an alky unit. Mistakes that one might get away with on another unit precipitate terrible corrosion rates on alky units. Operating errors that permit H_2SO_4 to escape from the reactor system can produce weak acid downstream of the reactor. Low-strength acid destroys carbon steel at an alarming rate.

The fundamental parameters of alkylation are isobutane concentration, acid strength and volume in the reactor, temperature, and mixing intensity. The engineer troubleshooting an alky problem should turn first to these fundamental precepts for help.

The Alkylation Process

In sulfuric acid alkylation, olefins and isobutane react to form a gasoline blending component. This reaction only occurs in the acid phase.

Olefins are extremely soluble in sulfuric acid; isobutane is only slightly soluble. Olefins will polymerize in the absence of isobutane. The polymerization reaction competes with the alkylation reaction:

$$olefin + olefin = polymer$$
$$olefin + isobutane = alkylate$$

As the percentage of olefin reacting to polymer increases, product octane drops and alkylate end point and acid consumption rise.

The object in running an alky unit is to suppress the competing polymerization reaction. This is done by:

- Diluting the olefin feed with large volumes of isobutane
- Maintaining high concentrations of isobutane in the reactor
- Mixing the isobutane liquid and acid phases vigorously
- Minimizing the reactor temperature to slow down the polymerization reaction relative to alkylation
- Dispersing the olefin feed in a large volume of acid

Olefins react immediately upon contacting acid. If the acid is cold and saturated with isobutane, alkylation operating problems will be minimized.

Process flow

A simplified sketch of an alkylation unit is shown in figure 6–1. Olefin and isobutane are charged to a refrigerated, stirred reactor. Acid in the reactor effluent is removed in a settler and recycled to the reactor.

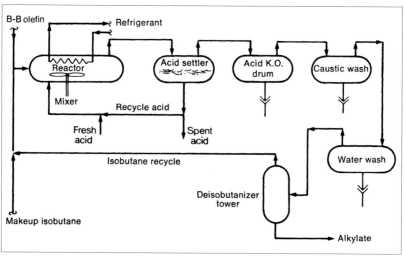

Fig. 6–1. Alkylation unit process flow.

Acidic components in the effluent are neutralized with caustic in the effluent treater. A deisobutanizer (DIB) tower distills off the isobutane from the alkylate bottoms product. The DIB overhead is recycled to the reactor.

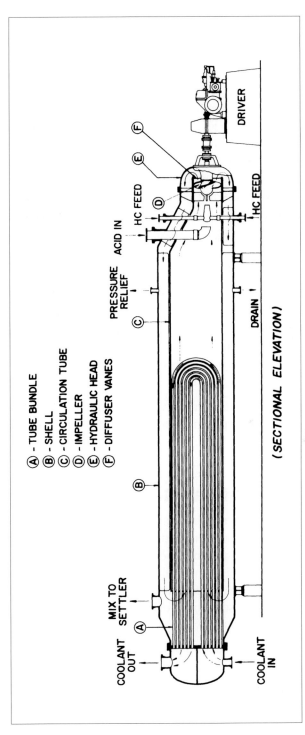

Fig. 6-2. Stratco horizontal alkylation contractor.

The most commonly used reactor in sulfuric acid alkylation service is the Stratco contactor. Figure 6–2 illustrates the improved eccentric impeller design of this reactor. The principal advantage of the Stratco contactor configuration is the inherently high isobutane concentration in the reactor effluent. This is achieved by operating the contactor at a pressure sufficient to suppress vaporization of the isobutane refrigerant recycle. The heat of the alkylation reaction is removed indirectly by partially vaporizing the settler effluent through the tube bundle shown in figure 6–2. Acid recirculation is achieved by gravity (i.e., density difference between hydrocarbon and sulfuric acid) and the action of the impeller. This eliminates the need for an acid recycle pump.

The older concentric type of Stratco contactor is subject to acid stratification in the contactor. The acid settles out in the bottom of the contactor and hence does not intimately mix with the olefin feed or circulating isobutane. To minimize this stratification, one refiner has:

1. Connected a 4-in. line to the "drain" connection shown in figure 6–2
2. Routed this 4-in. line back to the acid settler
3. Installed a globe valve on the "mix to settler" line shown in figure 6–2

Adjustments to this globe valve will force stagnant acid out of the 4-inch drain connection up into the acid settler for recirculation back to the contactor.

What Can Go Wrong

Operating deficiencies of sulfuric acid alkylation units can result in a unit shutdown. As the reaction environment deteriorates, acid is carried out of the reactor to downstream vessels. The weak acid, if not properly neutralized, is very corrosive and will cause hydrocarbon leaks. Early signs of reactor trouble are related to acceleration of the polymerization reaction. A unit may have some or all of these symptoms:

- High acid consumption
- Low octane and high end point of alkylate
- Red tinge to alkylate
- Increased caustic consumption in effluent treater
- Wild-looking spent acid (continues to fume and foam after weathering)

- Loss in DIB reboiler duty
- Plugging of flow meter taps
- Olefins in DIB overhead

In addition to difficulties related to alkylation chemistry, acid may be physically carried out of the reactor due to poor level or pressure control. Troubleshooting requires the engineer to differentiate between physical- and chemical-initiated carryovers.

Acid Carryover

There are two distinct types of alky process equipment: wet and dry. Equipment in dry service is acidic. Even if no free acid is present, hydrocarbon systems that have contacted sulfuric acid should be treated as sulfuric acid. If you can keep all moisture out of such systems, they will be noncorrosive.

Equipment in wet service must be kept above 6 pH. Even if no free water is present, hydrocarbon systems that have contacted water will be saturated with moisture. Keep acid out of such systems to minimize corrosion; an acid carryover mixes acidic components with the wet contents of the DIB tower.

The operating engineer should be able to distinguish between the two types of acid carryovers. Strong acid spillover from the settler is a physical acid carryover. Weak acid, containing a high concentration of acid esters, is the well-known phenomenon of acid runaway. It may also be called a chemical acid carryover.

The superficial result of an acid carryover is a sudden increase in corrosion and fouling. This is true whether the cause is of chemical or physical origin. The first step in troubleshooting such an incident is to check the following:

1. Has acid consumption recently been unusually high?
2. Does alkylate have a reddish tinge?
3. Has the alkylate end point suddenly increased by 30°F to 50°? Has alkylate octane dropped 1–2 research octane numbers?
4. Does an acid sample drawn from the acid settler have a strength of less than 80–85 wt%; does it have a wild, frothy appearance?
5. Are there olefins in the DIB overhead?

A positive response to these questions indicates the unit has experienced a chemical acid carry-over or an acid runaway. Indications of a physical acid carry-over are acid found in the settler is still strong (88+ wt%) and the alkylate properties are normal.

Physical Acid Carry-over

A massive carryover of strong acid from the settler may be caused by a number of unrelated factors. On start-up, many alky units have a tendency to carry over acid. With all other operating parameters normal, a large volume of acid appears in the effluent treater. A temperature rise is observed in the caustic wash drum. Several shifts later, the problem disappears. Particulates are the likely culprits. Finely divided solids, left in process equipment during the unit turnaround, cause the acid to foam by stabilizing the acid-hydrocarbon emulsion in the reactor.

The operating engineer can prevent foamovers of strong acid on start-up. After acid circulation has been established, obtain a completely weathered sample of the circulating acid. Use a small, stoppered bottle. Then, draw a similar sample of fresh sulfuric acid. Finally, shake both bottles vigorously.

Foaming of the samples drawn from the unit indicates dirty circulating acid. To prevent an acid carryover, change the acid prior to charging the olefins.

After start-up, there are many other factors that can cause a physical acid carryover:

- **Cold acid.** Alky operations are generally favored by low temperature. However, at temperatures below 40°F, the settling rate of acid in hydrocarbons is diminished. Operating below 30°F to 40°F has occasionally resulted in acid carryovers.

- **Acid settler levels.** Spent alkylation acid contains a sludge consisting mostly of corrosion products. The acid sludge accumulates in the settler. There it plugs the level gauge glass taps. This is one of the most frequent causes of physical acid carryovers.

Finding the acid-hydrocarbon interface level in the settler gauge glass is essential for the operators to control the rate of spent acid withdrawal. A plugged gauge glass fools the operators into believing that the settler acid level is too low. They will, therefore, reduce acid spending to raise this level. Not infrequently, the entire settler is filled with acid before the mistake is discovered.

Determine if a gauge glass tap is plugged by draining the glass down. If the acid level comes up very slowly, the tap is plugged. Blowing down an acid-filled gauge glass may be hazardous. To overcome acid carryovers resulting from plugged level taps, the gauge glass assembly should be modified, as shown in figure 6–3.

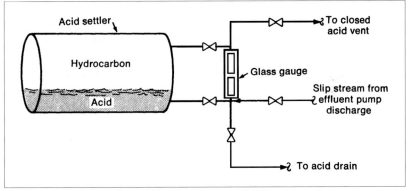

Fig. 6–3. Locating settler acid level (courtesy *Oil & Gas Journal*).

The features detailed in this sketch will permit operating personnel to drain down acid-filled gauge glasses safely and blow out plugged taps. High-pressure liquid from the discharge of the alky effluent pump is used to wash acid sludge out of the taps.

- **Reactor pressure control.** Unusually low reactor operating pressure can cause entrainment of acid in many types of reactors. A high propane content in the system or a high reactor temperature may also lead to acid carryovers. Poor control of these parameters will cause the contents of the acid settler to begin boiling. The resulting agitation inhibits acid settling. To determine if acid carryover is caused by boiling in the settler, increase the setting a few pounds per square inch on the settler pressure controller. This should stop the boiling and thus the acid carryover. Erratic operation of this controller may also precipitate an acid carryover.

- **Excessive acid recycle rate.** Intense mixing and high acid recycle rates improve alkylate octane. Very high acid circulation rates may cause a physical acid carryover. On the other hand, operators too often start reducing acid circulation at the first sign of a carryover. If the difficulty is due to a chemical acid carryover, cutting the acid recycle rate will make the problem worse by degrading the alkylation environment.

Chemical Acid Carryover

Acid esters are a product resulting from a poor alkylation reactor environment. Sulfuric acid can react with the olefin to produce a hydrocarbon-soluble compound called an acid ester. A small amount of esters is always coproduced with alkylate. When the environment for alkylation becomes less favorable, ester formation will accelerate. Conditions promoting the polymerization reaction are the same ones that lead to ester production:

- Low acid recycle rate
- Low isobutane concentration
- High temperature in the reactor
- Acid deficiency in the reactor
- Low acid strength
- Poor mixing in the reactor
- Too much propylene in the olefin feed
- Poor dispersion of the olefin feed in the reactor

If an acid carryover has been preceded by a period of deteriorating alkylate quality, the operating engineer can be assured that he is faced with a chemical acid carryover.

When no data are available, draw a sample of hydrocarbon from the reactor effluent. Allow the butane to weather off. If the residual alkylate has a reddish tinge, reactor conditions are conducive to an acid runaway or carryover.

Loss of acid strength

Once the sulfuric acid concentration drops below 85 wt%, acid strength will often continue to drop. This happens because the olefins are reacting to form long chain polymers and acid esters—both of which reduce acid strength. This is called an acid runaway.

Runaway acid has a wild appearance. It is foamy and continues to evolve acid fumes long after it is withdrawn from the reactor.

Frequently, an acid runaway is precipitated by a slug of contaminated hydrocarbon feed. The most common culprits are the following:

Butadiene, which per pound consumes 10 pounds of acid (really degrades sulfuric acid from 98% to 90%). Check with the operators of the FCCU to see if they have increased riser temperature. Figure 6–4 shows how butadiene concentration in the FCCU's butane-butylene (B-Bs) increases with riser temperature.[1]

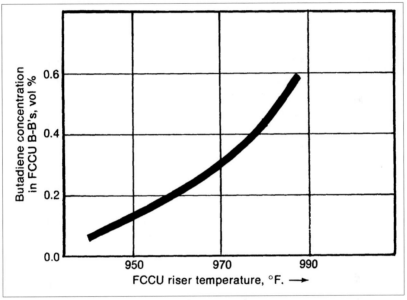

Fig. 6–4. FCCU can indirectly up acid consumption.

Water, which per pound degrades 10 pounds of acid. Most alky units have feed coalescers. Check the water level in the coalescer. Verify that the interface level controller is functioning properly.

Caustic, which per pound consumes 24 pounds of acid. Many acid runaways are caused by slugs of caustic carried into the alky reactor from upstream caustic treaters. Check the water drawn off the feed coalescers. A soapy feel indicates caustic in the feed (see chapter 9).

Mercaptan, which per pound consumes 20 pounds of acid.[2] Because mercaptan levels (even in untreated butane) rarely exceed several hundred ppm, this is an infrequent cause of acid runaways.

H_2S, which consumes large quantities of acid. If the concentration of propylene in the B-Bs has been increased, check the olefin feed for H_2S. This is conveniently done with easy-to-use lead-acetate paper.

Ethylene, which will also have a negative effect on acid consumption. This is a common problem with units alkylating propylene. Ask the FCCU operators to check for a low bottom temperature of their deethanizing stripping tower.

Low isobutane concentration

Chemical acid carryovers are frequently initiated by insufficient isobutane in the reactor effluent. As the isobutane in the effluent drops, formation of acid esters is favored. The minimum safe iso in effluent is roughly 35 to 50 mol%, depending on reactor configuration and other process variables.

Most large alky units have several reactors operating in parallel. Poor distribution of recycled isobutane between reactors is a common cause of acid runaways.

On one unit faulty flow meters caused the isobutane in effluent to be 32% and 54% from two parallel reactors. The reactor with the low isobutane in effluent experienced an acid runaway. This could have been prevented by sampling individual reactor effluents. In this case the operators had fooled themselves by only sampling the combined effluent for isobutane.

Reduced acid circulation

During an acid carryover crisis, operators too often cut acid recycle rates. If the unit is in an incipient state of chemical acid carryover, this may precipitate a full-blown acid runaway. The operating engineer can make a quick field decision whether to cut acid recycle by determining if:

Caustic strength is dropping rapidly in the effluent treater but little free acid is accumulating in the effluent acid KO drum (see figure 6–1). This is a chemical acid carryover. Do not reduce acid circulation or reactor mixing.

Caustic strength is dropping rapidly in the effluent treater, and a large, distinct acid phase is building in the effluent acid KO drum. This is likely a physical acid carryover. Reducing acid circulation and reactor mixing will help. Check the strength of the acid in the settler to make sure the unit is not experiencing an acid runaway.

The acid recycle is warmer than the reactor effluent. This is an indication that esters are continuing to react in the acid settler. The resulting heat of reaction increases the temperature of the acid recycle, while the hydrocarbon reactor effluent remains cool. The reaction of acid esters in an isobutane deficient environment increases the rate of acid consumption and lowers acid strength. Reduced acid strength produces more esters to fuel the after-reaction in the acid settler. This self-escalating production of esters is an acid runaway situation and is indicative of a chemical acid carryover requiring increased acid circulation. A 15°F temperature difference between the acid recycle and hydrocarbon effluent is an indication of an incipient acid runaway situation; a 3°F difference, while undesirable, is normal.

The alkylation reaction only takes place in the acid phase. The higher the acid circulation rate, the greater is the volume of acid available for dispersion of reacting olefins. Olefin dispersion discourages polymerization and acid ester formation. It is the author's experience that long-term, high-acid consumption problems are caused more often by low-acid recycle rates than any other single problem.

Poor mixing

Alkylation quality is adversely affected by poor mixing or unbalanced olefin dispersion in the reactor. However, field observations have shown mixing and feed dispersion failures typically do not initiate acid runaways.

For example, on one unit with a Stratco reactor, the impeller turbine (i.e., the mixer) was shut down for several hours. Nothing dramatic happened—even with zero mixer horsepower. Similar observations have been made for Exxon cascade reactor type units. Low pressure drop through the sparger rings of cascade reactors will, however, contribute to high acid consumption.

Propylene alkylation

Acid ester formation is much more pronounced when alkylating propylene as compared to butylene. When alkylating a barrel of propylene-propane (P-Ps), perhaps five times as much acid can be needed as compared to B-B alkylation.

If P-Ps are charged to a unit that normally only handles B-Bs, the operators should be alerted to watch for drops in acid strength. Also, caustic consumption in the effluent treater will increase. Many times, charging a slug of P-Ps has initiated a chemical acid carryover because of low acid strength. The reason for this is the inability to draw off enough spent acid to keep up with the accelerating acid consumption. One refinery always increases acid strength 1 wt% before charging propylene.

During propylene alkylation, spending acid at the minimum safe acid spending strength (89%) will actually increase acid consumption. At 89% acid, depending on the percent of propylene alkylate produced, sulfuric acid consumption (in tons per day) will be quite a bit higher then when spending acid at 91% strength. Acid strengths may be continuously monitored by gravity with an on-stream acid analyzer.[3]

High temperature

Stratco contactor or reactor temperatures are usually kept below 65°F. Cascade reactor effluent is typically less than 55°F. Ester formation and polymerization are favored as temperature increases. However, the operation of Stratco reactors for several hours at temperatures exceeding 75°F has been observed. Other than high acid consumption, nothing disastrous happened—at least in the short term.

Like poor mixing and olefin dispersion, high reactor temperatures in the field do not often initiate acid runaways. However, on some alky units, high reactor temperatures have resulted in a physical acid carryover of immense proportions. Boiling hydrocarbons in the settler, due to elevated temperature, was found to be the cause. Settler pressure must be maintained 3–5 psi above the effluent bubble-point pressure.

Emergency Response to Acid Runaway

Having decided that an acid runaway or chemical acid carry-over is in progress, a quick response is appropriate. The signal to initiate action is when a 6–8 pH in the effluent treater water wash drum cannot be maintained. (Several units have installed a continuous pH meter on the water wash drum with excellent results.) The following steps should then be taken:

1. Block olefin feed out of unit. Continue to run the refrigeration compressor. Most alky units will produce

enough vapors to keep the compressor on-line for half an hour, even without olefin feed.

2. Continue charging make-up isobutane.

3. Continue isobutane circulation at the maximum rate.

4. Go to fresh acid addition and acid spending at a maximum rate.

5. Check the pH of the circulating caustic stream in the effluent treater. If this stream is now really weak acid, shut down the caustic circulation pump until the caustic is renewed. Circulating weak acid through carbon steel pipes will start leaks at the welds in just a few hours.

6. Check the DIB reboiler shell for acid. Drain out any acid sludge and flush through with butane.

7. Resume charging olefin feed when acid strengths are restored to 90+ wt%.

Improving Alkylate Octane

There are a few basic ideas that the operating engineer can use to formulate projects to upgrade poor alkylate quality. The most common of these is to increase the isobutane concentration of the reactor effluent. A 10% increase in this concentration is worth 0.5 motor octane number.[4] Often, optimizing the isobutane make-up feed point to the DIB tower will help.

Increased sulfuric acid circulation will also increase product octane. On one unit a filter was installed on the acid recycle stream. By removing particulate matter from the circulating acid, the tendency of the acid to foam and carry-over was eliminated. This permitted an increase in the acid recycle rate. A large increase in alkylate octane and a decrease in acid consumption were observed.

For units operating above a 50°F reactor effluent temperature, debottlenecking the refrigeration section is worthwhile. If the compressor is horsepower limited, increasing the refrigerant condenser surface area will help. Changing the condenser bundle from bare tubes to low-fin tubes is a cost-effective way to gain condensing surface area. On one unit, an 80% increase in heat transfer was observed after a fin tube bundle was installed.

Upgrading effluent treating

Figure 6–1 shows an effluent treater consisting of a strong caustic followed by a water wash. Many refiners have modified this system as follows:

1. Both vessels are retrofitted with electrostatic precipitators.
2. The fresh acid is injected into the mix valve upstream of the former caustic wash. This extracts both acidic and neutral esters from the reactor effluent.
3. The resulting acid phase is pumped back to the contactors. This reduces acid losses, saves caustic, and increases alkylate yield by the recycled esters reacting with isobutane in the contactor.
4. As the neutral esters have been largely extracted in the acid wash, the former water-wash vessel only acts as a minor clean-up step for traces of acid carryover. A small amount of caustic is added to maintain a 10–12 pH in the circulating water.

The key to successful operation of acid-effluent treating is to charge all the make-up acid to the acid effluent mix valve and maintain a 10–15 psi ΔP across the mix valve.

Additives

There are several proprietary alkylation additives available that purport to improve alkylate octane and yield while reducing acid consumption. The value of these additives is a controversial subject.

In general, the use of additives in the refinery industry does little to overcome fundamental process operating problems. This is true of alkylation.

If your alky operation is badly deficient due to low isobutane and acid circulation, an additive will not retrieve the situation. Operating engineers should concentrate on the process fundamentals and not expect an additive to miraculously pull them out of trouble.

Hydrofluoric Acid Alkylation

A typical process flowsheet for an HF alky unit is shown in figure 6–5. The following five troubleshooting examples are based on problems encountered on large HF alky units along the Gulf Coast and in Oklahoma.

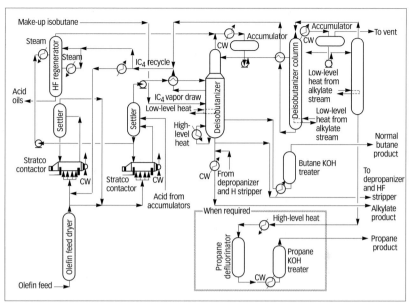

Fig. 6–5. HF alkylation-unit flow diagram (courtesy *Oil & Gas Journal*).

Propane stripper

The first case involved the operation of an alkylation unit propane stripper. The alkylation unit operators were experiencing great difficulty in controlling the propane content of the isobutane recycle loop.

In an attempt to keep the depropanizer pressure under control, the operators were venting vapor from the depropanizer reflux drum through the potassium hydroxide (KOH) vent gas scrubber. As a result of depropanizer venting, the alkalinity of the scrubber had declined, while the depropanizer pressure was increasing.

The operators felt that operating conditions were unsafe, and that preparations for a unit shutdown should be made. It is unsafe not to be able to neutralize the HF-rich vent gas.

A field analysis revealed the cause of the excessive propane buildup. The FCCU gas plant had encountered an upset, causing the ethane content of the olefin feed to increase.

The resulting composition of the propane-propylene feed being charged to the HF alkylation unit was:

C_2 2%

C_3 61% (propylene)

$$C_3 \quad 35\% \text{ (propane)}$$
$$C_4 \quad 2\%$$

The only outlet for the additional ethane in the olefin feed (except for nonroutine venting), was in the propane product (propane stripper bottoms).

As can be seen in figure 6–5, the propane stripper is charged from the depropanizer overhead accumulator. The purpose of the propane stripper is to remove trace amounts of HF acid from the propane product.

The operating procedure for this tower was to hold the stripper pressure constant and maintain a minimum tower-bottoms temperature. As the stripper bottoms product was essentially propane (with 1%–2% isobutane), holding the bottoms temperature and pressure constant fixed the ethane content of the propane product at approximately 3%. It was determined by an overall alkylation unit material balance that the propane product should have been about 5% ethane.

This discrepancy resulted in an accumulation of ethane in the depropanizer reflux drum. This, in turn, increased the ethane content of the propane stripper feed.

To maintain the stripper bottom temperature, the reboiler steam control valve kept increasing the heat input to the tower. Because the feed rate to the stripper was fixed to avoid tray flooding, the propane product rate declined, and both propane and ethane accumulated in the depropanizer. Eventually, the operators were forced to begin venting the depropanizer reflux drum to the KOH scrubber.

To prevent the alky unit from shutting down, we suggested that the propane stripper pressure be increased or that the bottom temperature be lowered. Note that a propane stripper removes traces of HF from the propane product by stripping and not by decomposing hydrofluoric-hydrocarbon chemical bonds.

Thus, there is no minimum column bottom temperature or maximum pressure that must be maintained. Only the stripping factor through the fractionation trays is critical to prevent HF acid from remaining in the propane product. A lower reboiler outlet temperature was targeted from vapor pressure charts. The new target was selected to obtain 6% ethane in the bottom product. Within two shifts, the propane balance in the alky unit had been restored to normal, and a needless shutdown was avoided.

The unit supervisor subsequently observed that the 6% ethane content of the propane product exceeded the maximum permissible specification. Of course, this was not a consequence of the stripper operations but solely a function of the amount of ethane in the alkylation unit charge, which had to be corrected at the FCCU plant.

Reduced alkylate quality

After a recent alkylation unit maintenance turnaround, a Southwest refiner was faced with a low-octane, high end-point alkylate product. In addition, HF acid catalyst consumption was higher than normal and was approaching the physical limits of the HF regeneration equipment. The refiner felt that the problem must be in the contactor reaction section. We checked all flow rates, temperatures, and pressures around the contactors. Special laboratory analyses of hydrocarbon and HF acid catalyst samples were performed. All process and mechanical variables around the contactors were acceptable, but the quality of the alkylate product was still poor.

Additional field observation and inspection revealed that some minor piping changes had been made during the turnaround. These piping changes did not allow the recycle isobutane to mix with the olefin feed stream before entering the contactors (figure 6–6).

Because the alkylation reaction occurs instantaneously, olefins will react with themselves to form low-octane polymers if not properly mixed with isobutane upon exposure to the acidic HF catalyst. It was suspected that the recent piping modifications were the cause of the poor alkylate quality.

The process configuration was returned to normal, allowing the recycle isobutane and olefin to mix externally to the reaction zone. This allowed for a homogeneous mix of olefin and isobutane before contact with the HF catalyst. In approximately eight hours, the quality of the alkylate leaving the alkylation unit had returned to normal.

Depropanizer reboiler limits

In another situation, the operating personnel at the alkylation unit needed to increase the propane-propylene feed rate, but they were experiencing depropanizer operating problems. The propane product was not within specification because of high isobutane concentration.

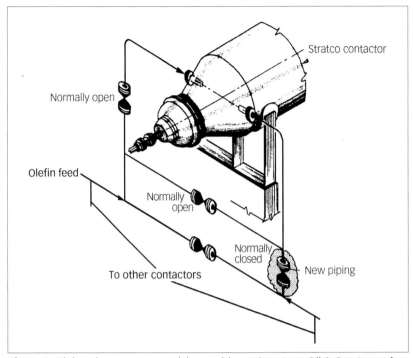

Fig. 6–6. Piping changes cause mixing problems. (courtesy *Oil & Gas Journal*).

Increasing the propane-propylene feed rate would aggravate the problem. To bring the propane product within specification, the depropanizer reflux rate was increased, but this caused the reboiler control-point temperature to fall. The operators noted that the reboiler steam inlet control valve was wide open, thus indicating that the depropanizer was limited by reboiler capacity.

We calculated the actual heat-transfer coefficient of the reboiler and found that it was quite close to the design value. The liquid level in the steam condensate pot was checked to make sure it was below the bottom of the channel head.

Steam pressure was at maximum and the recirculation to the thermosiphon reboiler appeared to be adequate. That is, the reboiler outlet temperature was only slightly higher than the depropanizer bottom temperature.

Apparently, the reboiler was functioning properly. Operating personnel were simply attempting to push the reboiler's capacity well beyond its design duty.

Further analysis indicated that enhanced reboiler capacity could be achieved by dropping the tower pressure. As the tower pressure dropped, the tower bottom temperature (at constant butane in the bottom product) also fell. This in turn permitted an increase in the reboiler heat duty:

$$Q = U{\cdot}A{\cdot}(T_2 - T_1)$$

where:

Q = Reboiler duty

U = Boiling heat-transfer coefficient

A = Exchanger surface area

T_2 = Condensing steam temperature

T_1 = Average reboiler process-side temperature

A drop in the depropanizer pressure of about 15 psi permitted the desired increase in reflux rate of 10%.

One of the unit operators inquired as to whether the combination of the lower tower pressure and higher reflux rate would promote vapor flood and reduced fractionation efficiency due to entrainment. We explained to the operators that as long as the temperature difference between the top and bottom of the depropanizer increased with increasing reflux rate, the trays would not suffer from excessive entrainment.

Methyl-tributyl Ether (MTBE) unit carry over

The operators at a Gulf Coast refinery noted that the alkylate product from the HF alkylation unit had a slightly off-color appearance. The operators obtained an HF acid sample which indicated a 4% drop in acid strength since the previous day's sample.

- The olefin feed was stopped, and the unit was put on isobutane circulation.
- The acid regenerator operations were revised to increase acid strength back to normal levels.

Knowing that this alkylation unit had an MTBE unit operating upstream, we suspected that excess carry-over of reaction products (oxygenates) caused the sudden loss in acid strength. Samples of the olefin feed indicated that the level of dimethyl ether had risen from the normal 40–60 ppm range to greater than 150 ppm. In the HF alkylation unit, oxygenates convert to water, reducing the system's acid strength.

It was determined that switching the MTBE guard reactor beds had caused a slug of oxygenates to enter the alkylation unit. To prevent shutdowns caused by low acid strength, we recommended revising the lab schedule for more frequent acid analyses, and a new HF acid sample location for obtaining more representative samples.

Isostripper turndown

Due to a reduction of FCCU charge, the olefin feed rate to the HF alkylation unit had dropped by 50%. As a consequence, operating personnel had reduced isobutane recycle by 50%.

At that time, the isobutane content of the normal butane product had increased from 5% to 60%. The normal butane was produced as a vapor side-cut from the isostripper (figure 6–5). Because an isostripper tower is not refluxed, the fractionation tray vapor and liquid loadings change proportionally to the tower's feed rate.

We advised that the severe reduction in the stripper's fractionation efficiency had resulted from the reduced tray loadings and concluded that the dry pressure drop per tray was slightly less than the equivalent height of liquid on the tray deck.

The dry-tray pressure drop is essentially a function of the vapor flow up the tower, while the equivalent height of liquid on the tray is largely determined by the height of the outlet weir. Also, the tray decks were equipped with valve caps. The trays were dumping due to the above factors.

While tray vendors state that valve trays maintain their fractionation efficiency even at lower vapor rates, this often proves not to be the case. Relatively small, out-of-level problems involving the tray deck installation, weir height, downcomer clearance, and so on, may cause severe vapor-liquid channeling at low dry-tray pressure drops. Because the isostripper had fractionated quite well at normal feed rates, it was concluded that the current problem was due to low vapor flow through the tray decks.

The isostripper was operating at its design pressure of 120 psig. The minimum operating pressure of any tower producing a net liquid distillate product is determined by the condensing temperature of the fluid in the overhead accumulator.

Because this problem occurred in December, it was possible to operate the isostripper well below the design pressure. To increase the volumetric vapor flow (but not the mass flow) through the tray decks, we suggested that the tower pressure be lowered from 120 psig to 65 psig. The objective was to increase the dry tray pressure drop by about one inch of liquid per tray.

The result of this pressure reduction was quite remarkable. The isobutane content of the alkylate product fell from 6% to 1%. The isobutane content of the normal butane sidecut dropped from 60% to 3%. All this was achieved without any increase in reboiler duty on the isostripper.

Troubleshooting Checklist for Alkylation

Problem Indicator
Low water wash pH
Low DIB reflux water pH
Acid in effluent KO drum
Fouled DIB reboiler
Acidic alkylate
Low octane and high end
  point alkylate

Physical Acid Carry-over
Strong acid in effluent KO drum
Alkylate octane normal
Dirty acid on start-up
Cold acid
Plugged settler level glass
Erratic reactor pressure control
Boiling in settler
Excessive acid recycle
Surfactants in feed
Temperature rise in caustic
  effluent wash drum

Acid Runaway
Low isobutane in effluent
Low acid circulation
Sulfur in feed
Water or caustic in feed
Fresh acid addition stopped
Butadiene in feed
Sudden increase in propylene
Olefins in DIB overhead
High ester concentration in acid

Emergency Response
to Acid Runaway
Block out olefin feed
Continue isobutane feed
Circulate maximum isobutane
Maximize fresh acid flow
Check pH of caustic treater
Drain acid from DIB reboiler

Safety Note

Strong sulfuric acid burns skin. However, as long as it is kept out of the eyes, most other accidents can be handled without injury.

If strong acid is spilled directly on your skin, wash it off with a large volume of water. You have several minutes before the acid starts burning, so keep calm. Most likely, just a slight reddening of the skin will result.

A large amount of acid spilled on your clothes is best dealt with by removing the affected garment before washing down in the safety shower. Bad acid burns have resulted from operators getting wet while wearing acid-soaked clothes. Again, one has several minutes to disrobe and get to the safety shower.

Exposing the eyes to sulfuric acid is another matter entirely. Alky unit personnel must always wear goggles because acid and caustic cause blindness.

References

1. NPRA Question-and-Answer Session on FCCU Operation, 1974.

2. NPRA Question-and-Answer Session, 1961.

3. *Continuous Sulfuric Acid Alkylation Analyzer* Bulletin 88–2. Drushal Design & Development, Baton Rouge, La., 1988.

4. R.C. Cupit et al., "Catalytic Alkylation," *Petro/Chem. Engineer,* January 1962, p. 208.

5. Liolios, G., and Lieberman, N. "HF Alkylation Operations Improved." *Oil & Gas Journal* June 20, 1988.

Fluid Catalytic Cracking Units

f you were to divide all refinery knowledge into three equal parts, fluid catalytic cracking unit (FCCU) technology would encompass one full portion. Not only is an FCCU the economic heart of most refineries, but advances in catalyst and mechanical design constantly expand the scope of the troubleshooter's problem.

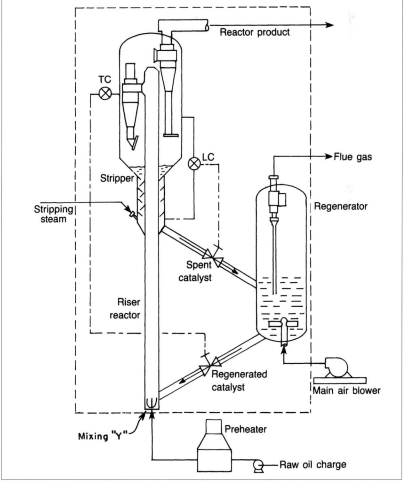

Fig. 7–1. A typical fluid cracking unit. The catalyst flow is controlled with slide valves.

Figure 7–1 summarizes the components of a modern FCCU. Conversion of FCCU feed takes place in the riser, while the "reactor" is little more than a vessel to contain the cyclones used to separate the catalyst from the vaporized hydrocarbon. The circulating catalyst promotes catalytic cracking of the feedstock. Also, it transfers a portion of the heat of combustion of the coke formed in the cracking reaction from the regenerator to the riser. The flow of catalyst is usually controlled by slide valves, and the operating parameters affecting the physical condition of the catalyst make earning a living as an FCCU troubleshooter a real challenge.

Catalyst Steam Stripping

Spent catalyst flows back to the regenerator through the spent slide valve. The spent catalyst is coated with carbon and heavy hydrocarbons. The purpose of the steam stripping section shown in figure 7–1 is to remove a portion of the heavy hydrocarbons adhering to the catalyst.

The hydrocarbons stripped off the catalyst are returned to the reactor and eventually recovered as products. The hydrocarbons left on the catalyst are burned in the regenerator with the following debits:

- Increased combustion air requirements
- Higher regenerator temperatures
- Loss in product yield
- Greater hydrothermal and thermal catalyst deactivation

As the process engineer, how do you know if the correct amount of stripping steam is being used in the catalyst stripper? Depending on the ratio of CO to CO_2 in the regenerator flue gas, there are a number of different answers to this question.

If the regenerator is operated with low concentrations (less than 3%) of CO in flue gas or in a complete CO burn mode, an increase in stripping steam to the catalyst stripper will lower the regenerator temperatures. This happens because the amount of fuel flowing into the regenerator, in the form of hydrocarbons adhering to the spent catalyst surface, is diminished.

As the regenerator temperatures drop, the regenerated catalyst flowing back to the reactor through the regen slide valve cools. This in turn drops the riser outlet temperature, unless the catalyst circulation rate is increased. The cooler riser results in less conversion and thus less coke is made. This effect further reduces the coke on the spent catalyst and lowers the regenerator temperature.

But suppose the stripping steam rate to the catalyst stripper is increased and no appreciable drop in regenerator temperatures is observed? Therefore, a proper or an excessive quantity of stripping steam is already being used. When this happens, try reducing the stripping steam rate below its initial level. If an excessive amount of stripping steam had been used, the regenerator temperature would not be affected. Of course, if the regenerator heats up, you have cut the stripping steam rate below the proper level.

Riser temperature control with stripping steam

One of the primary variables that an FCCU operator can use is riser temperature. Increasing riser temperature usually results in increased conversion, though not necessarily greater yields of gasoline. For units not on automatic reactor temperature control, riser temperature can be increased by:

- Increasing feed temperature
- Reducing recycle of FCCU cycle oils
- Cutting the fresh feed rate
- Increasing catalyst circulation rate
- Increasing the regenerated catalyst temperature

Left to his own, the shift worker operating the FCCU control panel often chooses the simplest method to hold riser temperature constant. On many units, the control panel operator has discovered that changing the steam rate to the catalyst stripper is a fast, easy route toward altering the riser temperature. Unfortunately, this mode of control results in the destruction of FCCU feed in the regenerator.

One cannot, and should not, generalize as to specific steam rates. However, for orientation, on one large FCCU, field tests determined that beyond a stripping steam rate of 6 lb of steam per ton of catalyst circulated, there was no significant reduction in regenerator temperature.

Observing optimum stripping steam rates with high concentration of CO in regenerator flue gas

If there is little or no excess oxygen in the regenerator flue gas, a change in the catalyst stripping steam rate will probably not affect the regenerator temperatures. The amount of combustion is limited by the availability of oxygen and not the amount of fuel. For most regenerators, if the CO in the flue gas exceeds 3%, there will not be any measurable amount of oxygen in the flue gas. So if the catalyst stripping steam rate is reduced and an increased amount of hydrocarbon is left on the catalyst, the amount of oxygen consumption in the regenerator will not increase. As a consequence, no appreciable change in regenerator heat release will occur, and the regenerator temperature will not rise. Instead, the ratio of CO_2 to CO will decrease. That is, more CO and less CO_2 will appear in the regenerator flue gas. There will also be an increase in the carbon on the regenerated catalyst. This will be particularly evident if the CO_2/CO ratio is near unity. The adverse effects of increased residual carbon are described in the section "Catalyst Regeneration Problems" in this chapter.

Plugged Spent Cat Steam Stripper Distributor

The stripping steam shown in figure 7–1 is distributed through a pipe grid. The orifices of this distributor can easily plug, if the stripping steam is wet. The problem is that a small amount of catalyst backs into the distributor and forms a cement-like deposit with the water entrained in the steam. The result is increased carbon on catalyst and a high stripping steam distributor pressure drop.

The one time I had this problem was due to water carry-over from a slurry pump around waste heat boiler in Delaware City. I lowered the water level set-point on the boiler and over a period of several days the steam distributor ΔP dropped back to a tolerable level and the excessive carbon on spent catalyst dropped to normal.

Catalyst Poisoning Affects Wet Gas Compressor Performance

The single factor that makes troubleshooting process operations so difficult is that the cause of a problem can be so very remote from the effect. The following story, describing how a minor incident at a crude unit caused a dramatic flow reversal on an FCCU, is instructive.

Many years ago, as a young process engineer, I was assigned to follow the operation of the refinery's crude unit. After carefully analyzing the economics of crude distillation, I concluded that the most important objective was maximizing FCCU feed production and minimizing residue product. Accordingly, I instructed the operating staff to curtail the use of vacuum tower wash oil by 50%. This step increased FCCU feed by 1,500 B/SD and reduced vacuum tower tar production by a corresponding volume.

A week later, having returned from a training course in computer technology, I chanced to be driving past the refinery. It was a crystal clear night in South Texas and the lights from the refinery sparkled attractively. Suddenly, a black cloud belched forth from the FCCU regenerator stack. Blacker than the night sky, the cloud spread at an alarming rate. I was witnessing a flow reversal of staggering proportions. Mercifully, at the time I was not aware that I was the cause of this titanic display of partially burnt hydrocarbon.

What is a flow reversal?

The hydrocarbon feed to an FCCU enters the unit at the *Y*. The FCCU feed can flow up the riser and be converted to gasoline and distillate or it can flow up the regenerator catalyst standpipe and be converted to black smoke. When the FCCU feed takes the latter route, we call it a *flow reversal.*

Now the differential pressure across the regenerator slide valve (regen ΔP) is normally kept high enough to prevent the FCCU feed from flowing into the regenerator. This differential pressure is nothing more than the pressure drop created by the catalyst as it flows across the restriction created by the regenerator slide valve. The pressure drop is usually kept at 3–5 psi.

If the regenerator slide valve pressure drop is low enough, the FCCU feed, which is supposed to flow up the riser, can take the path of least resistance and flow into the regenerator. This is called a flow reversal and results in the emission of a staggering volume of heavy black smoke from the regenerator stack. The factors that precipitate a flow reversal are:

- Sudden drop in the regenerator pressure
- Regenerator slide valve sticking open
- Sudden increase in the riser pressure

This last factor, an unexpected increase in the riser pressure, caused the flow reversal I witnessed years ago in Texas. But how could it be that I, the crude unit process engineer, was responsible for this increase in riser pressure?

Surging wet gas compressor

I drove through the refinery gates and arrived at the FCCU just in time to see the operators blocking in the main FCCU feed valve. The chief operator informed me that the wet gas compressor had "gone into surge," and that as a consequence they had "lost the regen ΔP." How can a problem with the wet gas compressor affect the regenerator slide valve differential pressure? Figure 7–2 shows that the pressure in the riser follows changes in pressure at the suction of the wet gas compressor. A sudden increase in compressor suction pressure raises the riser pressure and thus reduces the regen slide valve ΔP.

What did the chief operator mean when he said the compressor had gone into surge? Figure 7–3 illustrates a typical centrifugal compressor performance curve for a fixed speed (i.e., motor-driven) machine. Note that this figure actually shows two curves: One curve is for a gas of 34 mol wt, while the second curve is for a gas of 29 mol wt. Normally this compressor operated on point "A" shown in figure 7–3. That is, the compressor operating parameters were:

- 17,000 actual cubic feet/minute (ACFM)
- 34 mol wt gas at suction
- Suction volume, ACFM
- 20 psig suction pressure
- 200 psig discharge pressure

The compressor was putting up a differential pressure of 180 psig when handling 34 mole wt gas. The compressor discharge pressure was fixed by the operating pressure of the gas plant absorber. The compressor suction pressure was automatically controlled by the

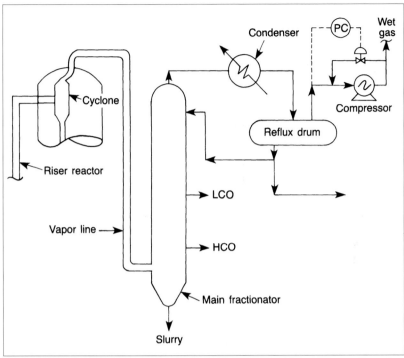

Fig. 7–2. The pressure in the riser follows pressure changes at the suction of the compressor.

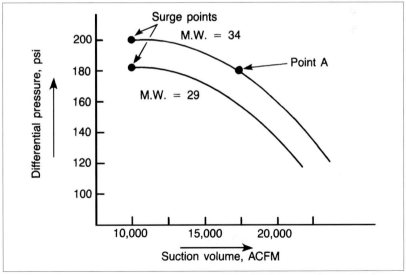

Fig. 7–3. A centrifugal compressor can start surging because the molecular weight of the gas is reduced.

pressure control valve (PC) shown in earlier figure 7–2. Normally, the flow of gas to the suction of the compressor consisted of:

- 10,000 ACFM of gas flowing from the reflux drum
- 7,000 ACFM of gas recycled from the compressor discharge

If the reflux drum pressure started to rise, the pressure control valve would start to close. The compressor would then move up (i.e., to the left) on its operating curve. This would automatically pull the reflux drum pressure back down to 20 psig. In this way the pressure of the riser also would be stabilized, and the regen slide valve differential would be held constant.

Unfortunately, this happy situation was disturbed by a change in the molecular weight of the gas. Specifically, the molecular weight started to drop. Of course, the FCCU operating personnel didn't notice this. All they observed was that the pressure control valve shown in figure 7–2 gradually closed over a period of a week. Now it transpires that for a constant-speed centrifugal machine the following applies:

$$\Delta P_f = \left(\frac{MW_f}{MW_i} \right) \times \Delta P_i$$

where:

ΔP = Differential pressure developed by the compressor, psig

MW = Molecular weight, lb/mole

i = Initial condition

f = Final condition

For a period of one week, unknown to the FCCU operating crew, the molecular weight of the gas from the reflux drum had dropped from 34 to 29. The compressor was slowly pushed back on its curve by the automatic closing of the pressure control valve, which was trying to maintain the 180 psig differential pressure between the reflux drum and the gas plant absorber. When this pressure control valve was shut, the flow of gas to the suction of the compressor was only 10,000 ACFM. This was also the absolute minimum flow that the compressor needed to prevent surge.

The chill wind that had cleared the sky of clouds had also dropped the reflux drum temperature by 10°F. This reduced the molecular weight of gas flow to the suction of the compressor from 29 to 28. The compressor could no longer put up a 200 psig discharge pressure because of the effect described by the equation above. The lower molecular weight gas was still trying to push the compressor further up on its curve. But the machine could not operate to the left of its surge point shown in figure 7–3. Hence, the compressor began to stall or surge.

Compressor surge affects regen slide valve

The erratic operation of the compressor caused the pressure in the reflux drum, the main fractionator, the vapor line, and of course, the riser to bounce between 20 and 25 psig. The pressure differential across the regenerator slide valve was lost. The FCCU feed entering the "Y" at the base of the riser began to take the path of least resistance—that is, the oil began to flow backward up the regenerated catalyst standpipe into the regenerator. This huge amount of oil thermally cracked to coke and gas in the 1,400°F regenerator. The resulting black smoke billowed out of the flue gas stack. The operators responded to this emergency by removing feed from the unit, diverting wet gas to the flare at the suction of the compressor, and partially closing the regen slide valve. My response was to rush over to the lab and examine the composition of the wet gas for the previous week.

Causes of increased hydrogen production

My heart raced and my hands went cold as I thumbed through the recent gas chromatographic analysis of the FCCU wet gas. To my dismay the data confirmed my worst fear: The hydrogen content of the wet gas had risen from 9% to 19% in just one week! Here was the reason that the compressor had been forced into surge. The tremendous increase in hydrogen content and a lesser increase in methane composition had lowered the molecular weight of the wet gas from 34 to 29.

With a sinking feeling I turned to the FCCU feed quality data sheets. I was especially interested in the analysis of vacuum gas oil originating at the crude unit. This stream made up more than half of the total FCCU feed. I compared the most recent analysis with one two weeks old:

	Previous Analysis	Recent Analysis
Initial boiling point, °F	401	405
Final boiling point, °F	981	1,008
Percent sulfur	0.41	0.43
Conradson carbon residue	0.3	0.8
Watson K factor	11.9	11.8
Gravity, °API	27.0	26.3
Vanadium, ppm	0.9	3.1
Nickel, ppm	0.5	2.0
Iron, ppm	0.2	0.2
Sodium, ppm	nil	nil

Nickel and, to a lesser extent, vanadium promote the production of molecular hydrogen during catalytic cracking of gas oil. The metals are deposited on the catalyst, and unless they are deactivated—for example by adding antimony—hydrogen production from FCCU feed is enhanced. This effect is quantified in figure 7–4.[1] One commonly accepted equation for calculating equivalent nickel is:[2]

$$\text{Equivalent Ni} = \text{Ni} + \text{V}/4.8 + \text{Cu}/1.2 + \text{Fe}/7.1$$

The reason for the reduced molecular weight of the FCCU wet gas was now transparent.

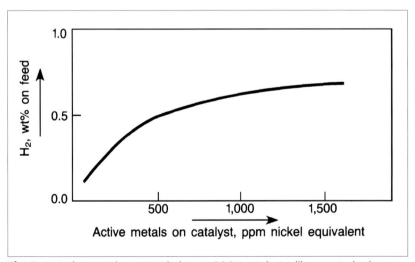

Fig. 7–4. Active metals accumulating on FCCU catalyst will promote hydrogen production from FCCU feed. (after Habib et al., courtesy *Ind. Eng. Chem.*, 1977).

Resid entrainment into FCCU feed

When I had instructed the crude unit operators to reduce vacuum tower wash oil by 50%, the entrainment of resid or tar into the gas-oil FCCU feed had substantially increased. As shown in the above tabulation of laboratory data, the entrainment had caused an increase in Conradson carbon residue of the vacuum gas oil. This was of no consequence. The extra coke made due to the higher Conradson carbon was compensated for by cutting the FCCU feed preheat temperature to hold the regenerator in heat balance. Of greater importance was the increase in nickel content from 0.5 ppm in vacuum gas oil to 2.0 ppm. This fourfold multiplication in nickel content was reflected by a concurrent increase of nickel accumulation of the circulating catalyst. As the nickel content of the catalyst increased, the hydrogen content of the FCCU wet gas also rose while the molecular weight of the wet gas dropped. The falling molecular weight caused the FCCU's centrifugal compressor to produce a greater polytropic head. As the polytropic head rose, the machine operated closer to its surge point. When the compressor started to surge, the reactor pressure rose, forcing the FCCU feed out of the riser and into the regenerator, thus precipitating the flow reversal I observed.

Life is replete with injustice. However, whenever I am erroneously blamed for a mistake, I mentally balance the injustice against the time I wasn't terminated for causing that flow reversal in South Texas.

Catalyst Regeneration Problems

"Coke makes coke." This oft-repeated quote is the hallmark of FCCU operations. It means that, if catalyst is inadequately regenerated and the FCCU is operated to maintain a constant conversion, the weight percent coke yield on FCCU feed will increase. Usually, a parameter that acts to produce more coke also increases the light hydrocarbon gas make. So inadequate catalyst regeneration increases the load on the wet gas compressor and also, by making more coke, increases the regenerator air requirements.

Some of the factors that may lead to deficient catalyst regeneration are:
- Damaged air grid
- Insufficient air
- Excessive regenerator velocity
- Poor spent catalyst initial distribution

Air grid troubles

No two air grids are ever designed the same. The operator may wonder if this means air grid design is approaching perfection or that the designers do not know what they are doing. Perhaps the difficulty with designing air grids resolves itself into two contradictory objectives:

- The need for uniform air distribution across the regenerator
- The requirement to permit regenerated catalyst to pass into the regenerated catalyst standpipe

To achieve this latter objective, air flow from the grid is usually restricted from the part of the grid immediately above the entrance to the standpipe.

The troubleshooter can suspect an air grid deficiency—either by design or mechanical failure—if one of the following problems develops:

- Regenerated catalyst takes on a salt-and-pepper appearance. A sample of regenerated catalyst appears to be a mixture of white and black particles as opposed to a uniform shade of grey.
- Regenerated catalyst is poorly regenerated (i.e., several tenths of 1 wt% carbon) while the regenerator flue gas contains 0.5 vol% or more of oxygen.
- A large amount of afterburn in one section of the regenerator. For example, the following regenerator temperature distribution would indicate afterburn:

Quadrant Number	Dense Bed Temperature, °F	Cyclone Outlet Temperature, °F
1	1,320	1,360
2	1,340	1,370
3	1,360	1,480
4	1,330	1,370

The data above indicate severe afterburn taking place in the cyclone above quadrant 3. Such a temperature rise might indicate that a disproportionate amount of air is entering this cyclone, perhaps from a ruptured air grid lateral in quadrant 3.

Excess air in any portion of a regenerator reduces the carbon on catalyst to low (0.05 wt%) levels, while a localized air deficiency results in under-regenerated catalyst with 0.3 wt% or more carbon on catalyst. This mixture of relatively low carbon on catalyst—which appears light grey—and relatively high carbon on catalyst—which appears charcoal grey—accounts for the salt-and-pepper look of the catalyst. Naturally, a damaged air grid can cause such localized air-distribution problems.

For a regenerator operating in a full CO combustion mode, one sign of maldistribution of air or catalyst inside the regenerator is a requirement for large amounts of excess oxygen in the flue gas (i.e., more than 2%) to prevent hot spots from occurring in the dilute phase or cyclones. For instance, on one FCCU the amount of air required for complete CO combustion was observed to increase during the last six months of operation prior to a unit turnaround. Also, increasing quantities of oxygen were noted in the flue gas. Internal inspection of the regenerator showed that the center hub of the air grid had eroded out. Note that many units require a CO combustion promoter to complete the CO burning in the dense bed. These units—usually shallow-bed or high air-velocity designs—show high dilute-phase or cyclone temperatures because the CO combustion catalyst activity is low. Consequently, in the case of a regenerator operating in full CO combustion, air grid failures cannot always be readily identified by excessive afterburning in the cyclones.

Insufficient air

FCCU operating personnel often determine the correct flow of combustion air to the regenerator based on the shade of the regenerated catalyst. A sample of catalyst is drawn every few hours and visually compared against a standard. Typically, a light grey shade indicates proper regeneration, while dark grey shows that more air is needed. Sometimes operators adjust combustion air rates on the basis of regenerator bed temperatures (i.e., to a constant ΔT between dense bed and the dilute phase) or CO to CO_2 ratios in the flue gas. However, only the shade of grey of the catalyst gives a completely reliable indication of the carbon on catalyst.

A process engineer who observes a 15% increase in FCCU wet gas flow over a two-hour period can be pretty sure that the FCCU control board operator has allowed the unit to get behind in burn. This means the regenerator has been operating with an air deficiency for some time, and coke has built up on the circulating catalyst. The increased gas make is a symptom of lower catalyst activity and selectivity due to carbon on catalyst. Correcting this situation is called catching up in burn. Trying to catch up too fast causes excessive regenerator temperatures.

Regenerator size and spent catalyst distribution

Uneven catalyst regeneration and excessive afterburn are often observed together. These difficulties may not, however, be related to mechanical failure. Regenerating large flows of catalyst in a relatively small combustion zone has resulted in uneven burning. This phenomenon is indicated by an increasing flue gas temperature spread between the cyclone outlets as the catalyst circulation rate rises.

Poor initial distribution of the spent catalyst can account for an uneven temperature rise through the cyclones. For instance, in a relatively small regenerator (i.e., a regenerator where the flue gas superficial velocity exceeds 4 ft/sec or with an inventory of less than five minutes of catalyst circulation), catalyst flowing from the spent catalyst stripper may flow clear across the regenerator and accumulate on the side opposite the spent catalyst inlet. The high concentration of spent catalyst, and thus coke in this portion of the regenerator, may cause a local deficiency of oxygen. The resulting flue gas is rich in CO. In another portion of the regenerator, the deficiency of coke-rich spent catalyst causes residual oxygen to be present in the flue gas. These two flue gas streams—one rich in CO and the other containing O_2—may be drawn into a cyclone together. The resulting combustion of CO is observed as afterburn in the cyclone.

Identifying air grid damage

The above discussion illustrates the difficulty of differentiating between an air grid failure and a spent catalyst distribution problem based on a regenerator's operating characteristics. On more than one occasion I have seen FCCUs shut down to repair anticipated regenerator air grid damage. Management had concluded, based on observed afterburn data in the cyclones and the salt-and-pepper appearance of regenerated catalyst, that the air grid had failed. However, internal inspection of the regenerator did not reveal extensive damage to the grid.

The best way to avoid such an embarrassment is to monitor the air grid pressure drop carefully. Figure 7–5 shows that the pressure drop measurement should be made between the regenerator and the air inlet line. For these data to be meaningful, they must be collected when the air grid is first put in service, as well as when damage is suspected. If adequate data are thus assembled, a plot as shown in figure 7–6 can be kept. The data points that correspond to low air grid pressure drop and high afterburn temperatures are signs

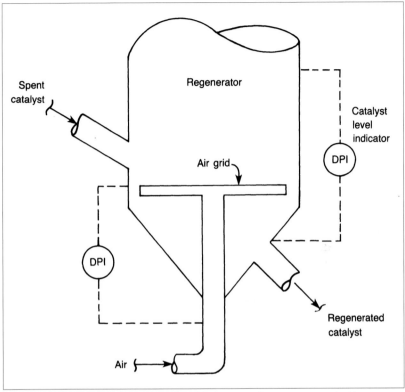

Fig. 7–5. An inexplicable drop in air-grid pressure drop accompanied by an increase in afterburn indicates air-grid damage.

of a damaged air grid. Note that for the pressure-drop data to be consistent, they must be normalized for air flow rate:

$$\Delta P_n = \Delta P_o \left(\frac{V_b^{\,2}}{V_o^{\,2}} \right) \left(\frac{P_o}{P_b} \right)$$

where:

ΔP_n = Normalized air grid pressure drop, psi

ΔP_o = Observed air grid pressure drop, psi

V_o = Observed actual volume of air, ft^3

V_b = Volume of air based on design condition, ft^3

P_o = Observed regenerator pressure, psia

P_b = Design regenerator pressure, psia

Note that V is not the scf of air, but the actual volume of air flowing to the grid.

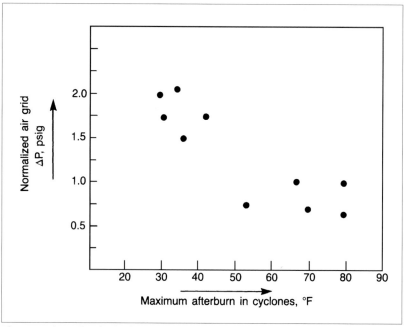

Fig. 7–6. The points in the lower right-hand corner are a symptom of air-grid failure.

Catalyst Deactivation

A loss of conversion or selectivity in an FCCU is a major economic debit to a refinery. The vendor supplying the FCCU catalyst is best equipped to troubleshoot this problem. However, a few of the difficulties you could look for are:

- **Thermal deactivation.** Caused by excessive regenerator temperatures such as severe afterburn in one of the cyclones. Note that if one out of five sets of cyclones in a regenerator is experiencing excessive (i.e., 1,400+°F) temperature, almost all of the catalyst, not just 20%, is being exposed to a 1,400°F temperature.

- **Hydrothermal deactivation.** Caused by exposing catalyst to a steam atmosphere at high (1,300+°F) temperature. Since most of the water vapor in the regenerator flue gas is produced from coke combustion, the presence of steam in the regenerator is unavoidable. Consequently, excessive overall regenerator temperatures cause hydrothermal deactivation of the catalyst. A potentially more serious area of hydrothermal deactivation is the regenerated catalyst standpipe. Small amounts of steam used to purge pressure taps or slide valve guides can create a very high steam partial pressure in the regenerator standpipe, which is just as hot as the catalyst in the regenerator dense bed. Substituting inert gas (nitrogen) for steam as a purging medium eliminates this problem.

- **Loss of antimony flow.** The effect of nickel accumulation on the catalyst has already been reviewed. To offset the effect of metals, an antimony-containing solution may be injected into FCCU feed. If you observe a loss in catalyst selectivity, such as a sudden increase in hydrogen in the wet gas, check the antimony pump.

- **Sodium and other metal contamination.** Caustic and sodium chloride are the most common reasons for the unanticipated loss of catalyst activity due to metal contamination. Metals such as vanadium and nickel are uniformly dispersed in FCCU feed. Caustic and NaCl are most likely contained in a separate aqueous phase. Checking a sample of feed for metals may not reveal the presence of sodium ions dissolved in water that is not uniformly dispersed in the hydrocarbon phase. Also, such sodium ions may appear from unexpected sources.

For example, one integrated Midwestern refinery had a terrible incident of FCCU catalyst poisoning due to caustic carry-over from an upstream unit. Figure 7–7 tells the story. The scenario was as follows:

- Mercaptans were extracted from virgin butanes with a circulating caustic stream.

- The caustic was regenerated with air to convert the mercaptans—which are slightly soluble in caustic—to disulfides—which are very much less soluble in caustic.

- The caustic was scrubbed to remove entrained disulfides with a small stream of naphtha.

- The naphtha, now containing the disulfides, was supposedly separated from the circulating 10% caustic stream in the disulfide separator vessel.

- The naphtha was charged directly to the FCCU to convert the hard-to-handle, toxic disulfides to hydrogen sulfide.

- Operating personnel at the mercaptan extraction unit did not effectively control the naphtha-caustic interface. Caustic carried over direct to the FCCU feed line.

It took a week for FCCU technical personnel to identify that sodium was the contamination deactivating their catalyst and another week to track down the source.

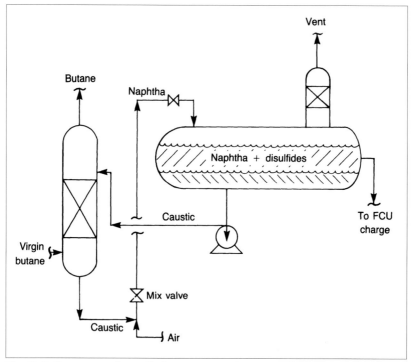

Fig. 7–7. Carry-over of caustic from a mercaptan-extraction unit resulted in sodium poisoning FCCU catalyst.

Catalyst deactivation versus refractory feed

Catalyst activity can be measured in a sophisticated laboratory with decent reproducibility. An activity of 70+ is good; less than 50 is very bad. Since the lab analysis typically is a service provided by catalyst vendors, turnaround time for results may be one week. The troubleshooter faced with a sudden loss in feed conversion may not have a week to wait. The plant manager may need to know right away if the loss in FCCU gasoline production is due to a more refractory (i.e., harder to crack) feed or poisoned catalyst.

To make this determination without waiting for lab data, have the operators load several tons of fresh catalyst over a few hours. If gasoline production suddenly improves by 5%–10% and then dies off again, you are circulating low-activity catalyst. On the other hand, if the shot of fresh catalyst does not have a dramatic effect on gasoline production, your feed has probably become more refractory due to increased aromatic content. A check of its Watson K characterization factor will verify this.

Suppose you have an FCCU with a catalyst inventory in the reactor-regenerator of 200 tons. Does adding two tons of fresh catalyst (1%) significantly affect the apparent cracking activity of the circulating catalyst? Field observations indicate that the answer is yes. If 45 microactivity catalyst is being circulated, a 1% addition of fresh, high-activity catalyst can make a big difference—for a little while.

Reducing Regenerator Temperature

High regenerator temperatures (i.e., greater than 1,380°F) may be caused simply by excessive heat release. Other than the obvious step of reducing CO combustion by cutting the air flow, you might try:

- Lower feed preheat. This will only work if the catalyst circulation also can be increased to maintain the riser temperature.
- Inject water in the regenerator torch oil nozzles. Of course, this accelerates hydrothermal deactivation of the catalyst.
- Reduce the air-blower discharge temperature. This temperature can be 300°F or more. An air cooler on the blower discharge can cut the regenerator temperature by 150°F.
- Inject water to the riser along with the feed. This has the same effect as reducing feed preheat. Care must be taken to use only water that is free of sodium chloride. Incidentally, relatively small amounts of water or steam in the feed can sometimes accomplish wonders in improved yields by promoting dispersion in the riser.
- Recycle naphtha or LCO to the riser. Again, this will only be effective in removing heat from the regenerator if the catalyst circulation can be increased.

Troubleshooting Cyclone Malfunctions

Reactor cyclones are used to separate cracking catalyst from vaporized reaction products; regenerator cyclones perform the same function for flue gas. In both services the erosive nature of the catalyst, combined with rapid gas velocities, may be highly destructive to the steel cyclones. Typically, the reactor cyclone is exposed to 950°F–1,000°F temperatures, while the regenerator cyclones must handle flue gas from 1,250°F–1,500°F. Due to the temperature, the regenerator cyclones are especially prone to failure.

A sketch of a typical two-stage cyclone is shown in Figure 7–8. Some of the more common causes of cyclone operational failures are:

- Dipleg unsealed
- Dipleg plugged
- Dipleg failure
- Cyclone volute plugged
- Hole in cyclone body

Cyclone malfunctions result in catalyst loss. A deficient reactor cyclone may be identified by high BS&W levels in the slurry oil product. Regenerator cyclone problems are visibly identified by the increased opacity of the regenerator flue gas or by reduced rates of spent catalyst withdrawal.

Dipleg unsealed

Note the flapper valve in figure 7–8. The flapper functions as a check valve, permitting the catalyst to flow out of the dipleg but preventing gas from backing up into the dipleg. If this flapper becomes jammed in an open position or if it breaks off, flue gas can flow into the bottom of the dipleg. This is a particularly significant problem in the second stage of a two-stage cyclone. For this case the flue gas bypasses the first-stage cyclone through the second-stage cyclone's dipleg. Catalyst flow out of the dipleg is inhibited, and as a consequence, there is an increase in the catalyst lost from the regenerator stack.

You can verify that an observed increase in catalyst loss is due to an unsealed dipleg by carefully raising the catalyst level in the regenerator. This stops the flue gas from backing up into the dipleg.

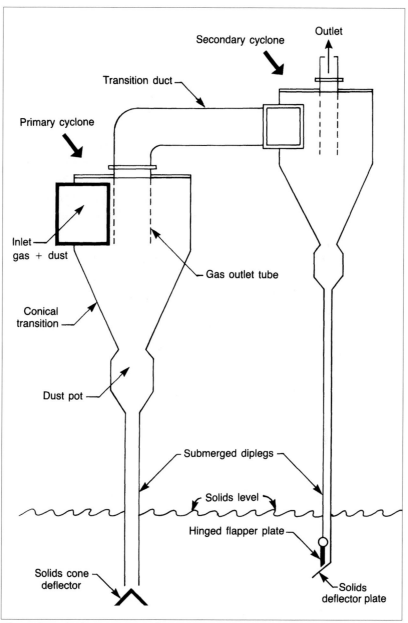

Fig. 7–8. Typical two-stage cyclone.

Plugged dipleg

If a cyclone dipleg plugs off entirely, the cyclone will flood with catalyst and become nonfunctional. Broken refractory is the most common cause of a plugged dipleg. Occasionally, an excessively high catalyst level tends to restrict catalyst drainage from the dipleg. I have observed some really impressive losses of catalyst in the regenerator stack due to high catalyst level in the regenerator.

Note that the catalyst levels indicated in the control room are not measured catalyst levels. They are calculated based on the pressure differential, as shown earlier in figure 7–5, and an assumed dense bed apparent catalyst density. If something happens to the catalyst to make it less dense, the indicated catalyst level will be lower than the actual catalyst level.

Dipleg failure

Dipleg failure has the same effect as unsealing a dipleg. That is, gas bypasses the normal cyclone inlet, and catalyst drainage from the cyclone volute is restricted. Damage to the cyclone body is fairly common in regenerators, but rare in reactors. Pluggage inside the cyclone volute is typically caused by refractory failure.

Other than altering the catalyst level, no operational changes pinpoint a particular type of cyclone failure. However, the troubleshooter can differentiate between catalyst attrition and cyclone damage as a cause of excessive catalyst losses. By collecting a sample of catalyst entrained in the regenerator flue gas, the amount of catalyst fines can be measured. This value is then compared to the percentage of fines in a sample of regenerated catalyst. A low relative percentage of fines in the sample obtained from the flue gas indicates cyclone damage. A high percentage of fines in the flue gas indicates catalyst attrition.

Air-Blower Problems

Many FCCUs are coke-burn limited, which usually means they are limited by the regenerator combustion air blower. On most units the air blower is a low-head centrifugal compressor. Factors to look for in defining the cause of reduced blower capacity are:

- Plugging of the air intake filters. Anything over a few inches of mercury pressure drop on these filters is excessive.
- High pressure drop on the compressor discharge. For a machine operating far out on its compressor curve, an increase of a few psig in the discharge pressure can result in a large decrement in air flow. A combined ΔP of 3–4 psi across the air grid and blower discharge piping is typical.
- Increased ambient temperature. The capacity loss between a 40°F day and a 90°F day is about 10%.
- Dirt in the compressor rotor. This can be a big problem in dusty environments. Compare the observed head-capacity characteristics of the compressor against the performance predicted by the compressor curve.
- Slow speed. A big problem for turbine-driven machines.
- High wet gas compressor suction pressure. An increase in the pressure at the suction to the wet gas compressor backs up through the fractionator overhead condensers and trays, the vapor line, and the reactor cyclones to the riser. Now, for the operator to maintain a proper regenerator slide valve ΔP (i.e., at least 3 psi), he must pinch back on the flue gas slide valve to raise the regenerator pressure. This, of course, translates directly into an increase in air-blower discharge pressure.
- Pluggage of fractionator overhead condensers. The resulting pressure drop has the same effect as raising the wet gas compressor suction pressure. Ammonium chloride salts may be the culprit.
- Excessive regenerator slide valve ΔP. Each psi of pressure lost across this slide valve translates into one more psi of air-blower discharge pressure, unless the flue gas slide valves are already wide open.

The above discussion emphasizes an interesting and vital aspect of troubleshooting an FCCU: the hydraulic relationship between the wet gas compressor, the combustion air blower, the flue gas slide valve, the spent catalyst slide valve, and the regenerated catalyst slide valve. A schematic diagram summarizing this relationship is shown in figure 7–9. There is no way to understand a "cat" unless the process engineer firmly grasps this concept.

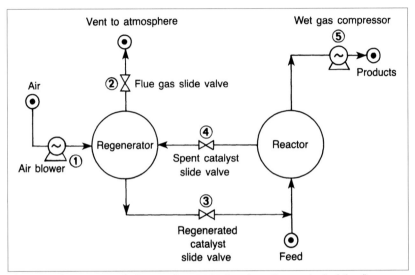

Fig. 7–9. The pressure balance of an FCCU involves the control of the five parameters shown in this figure.

Catalyst and Feed Mixing

Referring to figure 7–1, we can see that the FCCU feed mixes with the catalyst in the Y. Let us suppose that our target riser temperature is 990°F. Our FCCU feed might be 500°F and still a liquid as it flows into the Y. The catalyst might be 1,380°F. The hot catalyst cools down to 980°F and in so doing supplies energy to:

- Vaporize the feed.
- Heat it to 980°F.
- Provide the endothermic heat of cracking.

But suppose the oil and catalyst are not well-mixed in the base of the riser. Suppose the oil and catalyst do not come to thermal equilibrium until they reach the top of the riser. How can one identify this problem and how does it affect FCCU yields?

A West Coast refinery had recently revamped its cracker to enhance product yields. However, the yield structure of the revamped unit was worse than before the expensive modifications were made.

The first abnormality I observed on this unit was that the riser outlet temperature was 13°F hotter than the reactor product (i.e., second-stage cyclone) outlet temperature. This could only be due to continuous cracking as the riser effluent flowed through the cyclones. This reaction, being endothermic, lowered the riser effluent temperature.

If cracking was continuing after the catalyst-oil mixture had passed through the long riser reactor, however, certainly the oil and catalyst could not be mixing very efficiently in the Y. This meant that some of the oil was mixing with very little catalyst and some of the oil was contacting much more catalyst and was consequently being heated to a temperature far greater than the riser outlet temperature. Such localized overheating of the FCCU feed would produce an excessive amount of light gas. This was exactly the problem that my client reported: abnormally high gas make and low molecular weight wet gas. Of course, the gas production was at the expense of gasoline and distillate yields.

The usual cause of poor mixing in the Y is uneven distribution of the oil. Figure 7–10 shows two of the 12 injection nozzles used to disperse the oil into the flowing catalyst. At first glance there appears to be a 22-psi ΔP to distribute the oil from the FCCU feed header into each injection nozzle. My client therefore reported to me that this 22 psi ΔP was considered insufficient to distribute the oil evenly through the Y. However, take a closer look at the data presented in figure 7–10.

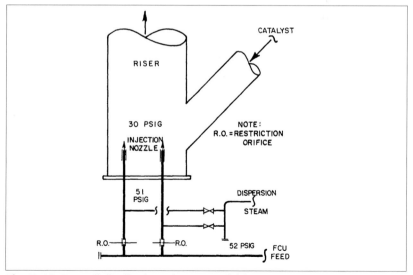

Fig. 7–10. Poor oil distribution ruins FCCU yields.

The pressure drop through the injection nozzles is certainly 21 psig. But this ΔP is all due to the dispersion steam. The pressure drop through the FCCU feed restriction orifices was only one psi. To prove the point, we recalculated the ΔP for the orifice size and found that the calculated orifice ΔP was also about one psi.

My client replaced the restriction orifices with orifices of one-half the diameter. The pressure drop went from 1 psi to 16 psi. The temperature loss between the riser and second-stage cyclone outlets was reduced from 13°F to 7°F, and a substantial reduction in the light wet gas production was observed. This is another good example of the importance of making personal field observations, even when obtaining simple pressure measurements.

Troubleshooting Checklist for Fluid Catalytic Cracking Units

Note: This list encompasses only a small portion of FCCU problems.

Catalyst Steam Stripping
Remove oil of catalyst
Increased regenerator air
Higher regenerator temperature
Loss product yield
Hydrothermal deactivation
Alter steam rate to test
Change riser temperature
  with steam
Check CO/CO_2 ratio

Raise Riser Temperature
Higher feed temperature
Lower cycle oil recycle
Reduce feed rate
Increase catalyst circulation
Raise regenerator temperature
Poor mixing in Y

Catalyst Poisoning Effect on Gas Compressor
May cause flow reversal
Lowers regen slide valve ΔP
Lowers molecular weight
  and causes surge
Increased hydrogen make
Increased Ni content of feed

Catalyst Deactivation
Thermal deactivation
Hydrothermal deactivation
Metals poisoning (Ni + Va)
Loss of antimony flow
Caustic of salt
Differentiating between catalyst
  problems and refractory feed
Effect of fresh catalyst

Reducing Regenerator Temperature
Lower feed preheat
Inject water through torch
  oil nozzles
Cut air blower discharge
  temperature
Add water to riser feed
Recycle naphtha and LCO to riser

Cyclone Malfunctions
Unsealed diplegs
Plugged diplegs
Dipleg failure
Pluggage in cyclone volute
Hole in cyclone body

Catalyst Regeneration Problems
Coke makes coke
Damaged air grid
Insufficient air
Excessive regenerator velocity
Poor spent catalyst
  initial distribution
Check air grid pressure drop
Afterburn problems

Air Blower Problems
Air intake filters plugged
Pressure restriction of
  blower discharge
Increased ambient temperature
Compressor rotor fouled
Blower turbine running slow
Wet gas compressor problems
Fractionator condensers plugged
High regen slide valve ΔP

References

1. E. Habib et al., *Ind. Eng. Chem.*, Prod. Res. Div., 16, 291 (1977).

2. W. L. Nelson, *Oil & Gas Journal*, 59(43), 143 (1961).

FCCU Product Fractionation

During my first hour on the first day of my new job as technical manager of a large Gulf Coast refinery, the plant manager called me into his office. Without so much as a word of welcome, he informed me that something terrible had happened. The refinery had just completed a major expansion. Part of this huge project had been the construction of a new FCCU fractionator. This large tower had recently been commissioned, and unit operating personnel were reporting a serious problem: The light cycle oil (LCO) product was a light brown rather than the required transparent yellow. Furthermore, the operators reported that the only way the LCO product color could be improved was to reduce FCCU charge below the design feed rate. The dark color of the LCO was probably indicative of slurry oil entrainment to the LCO product drawoff tray.

I told the plant manager that there were three possible explanations:

- **Leaking LCO versus feed heat exchanger.** He informed me, however, that the LCO product rundown pressure was higher than the feed pressure, so the dark-colored feed could not be leaking into the LCO.
- **Insufficient LCO internal reflux.** The plant manager immediately telephoned the FCCU control center and instructed them to increase the internal reflux or LCO wash oil rate. Twenty minutes later the chief operator informed us that the LCO was no longer light brown-it was now dark brown.
- **Flooding.** Since this was a brand-new tower operating at its design throughput, it did not seem probable that it would flood.

Unfortunately, I knew from previous experience that one indication of flooding is that as the internal reflux rate is increased, fractionation is adversely affected. The plant manager became agitated when I mentioned this possibility. He instructed me to immediately investigate the problem in the field.

Once at the FCCU, I instructed the instrument mechanics to hook up a Magnehelic ΔP cell (i.e., a differential pressure gauge) as shown in figure 8–1. Note how the Magnehelic instrument is physically located above the upper pressure tap to prevent liquid accumulation in the pressure sensing lines. Not only must the pressure-sensing lines be self-draining, but they should all be one inch or larger. Smaller sensing lines will accumulate liquids due to the force of capillary attraction. The source of the liquid is hot vapor in the column which accumulates in the sensing lines. I then proceeded as follows:

1. Reduced the FCCU feed rate until the LCO appearance returned to a light, transparent yellow. The measured pressure drop across the tower internals indicated on the ΔP cell was 44 inches of water.

2. Increased the FCCU feed rate back to the design point. The internal reflux and pumparound rates were increased to hold a constant temperature profile.

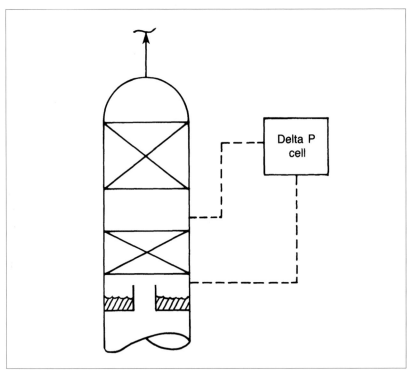

Fig. 8–1. When measuring pressure drops across tower internals, connect ΔP cell in a vapor space and elevate the cell above the top tap.

During a 20-minute period, the LCO color changed from a bright, clear yellow to a dirty blond and then to a translucent light brown. Coincident with this degradation of product color was a gradual increase in pressure drop between the LCO product drawoff tray and the reactor vapor inlet line. The Magnehelic ΔP cell leveled off at 79 inches of water-pressure drop. From these observations I could now draw two definite conclusions:

- The tower was flooding. Any time reduced fractionation coincides with increased tower pressure drop, incipient flooding is indicated. To confirm this conclusion, I increased the LCO internal reflux rate and observed that the Magnehelic reading increased and the LCO product color darkened.

- Flooding was being initiated from the bottom of the tower. I knew this because 79 inches of water-pressure drop represented a ΔP per tray of about 7 inches of water. Therefore, each of the trays below the LCO product draw tray was flooding. Put another way, the apparent fluid foamed specific gravity on a flooded tray deck in hydrocarbon service is typically 0.3. For trays spaced 24 inches apart, the observed pressure drop corresponding to a fully flooded tray is then:

$$\Delta P = 24 \text{ in.} \times 0.3 = 7 \text{ in.}$$

Trays may flood from the bottom up, but never from the top down. That is, flooding on a tray in the center of a tower causes all of the trays above to flood but does not flood the trays below. Of course, flooding from the lowest tray in a fractionator may simply be caused by a high liquid level in the bottom of the tower. In this case, I checked the liquid level in the sight glass at the base of the tower to eliminate that possibility.

I had concluded that the FCCU fractionator was entraining black slurry oil into the LCO product because the slurry pumparound section was flooding. But how could a brand-new tower, designed by a reputable engineering contractor, flood at design throughput?

Superheated Reactor Vapor Affects Tower Capacity

As shown in figure 8–2, the reactor vapor enters the fractionator at 980°F, while the vapor leaving the slurry oil pumparound has been cooled to 630°F. For a 50,000-B/SD FCCU charge, the weight of reaction vapor in the vapor feed to the fractionator is about 700,000 lb/hr. This amount of vapor must be cooled or desuperheated in the slurry pumparound section. What then is the maximum weight of vapor flow that passes through the slurry pumparound section?

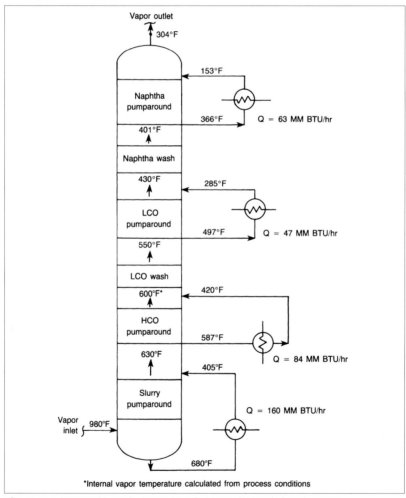

*Internal vapor temperature calculated from process conditions

Fig. 8–2. Observed heat balance in a fluid cracker unit fractionator (after Lieberman, courtesy *Hydrocarbon Processing*, April 1984).

Before answering this question the process engineer should recall a few facts:

- The FCCU feed rate at which the tower will flood is proportional to, among other parameters, the weight of vapor flow passing up through the slurry pumparound section.
- Process engineers who design towers that flood prematurely are fired.

The engineering contractor who designed the slurry pumparound rate for 700,000 lb/hr vapor flow was mistaken. He forgot that the superheated reaction vapors were cooled primarily by evaporating cycle oil and not through direct heat exchange with the relatively cool slurry oil pumparound.

The bubble effect

How many pounds of heavy hydrocarbon liquid must be vaporized to cool 700,000 lb/hr of vapor from 980°F to 630°F? Assuming a specific heat of 0.6 for the vapor and a latent heat of vaporization of 90 BTU/lb for the cycle oil:

$$\text{Pounds vaporized} = \frac{700,000 \times 0.6\,(980 - 630)}{90} = 1,633,000\ \text{lb/hr}$$

The maximum vapor flow up through the slurry oil pumparound was then:

$$700,000 + 1,633,000 = 2,333,000\ \text{lb/hr}$$

No wonder the slurry pumparound, which had been designed for 700,000 lb/hr vapor flow, was flooding. This extreme increase in vapor flow due to desuperheating is called the *bubble effect*. It is most noticeable in packed columns where there is intimate contact between vapor and liquid. Of course, the size of the bubble is proportional to the temperature change of the upflowing vapor.

The preceding calculation assumes an infinite number of contacting stages in the slurry oil pumparound. Detailed tray-to-tray computer simulations show that for three theoretical stages in the slurry oil pumparound, the maximum tray loading is about 65% above that calculated using 700,000 lb/hr vapor and 980°F. The computer simulation takes into account the reduced temperature, liquid flows, and vapor density.

It was difficult to explain the bubble effect to the plant manager. Logically, he could follow my explanation. But emotionally, he could not accept that the premature flooding of the new fractionator was an intrinsic process problem, rather than a mechanical difficulty that could be rectified.

Vapor-line quench

If the premature flooding problem was really attributable to excessive vaporization of cycle oil by the hot reactor vapors, then desuperheating the vapor external to the fractionator ought to mitigate the problem. I discussed the possibility with a colleague who suggested the modification shown in figure 8–3. Cooled slurry oil from the pumparound return line is sprayed into the reactor vapor line. The slurry oil does not start to vaporize until it is heated to 680°F (i.e., the tower bottoms temperature). So if 50,000 B/SD of 440°F slurry oil is well mixed with 700,000 lb/hr of 980°F vapor, the resulting mixed phase stream entering the fractionator will consist of:

- 700,000 lb/hr vapor at 700°F
- 50,000 B/SD liquid at 700°F

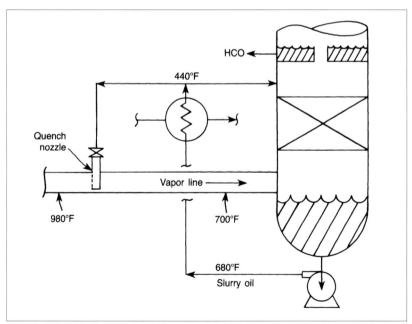

Fig. 8–3. Use of a vapor line quench nozzle is a handy way to de-bottleneck a FCCU fractionator.

The net effect of desuperheating the vapor from 980°F to 700°F is to prevent vaporization of about 1,600,000 lb/hr of cycle oil as calculated above. The plant manager, desperate for a way out of the fractionator flooding problem, decided to immediately install the nozzle shown in figure 8–3 in the reactor vapor line. The FCCU was shut down the following week. Four days later the unit was back online with the vapor-line quench nozzle in service. The FCCU charge was increased to 20% over design; the LCO product stayed a clear, bright, and very beautiful yellow.

Epilogue

Some months later a resourceful technical service engineer decided to use heavy cycle oil (HCO) instead of slurry oil in the vapor-line quench nozzle. I leave it to the reader to explain why the tower began to flood at its former rate, even though the vapor feed was desuperheated by the HCO in the reactor vapor line. The above calculations quantifying the bubble effect partially provide the answer.

Fractionator Difficulties

An FCCU fractionator is very much like a crude tower above the slurry oil section. Troubleshooting trayed crude towers are covered in chapter 1 and need not be repeated here. However, the process evaluation and operation of FCCU fractionators using structured—as opposed to dumped—packing is a topical subject and is discussed in detail. The process engineer engaged in the operation of a packed FCCU fractionator has three concerns:

- Tower vapor and liquid capacity (flooding)
- Packing fractionation efficiency
- Packing heat-transfer coefficient

In each case, one of the process engineer's main questions would be how the packed sections performed compared to trays.

Observed flooding

To realistically observe flooding in a packed column, a differential pressure cell is required across each packed section. For commercial facilities, the following rules of thumb appear generally applicable:

- Pressure drop less than ½ in. H_2O/ft of packed height—no flooding.
- Pressure drop of ¾–1 in. H_2O/ft of packed height—incipient flooding. For example, a tower had continuously operated at ¾ in. H_2O pressure drop and achieved reasonable fractionation. However, when the pressure drop increased to 1 in. H_2O, liquid from the packed section was observed in the upper product drawoff within 15 minutes.
- Pressure drop of 2 in. H_2O/ft of packed height—fully developed flooding. In a tower with multiple packed sections, observing a 2 in. H_2O pressure drop is not particularly informative. An upper section may be flooding because it is overloaded or because a lower section has flooded and backed liquid up the column.

Observed loads

Figure 8–4 shows the packed sections of an FCCU fractionator. Table 8–1 summarizes the maximum loads observed on the fractionator. The observed pressure drops reported for sections 4, 5, and 6 are at incipient flood for the individual sections. The low pressure drop reported for the top three sections is due to the fractionator being limited by the capacity of the tower by the lower sections.

Liquid loading sensitivity

The only pressure drop out of line with published correlations is section four.[1] This section flooded at low liquid rates, evinced by the rapid increase in pressure drop as the reflux to this section was raised. A likely explanation of this is excessive liquid entrainment from section 5, which may have raised the actual gpm in section 4 above the calculated value.

Due to the high percentage of open area of the Flexigrid packing, section 6 seems relatively insensitive to changes in liquid loads. However, increases in vapor rates of as little as 2% are sufficient to flood section 6. Within 30 minutes of a small increase in vapor flow, the LCO pumparound turns black.

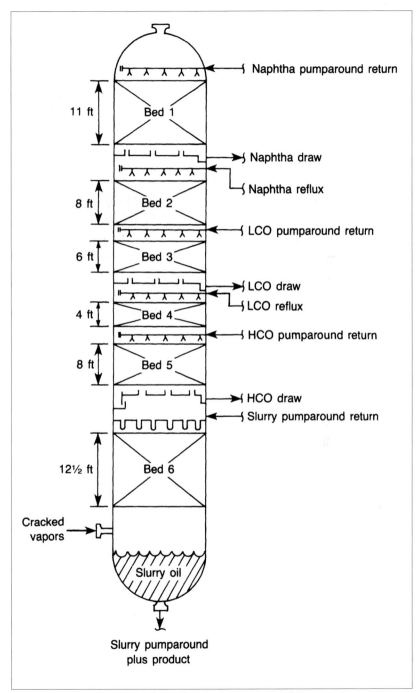

Fig. 8–4. Packed sections of a 14-ft-0-in. diameter fractionator
(after Lieberman, courtesy *Hydrocarbon Processing*, April 1984).

Capacity comparison

Table 8–2 compares the observed percentage of flood of the packed sections of the fractionator to the calculated percentage of flood for valve trays. The Flexipac type 4 sections are compared to trays on a 36-inch spacing, whereas Flexipac type 3 is compared to a 24-inch tray spacing. This is done because 2 feet of No. 2 packing or 3 feet of No. 4 packing typically provide roughly the same fractionation efficiency as a single tray.[2] The percentage of flood for the packed sections is determined from the "Generalized Pressure Drop Correlation" for packed columns, based on observed pressure drop data.[3] The percentage of flood for valve trays is calculated from standard industry correlations as based on the observed flows listed in table 8–1.[4]

Table 8–1. Tower loads and observed pressure drop on a 14-ft 0-in fluid cracking unit fractionator utilizing structured-type packing.

Section Number	Type of Packing	Service	Vapor Density, lb/ft³	Liquid Density, specific gravity
1	Koch #4 Flexipac	Naphtha pumparound	0.39	0.74
2	Koch #3 Flexipac	Naphtha wast	0.40	0.77
3	Koch #4 Flexipac	LCO pumparound	0.43	0.77
4	Koch #3 Flexipac	LCO wash	0.40	0.76
5	Glitsch Grid	HCO pumparound	0.42	0.77
6	Koch Flexigrid	Slurry pumparound	0.55	0.74

*Tabulated loads do not correspond to a single fractionator operation.
**F_s = Vapor superficial velocity multiplied by the square root of vapor density.
†Indicated pressure drops correspond to observed incipient flood conditions.

Table 8–2. Capacity comparison—Trays versus packing.

Section Number	Service	Observed Percentage of Flood of Packed Sections	Type of Packing
1	Naphtha pumparound	46	Koch #4 Flexipac
2	Naphtha wash	56	Koch #3 Flexipac
3	LCO pumparound	63	Koch #4 Flexipac
4	LCO wash	100	Koch #3 Flexipac
5	HCO pumparound	100	Glitsch Grid
6	Slurry pumparound	100	Koch Flexigrid

For the majority of the Flexipac sections, the anticipated percentage of flood is 60% higher than for comparable valve trays. Section 4 flooded prematurely, presumably because of excessive entrainment originating in section 5. Flexigrid packing exhibited a high capacity and maintained a reasonable heat-transfer efficiency in the bottom pumparound service.

Heat-transfer coefficients

Table 8–3 summarizes observed heat-transfer coefficients for commercial units. The coefficients are expressed in terms of cubic feet of tower volume devoted to pumparound service.

The observed heat-transfer coefficients per square foot of tray surface area are divided by 3 feet, which corresponds to the 36-inch tray spacing assumed in table 8–2. If an 18-inch tray spacing had

Vapor Flow, lb/hr	Liquid Flow, lb/hr	F_s**	gpm/ft^2	Observed P, in. H$_2$O/ft packing
820,000	580,000	2.37	10	0.1
810,000	150,000	2.31	3	0.3
1,190,000	650,000	3.29	12	0.3
920,000	100,000	2.63	2	0.9†
1,160,000	1,350,000	3.22	22	0.8†
1,830,000	1,850,000	4.46	33	1.1†

From: N. P. Lieberman. "Packing Expands Low-Pressure Fractionators," *Hydrocarbon Processing*, April 1984.

Calculated Percentage of Flood of Comparable Sections with Valve-Type Trays	Valve Tray Configuration Assumed
76	36" tray spacing, 4 pass
85	24" tray, 2 pass
100	36" tray spacing, 4 pass
95	24" tray spacing, single pass
136	36" tray spacing, 4 pass
194	36" tray spacing, 4 pass

From: N. P. Lieberman. "Packing Expands Low-Pressure Fractionators," *Hydrocarbon Processing*, April 1984.

Table 8–3. Heat-transfer efficiency comparison—Structured packing versus trays.

	Service	Contacting Device	Duty, MMBTU/hr
Section 1:	Naphtha pumparound FCCU fractionator	Koch #4 Flexipac	63
Section 3:	LCO pumparound FCCU fractionator	Koch #4 Flexipac	47
Section 5:	HCO pumparound FCCU fractionator	Glitsch Glid	84
Section 6:	Slurry pumparound FCCU fractionator	Koch Flexigrid	145
Coker fractionator		Valve trays**	34
Rerun still pumparound		Bubble-cap trays**	23
Combination unit pumparound		Bubble-cap trays**	26
Crude still pumparound		Bubble-cap trays**	67

*Heat-transfer coefficient is based on one cubic foot of tower volume. For purposes of comparison, coefficients for trays are based on 36" tray spacing.
**Observed heat-transfer coefficients are all for trays between 75%–90% flooding.

been assumed, the heat-transfer coefficients for the trays would have doubled, but then the capacity of the trays would be difficult to compare to the packed sections.

The data presented in table 8–3 are chosen for trays operating at roughly 75%–90% of flood. Data for trays operating at low vapor rates show a marked decrease in heat-transfer coefficients.[5]

Structured packing heat transfer

The observed heat balance for the FCCU fractionator is shown in figure 8–2. Referring to table 8–3, the Flexipac type 4 material had a heat-transfer coefficient somewhat better than trays. The Glitsch Grid section appeared about 20% better than the Flexipac type 4 packing. This difference may be due to the higher heat-transfer coefficients attributable to greater superficial mass velocities (see table 8–1) rather than to the difference in the packing.

The slurry pumparound section has a heat-transfer coefficient significantly lower than the other packed pumparound sections and lower than reported coefficients for trays. The duty in the section is, however, devoted largely to desuperheating the cracked vapors from 940°F to 700°F, whereas the major component of duty in the other pumparound sections is condensation. It is likely that heat transfer coefficients would be higher for condensation as opposed to desuperheating vapors.

LMTD, °F	Tower Diameter, ft.	Heat-Transfer Coefficient BTU/hr/ft³/°F
80	14	470
92	14	557
104	14	660
252	14	301
78	11	311
47	13	510
92	10	400
122	16.5	415

From: N. P. Lieberman. "Packing Expands Low-Pressure Fractionators," *Hydrocarbon Processing*, April 1984.

Height equivalent to a theoretical stage

A computer simulation of a thermal cracker fractionator pumparound section based on equilibrium flash vaporization calculations shows that the heat-transfer coefficient for a theoretical separation stage was 1,600 BTU/hr/ft²/°F. On this basis, the height equivalent to a theoretical stage of packing, such as the Flexipac type 4 in section 3 (see table 8–3), is:

$$\frac{1,600 \text{ BTU/hr/ft}^2/°\text{F}}{557 \text{ BTU/hr/ft}^2/°\text{F}} = \text{height of a theoretical stage} = 2.9 \text{ ft}$$

This value is in line with published commercial correlations.[6]

Fractionation efficiency

Table 8–4 summarizes the product distillations for the FCCU fractionator. These distillations are consistent with the vapor and liquid flows presented in table 8–1. The degree of separation between the tower overhead and LCO products is represented by the ASTM 5% to 95% gap of –2°F. Based on the Packie method,[7] calculations indicate that there are 10 effective trays above the LCO draw tray. This is equivalent to roughly six theoretical separation stages. Since each pumparound section is usually represented as a single stage, we can see by referring to figure 8–2 that the naphtha wash section is equivalent to four theoretical separation stages.

Table 8–4. Product distillations using structured packing.

ASTM Distillation, °F	Stabilized Overhead Product	Stripped LCO Product	Unstripped CSO Product
IBP	87	364	330
5%	110	416	520
10%	124	470	614
50%	242	550	—
90%	380	668	—
95%	418	690	—
E.P.	446	702	—
RVP, psi	10	—	—
Flash, °F	—	168	—

*These distillations are consistent with the internal flows listed in Table 8–1.
From: N. P. Lieberman. "Packing Expands Low-Pressure Fractionators," *Hydrocarbon Processing*, April 1984.

The naphtha wash section has a height of 8 feet. So for this particular service, section 2, consisting of Flexipac type 3 packing, has an apparent fractionation efficiency of 24 inches per theoretical separation stage.

The process engineer who draws the conclusion from the above data that structured-type packing is superior to trays is correct. However, he should also note that the performance of packing can be no better than the initial liquid distribution provided by the spray headers.

Reflux distribution

A Midwestern refinery had recently revamped the FCCU main fractionator. The top five trays, used to fractionate between naphtha and hydrocracker feed, were replaced with 8 feet of structured-type packing. The naphtha end-point specification was 420°F and the hydrocracker feed end-point spec was 800°F. The purpose of the revamp was to improve fractionation between naphtha and hydrocracker feed.

Unfortunately, when the tower was commissioned after the revamp, fractionation was observed to be worse, not better, than before the revamp. To meet the naphtha end-point spec, 35% of the FCCU naphtha had to be undercut into the hydrocracker feed product.

Coincidentally, with the revamped FCCU start-up, my client had appointed a new plant manager. The plant manager was outraged that the refinery, having spent almost $1 million to revamp the fractionator, had actually lost fractionation efficiency. The plant manager retained my services to troubleshoot the problem and asked me to address three questions:

1. Why was fractionation bad?
2. Whose fault was it?
3. What could be done to minimize the naphtha content of hydrocracker feed?

Figure 8–5 shows the configuration at the top of the tower. The reflux was distributed by an orifice plate chimney type, pan liquid distributor. Vapor flowed up through the chimneys; reflux was supposed to be evenly distributed by the ½-inch holes drilled in the pan across the top of the packed bed. To determine if the reflux was truly being equally distributed across the structured bed, I proceeded as follows:

1. I located the elevation on the side of the column a few inches above the top of the chimneys.
2. Next, at 90°F intervals around the tower, I cut 2-inch diameter holes through the insulation.
3. Using a surface temperature indicator, I measured the surface temperatures shown in figure 8–6.

These surface temperatures correlate with the temperature of the vapor exiting from the chimney tray, which in turn is a function of the reflux rate to a particular quadrant of the packed bed. (This assumes that the vapor flow to the bottom of the bed is well-distributed. Vapor distribution is much more easily achieved than liquid distribution. However, vapor channeling will also result in diminished fractionation efficiency.)

A low vapor temperature indicates a high liquid flow; a high vapor temperature indicates lack of liquid. Was it likely that the hottest quadrant indicated in figure 8–6 was running dry? An orifice plate

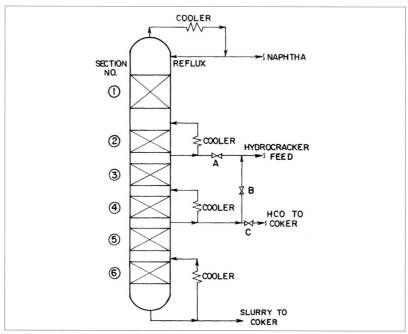

Fig. 8–5. Eliminating a redundant HCO draw-off.

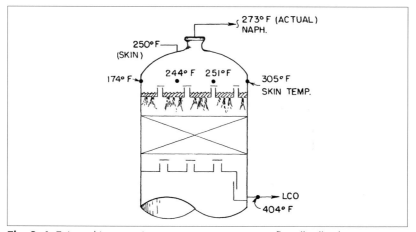

Fig. 8–6. External temperature survey proves poor reflux distribution.

will not properly distribute reflux when the plate is substantially out of level. For instance, this 14-foot ID tower would have a difference of 0.9 inch between the high and low side of the plate when the plate was installed one-half degree out of level. However, is this 0.9 inch "out-of-level" sufficient to cause significant liquid maldistribution?

To make this decision, we must calculate the average pressure drop of the liquid flowing through the orifice holes:

$$\Delta PL = (0.7)(V^2)$$

where:

ΔPL = Pressure drop of the flowing liquid, in. of liquid

V = Velocity of the liquid through the orifice holes, ft/sec

The 0.7 value is a typical orifice coefficient which will vary with plate thickness and whether the hole was punched or drilled. If ΔPL is smaller than the expected out of levelness of the plate, vapor will start to bubble up through the orifice holes on the high point of the plate, and the reflux flow will stop in that section of the plate.

Please note that ΔPL is not the liquid level on the plate. The liquid level is the sum of ΔPL and the pressure drop of the vapor flowing through the chimneys (ΔPV):

$$\Delta PV = \frac{(0.9)(pv)(V^2)}{pl}$$

where:

pv = Vapor density

pl = Liquid density

V = Velocity of vapor through the chimneys, ft/sec

The 0.9 value is a typical coefficient for vapor flowing through a chimney riser and hat.

The ΔPL value for this fractionator reflux plate distributor was 0.8 inches, that is, smaller than the "out-of-levelness" value of the plate assuming a maximum one-half degree level tolerance. Now a half degree leveling error is a pretty tight spec, so I felt confident in telling the plant manager that the reason the fractionation was bad was that the ΔPL value was too small. Put another way, the orifice holes were too big.

This was, then, a design error. The plant manager suggested the design engineer should have used a ⅜-inch hole instead of the ½-inch hole actually used. I replied that ⅜-inch holes were subject to plugging and that the designer should have used a "narrow trough tray distributor," and not an orifice plate. Each trough can be leveled quite exactly, and the troughs can be designed to avoid plugging. Of course, they are far more expensive and complicated to design than an orifice plate.

"So this was a design error which caused reflux maldistribution," summarized the plant manager. "Your peripheral external temperature survey above the top reflux distributor supports your theory. While I agree with your conclusion," he continued, "how can we improve fractionation without an extended FCCU shutdown?"

Eliminating HCO drawoff

I had noticed upon first reviewing the product ASTM distillation lab data that fractionation between HCO and slurry was quite good. As can be seen in figure 8–5, however, the HCO and slurry products were commingled as coker feed. Evidently, packed section 5 in this tower was not serving a useful purpose. With this in mind, we proceeded with the following (see figure 8–5):

1. Slowly closed valve A over a period of four hours.
2. Permitted the drawoff temperature from section 4 to fall by about 60°F.
3. The product flow to valve C was initially very dark. We waited until this color turned a transparent yellow. Surprisingly, this took almost six hours.
4. When the end point of the flow through valve C dropped to 800°F, we opened valve B, shut valve C, and completed closing valve A.

Sections 1, 2, 3, and 4 were now being used to fractionate between naphtha and hydrocracker feed. The loss in naphtha product to hydrocracker feed fell from 35% to 5%. The slurry oil pumparound draw temperature, while somewhat reduced, was still hot enough to maintain the fractionator heat balance. Although this change did nothing to enhance the top reflux distribution, it did bring all products back on spec without an FCCU shutdown.

This story serves to emphasize a vital principal during retrofitting of any large-diameter tower from trays to packing. The most important, but also the most difficult part of the job, is proper initial liquid reflux distribution onto each packed bed.

Feed versus Slurry Heal Exchanger Leaks

A well-designed FCCU uses a large proportion of the heat contained in the reactor effluent vapors to preheat fresh feed. The heat contained in the 950°F–1,000°F vapor is absorbed by the slurry oil pumparound as shown in figure 8–7. The hot slurry is then heat-exchanged against cooler feed in the slurry pumparound heat exchangers.

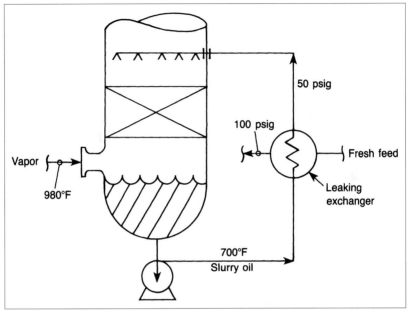

Fig. 8–7. Leaking fresh feed into slurry-oil pumparound can quickly coke the slurry-oil pumparound section.

The consequences of a leak in these exchangers can be very different, depending on whether the slurry or the fresh feed is higher in pressure. If the slurry is leaking into the feed to the riser, nothing particularly disastrous will transpire. The highly aromatic refractory slurry oil will pass through the riser and reactor cyclones without much change. At most, a slight regenerator temperature increase and a small decrease in slurry oil product degrees API may be observed.

On the other hand, leaking fresh feed into the slurry pumparound return can have some pretty dire consequences. In particular, the bottom section of the fractionator can coke up. As shown in figure 8–7, the fractionator bottoms temperature is typically 700°F. Slurry oil, being essentially condensed aromatic rings, is too refractory to thermally crack at 700°F. After all, the slurry oil is created in the riser at temperatures between 950°F–1,000°F. The fresh feed, being much more paraffinic than the slurry, is easily subject to thermal degradation, or coking.

But, one may ask, is the rate of coking of even very paraffinic FCCU feed appreciable at the relatively low temperature of 700°F? No, it is not. Unfortunately, the slurry pumparound liquid is exposed to temperatures far in excess of 700°F. The problem is that the cool slurry liquid and the hot reactor vapors are not necessarily evenly contacted in the pumparound section. A certain amount of lateral maldistribution is inevitable. Indeed, one set of plant data showed a 90°F spread in temperatures at the same elevations in a 14-foot diameter FCCU fractionator in the slurry pumparound section.

Exposed to unevenly distributed 980°F vapor, paraffinic components in the slurry pumparound may be converted to coke and lighter products. For every 10 pounds of heavy paraffinic FCCU feed that leaks into the slurry pumparound and cracks, about 1 pound of solid coke is formed.

Let's say a 50,000-B/SD FCCU is leaking 0.10 vol% of its feed into the slurry pumparound through a heat exchanger. Therefore, 70 lb/hr of coke can be forming in the slurry each hour, or 50,000 lb/month! (The coke is formed from the thermal degradation of the fresh feed, which produces approximately 10 wt% coke and 90 wt% hydrocarbon vapors.) Although most of the coke probably is purged out with the slurry product, the remaining coke, over a period of time, is sufficient to plug the bottom section of a large fractionator.

The process engineer can recognize a heat exchanger leaking fresh feed into the slurry pumparound by analyzing the slurry oil product for paraffins. An increase in paraffin content from near zero to several percent usually indicates the presence of fresh feed in the slurry. If feedstock and operating conditions are relatively constant, increases in slurry oil API gravity or aniline point may also indicate fresh feed in the slurry. An increase in metals in the slurry product also indicates a leak.

Incidentally, a leak of fresh feed into any of the main column pumparound circuits results in the same effect as a slurry exchanger leak. If fresh feed is found in the slurry oil, all the fresh feed/product exchangers must be checked. Start with the slurry exchanger, though, because it is the most likely culprit. Also, a leak in an upper pumparound circuit would increase the end point of LCO product.

Gasoline production

A reduction in FCCU gasoline product octane or rate is always a serious economic problem. A few rules of thumb that can assist the troubleshooter are as follows:

- Dropping reactor temperature by 20°F reduces gasoline research octane number by 1.0.
- An increase in feed Watson K characterization factor increases gasoline yield, as shown in figure 8–8.

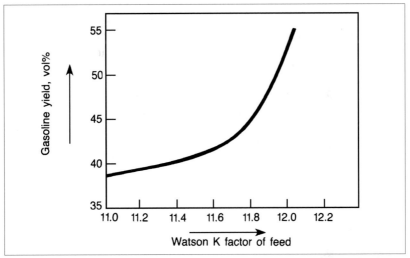

Fig. 8–8. Gasoline yield is primarily a function of the Watson characterization factor for virgin vacuum gas oil. Coke yield held constant (adapted from the *Davison Fluid Cracking Catalyst Handbook*, p. 53).

- A good-quality paraffinic gas oil (800°F ASTM 50% point and 35°API) yields 20% more gasoline at a given conversion than a poor-quality aromatic gas oil (800°F ASTM 50% point and 20°API).
- Adding 20% coker gas oil to FCCU feed can reduce gasoline production up to 15%.
- Changing the gasoline end point does not significantly affect octane but does have a large effect on yield.
- A decrease in LCO flash correlates with a decrease in gasoline production.

- A gas chromatographic analysis showing 2 mol% pentanes and heavier in the gas plant absorber off-gas indicate 1 to 1½ vol% of gasoline production is being lost to fuel gas.
- At constant operating conditions (riser temperature, catalyst to oil ratio, etc.), cracking virgin diesel oil yields 75%–80% as much gasoline as cracking virgin gas oil.

If none of these problems explain the observed loss in gasoline production, a reduction in catalyst activity or selectivity is indicated. Of course, one should not overlook less technical, but certainly not infrequent, causes for lost gasoline production. For example:

- Gasoline is being diverted to the refinery slop system through an open start-up connection.
- The riser or reactor temperature is reading too high. It's pretty easy to attribute the results of a faulty riser TI to inactive catalyst. Note that gasoline production does not always increase with riser temperature or conversion. Excessive riser temperature results in overcracking, and gasoline output suffers.

Vapor Recovery Unit

An FCCU gas concentration plant or vapor recovery unit typically consists of:

- Reboiled absorber-stripper
- Debutanizer
- Depropanizer
- Gasoline splitter

Only the reboiled absorber presents novel problems unique to FCCU operations.

Recontact drum overcondensation

Figure 8–9 shows an absorber-stripper combination. Note how both the liquid from the bottom of the absorber and the stripper overhead vapor flow into the recontact drum. This feature is intended to cut propylene losses to the absorber off-gas. Normally, this purpose is achieved. Unfortunately, the amount of condensation in the recontact drum can get out of control. On one Gulf Coast FCCU absorber I designed, this would happen regularly every winter.

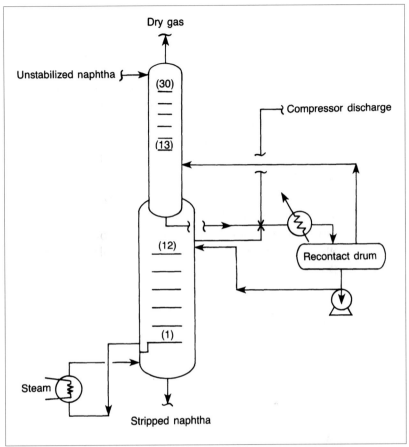

Fig. 8–9. A FCCU absorber-stripper using a recontact drum can flood due to overcondensation of ethane during cool weather.

The problem manifests itself by the escalating flow of liquid from the recontact drum outlet pump. A gas chromatic analysis of the gas vented from the drum showed an extraordinary concentration of ethane. This ethane buildup would continue until the stripper section flooded. To correct the problem, it was mandatory to educate each annual crop of FCCU operators to maintain a minimum recontact drum temperature by pinching off on the cooling water flow to the condenser feeding the recontact drum.

Stripper inefficiency

There is a tremendous variation in tray loadings between the top tray (tray 12) of the stripper section shown in figure 8–9 and the bottom tray (tray 1). Specifically, if the bottom tray is operating at 75% of flood capacity, the top tray of the stripper is running at about 25% of flood. This large spread in tray loading is a consequence of the necessity to heat the stripper feed from about 100°F to 280°F as the liquid cascades down the stripping trays. Most valve, grid, or sieve trays do not have sufficient turndown capacity to operate efficiently at 25% of flood. Liquid weeps through the tray decks, and the ethane content of the tower bottoms becomes more difficult to control.

To correct this problem, it is necessary to retray the upper 40% of the stripper with trays having a superior turndown ratio, or less vapor capacity. Some of the alternatives I have seen used with success are:

- Blank off half of the openings on the tray deck.
- Change the valve tray decks to trays having dual light and heavy valve caps. This type of cap only permits the valve to open halfway at low vapor rates (see figure 8–10). This is not as effective as the other two options listed.
- Replace the valve trays with bubble-cap trays. This is an expensive but effective solution.

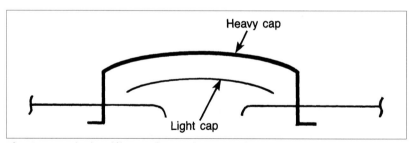

Fig. 8–10. A dual-weight cap for a valve tray enhances tray turndown ratio.

Hydrogen blistering

Thinning of the vessel wall of an absorber-stripper used in FCCU gas plant service is a common problem. The thinning occurs several trays down from the top of the stripper section. The cause of this metallurgical deterioration is hydrogen blistering. The sequence of events leading to the formation of hydrogen blistering are as follows:

1. Cyanides are formed as the hot catalyst contacts nitrogen compounds in the gas-oil feed flowing into the riser.

2. Due to inadequate water washing of the wet gas compressor discharge, these cyanides are carried over into the absorber-stripper.

3. Water entrained in the stripper feed is not properly removed at the water drawoff tray. (See figure 20–3 for details of a water drawoff tray.)

4. The aqueous-phase CN^- ions sensitize the steel walls of the vessel to permit atomic hydrogen to diffuse through the metal lattice structure.

5. The hydrogen atoms recombine into molecular hydrogen at an imperfecture inside the vessel wall. The accumulating hydrogen forms a bubble and then a blister.

The prevention of such blisters is achieved with proper water washing—preferably with steam condensate—of the wet gas after compression and withdrawing water that accumulates on the top trays of the stripper section. Polysulfide injection to the wash water is necessary for quantitative cyanide removal.

Low propylene recovery

A well-designed, high-pressure (300 psig), refrigerated, reboiled absorber can recover 97% of available propylene. High-temperature, low-pressure, and low lean-oil circulation all result in reduced propylene absorption. Overstripping the bottoms so that the molar ratio of ethane to propane is less than 6%, i.e., HD-5 propane specification, also inhibits propylene recovery. A less obvious cause of propylene carryover into fuel gas occurs when unstabilized naphtha from the FCCU fractionator reflux drum is used as the only absorption oil to the top tray of the absorber. This standard design, widely used for high-pressure absorbers, is illustrated in figure 8–11. Note how propylene absorption is improved through the use of a refrigerated intercooler at an intermediate tray, but the unstabilized naphtha is not chilled by refrigeration.

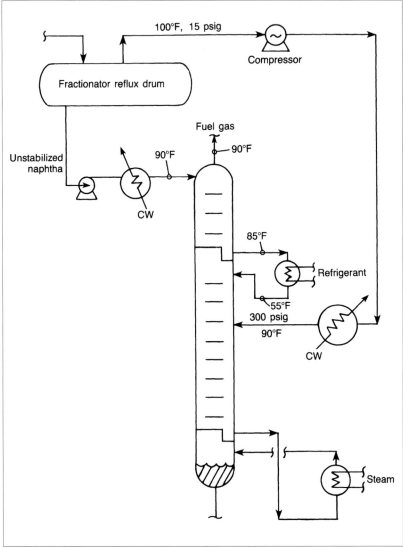

Fig. 8–11. This type of reboiled absorber can lose efficiency as absorption is improved in the fractionator reflux drum.

An absorber-stripper of this configuration can have the peculiar habit of rejecting more propylene to fuel gas when the temperature drops or the pressure rises in the FCCU fractionator reflux drum. This transpires because the propylene content of the unstabilized naphtha increases as the reflux drum temperature drops or the pressure rises. The unstabilized naphtha reaches vapor-liquid equilibrium with the absorber off-gas on the top tray of the absorber. If the partial pressure of the propylene dissolved in the unstabilized naphtha exceeds the partial pressure of the propylene of the vapor flowing to the top tray, then the propylene content of the vapor from the top tray will increase.

This rather nebulous concept can best be appreciated by performing a tray-to-tray computer simulation of the column. In the field, I first observed this phenomenon when the operators noticed that shutting down the intercooler shown in figure 8–11 did not affect the propylene content of the absorber off-gas.

Operational changes or equipment modifications that result in enhanced propylene recovery in an FCCU absorber-stripper will inevitably increase hydrogen sulfide recovery. Moreover, the percent recovery of hydrogen sulfide from the sour fuel gas may be an order of magnitude greater than the increased percent recovery of propylene from fuel gas.

For instance, increasing propylene recovery from 88% to 93% in one plant doubled the amount of H_2S in the downstream depropanizer overhead. This led to a doubling in the amine H_2S absorber acid gas loading and as a consequence a great increase in corrosion in the bottom of the amine absorber in the form of hydrogen blistering and hydrogen-assisted stress corrosion cracking. Eventually, this resulted in a catastrophic failure of the absorber. Sixteen people died as a result of this incident at the Unocal plant, south of Chicago, in 1984.

Formation of diolefins

Downstream units that process FCCU olefins are adversely affected by diolefins. A sudden increase in acid-soluble oils in an HF alkylation unit or a dramatic increase in catalyst consumption on a Dimersol polymerization unit is likely caused by an increase in the diolefins in olefin feedstock.

The effect of riser temperature on the diolefin content on FCCU butanes is correlated in figure 6–3. One refiner also reported that an increase in regenerator temperature from 1,280°F to 1,320°F doubled the diolefin content of the alkylation unit feed.[8]

Personally, I have noted an increase in the diolefin content of FCCU C_3-C_4 when an ill-advised modification of the feed-injection distributor at the base of the riser reduced feed dispersion up the riser. Incidentally, a selective hydrotreater to destroy diolefins is one of the more cost-effective investments that a refiner can make. It is not uncommon to observe a decrease in sulfuric acid consumption on a sulfuric acid alky unit of 15% when such a hydrotreater is put online to treat butylene feedstock.

Troubleshooting Checklist
for FCCU Product Fractionation

Dark LCO Product
Leaking LCO versus feed exchanger
Low internal reflux
Flooding of LCO wash section

Flooding of Fractionator Slurry PIA
Darker LCO with increased reflux
Check ΔP at unflooded conditions
Determine which section
    initiates flooding
Desuperheating reactor vapor
Bubble effect in slurry P/A
Reactor vapor line quench

Packed Fractionator Problems
Maximum vapor and liquid capacity
HETP
Heat-transfer coefficient
Liquid loading sensitivity
Structured packing
Reflux distribution

Feed vs Slurry Exchanger Leaks
Promotes coking of tower bottoms
Check slurry for paraffins
API gravity
Aniline point

Loss in Gasoline Production and Octane
Drop in reactor temperature
Decrease in Watson K factor
Coker gas oil in feed
Reduced product ASTM
    end point
Increase LCO flash
Pentanes in absorber off-gas
Catalyst activity or selectivity
Losses to refinery slop
Riser temperature reading high

Vapor Recovery Unit
Recontact drum
    over condensation
Stripper inefficiency
Stripper turndown problems
Cyanides cause
    hydrogen blistering
Low propylene recovery
    in absorber
Absorber top tray
    equilibrium problems
Low absorbtion (L/VK) factor
Diolefins in olefin products

References

1. Koch Flexipac, Bulletin KFP-2, Figure 8c, Wichita, Kansas.

2. K. J. McNulty and C. Hirsch, "Hydraulic Performance and Efficiency of Koch Flexipac Structured Packing," presented at the 1982 Annual Meeting of the AIChE, Los Angeles, California, January 1982, Figures 19 and 20.

3. F.A.1. Holloway, T. K. Sherwood, and G. H. Shipley, *Ind. Eng. Chem.*, 30765 (1938).

4. Koch Flexitray Design Manual, Wichita, Kansas.

5. Circa 1940 data on bubble-cap trays. Personal files of author.

6. McNulty and Hirsch, 1982.

7. R. N. Watkins, *Petroleum Refinery* Distillation (Houston: Gulf Publishing, 1982), Figure 2.6.

8. E. S. Johnson, NPRA Question-and-Answer Session 1979, Light Oil Catalytic Processing.

Saving Energy at Reduced Feed Rates

Every businessman knows that per-unit costs go down as volume goes up. In a refinery we have all observed that energy use per barrel of crude processed goes down as refinery throughput rises.

This is a paradox to the designer. He knows that at higher charge rates more heat-exchange surface and furnace convective tube area are needed to maintain thermal efficiency. Certainly a higher temperature driving force is required to transfer heat at increased duties when heat-transfer surface is fixed. And, as the second law of thermodynamics states, increased temperature levels always reduce the thermal efficiency of a process.

The answer to this paradox is explained by dirt, leaks, and human inertia. A partial list of the causes of increased per-barrel energy use at reduced throughput is:

- Exchanger fouling rates increase at reduced velocities.
- Fractionating trays leak more at lower vapor rates, decreasing tray efficiency.
- Operators don't bother to close secondary air registers on furnaces when throughputs drop. Thus, excess air is increased.
- Motor-driven pumps operate less efficiently when pushed back on their pump curves.
- Ambient heat loss stays constant regardless of unit throughput.
- Cold air in-leakage through holes in a furnace's skin remains the same when the firing rate is cut. Reduced heater efficiency results.
- Operators fail to slow down steam-driven pumps and compressors when the load drops.
- A leaking total trapout pan drops clean gas-oil product into black oil. The lost gas-oil will eventually be boiled out of the black oil again.

- Low throughput disrupts the refinery steam balance, forcing the venting of low-pressure steam.
- Distillation tower reflux rates are kept high. Overfractionation wastes energy.
- Vacuum ejectors require a constant flow of steam regardless of gas rate.
- Atomizing steam to oil burners is not reduced.
- Air blowers are vented to the atmosphere to keep the machine out of surge.

Operating practices that made money for the refinery when management objectives were to maximize throughput eat into profits when crude runs are cut back. The effect of leaking trays and exchanger fouling rates becomes more pronounced when process units are loafing. The specific examples cited below illustrate various methods to save energy at reduced throughputs.

Steam-Turbine Hand Valves

Hidden underneath the governor trip assembly and a maze of small steam piping at the front of the steam turbine are the hand or jet valves. Most medium-sized turbines have two or three of these valves. Their function is vital to saving energy at reduced unit throughput.

When steam enters a turbine case through the nozzle block, its pressure and enthalpy energy is converted to kinetic or velocity energy. This is an isoentropic expansion, in that the potential of the steam to do work is preserved. When a turbine runs too fast, it may be slowed down in one of two ways. First, allow the governor steam inlet valve to reduce the flow of steam. Not only is the steam flow reduced, but the pressure of the steam flowing into the nozzle block, and thus its ability to do work, are diminished. The other way is to reduce the nozzle block cross-sectional area by closing one or more hand valves. This simply reduces the flow of steam but does not impair the ability of the steam to do work.

Likely, one will find the hand valves on a turbine stuck wide open. They have probably not been used since someone opened them years ago to obtain the maximum horsepower from the machine. Ask the unit machinist to make the hand valves operable. Then proceed as follows:

1. Close one hand valve all the way (hand valves must be completely opened or closed to achieve their purpose).
2. Check the turbine speed with a tachometer. If the governor opens up enough to maintain speed, close the next hand valve.
3. Continue closing the hand valves until the governor is wide open.

This procedure will cut turbine steam requirements by roughly 10% to 20%. The horsepower versus steam curves supplied by the turbine vendor can be used to calculate exact savings.

The next step is to slow the turbine to the minimum speed that is required to produce the required pump discharge pressure. Assume that a steam turbine utilizing 5,000 lb/hr of steam is driving a centrifugal pump. The process flow control valve downstream of the pump is only open by 25%. By slowing the turbine (turn the governor speed control valve knob counterclockwise), the flow control valve will open to 70% as the pump discharge pressure declines. The turbine speed falls, for example, from 3,300 rpm to 3,200 rpm. The rule of thumb is for every 1% reduction in speed, 3% of the turbine motive steam is saved. So slowing this turbine saves 500 lb/hr of steam.

Pump Impellers

Whereas energy can be saved by running steam pumps at a reduced speed, conventional (without motor frequency control) motor-driven centrifugal pumps run at a constant speed. The electricity used by a motor does not depend on the motor horsepower rating but is a function of the head and flow developed by the pump. The pump curve supplied by the vendor (see fig. 9–1) fixes this relationship for a particular-sized impeller.

Installation of a smaller impeller is a simple mechanical procedure. When a motor-driven pump puts up excess head, as indicated by the downstream control valve position, switching to a smaller impeller will significantly cut motor horsepower requirements. Table 9–1 relates the effect of changing impeller size to flow, head, and horsepower.

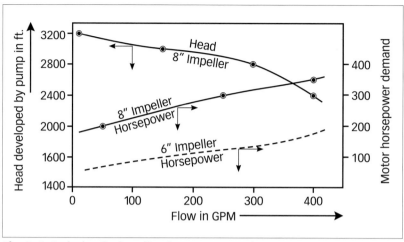

Fig. 9–1. Reducing the impeller size saves motor horsepower.

Table 9–1. Effect of reducing a centrifugal pump's impeller diameter.

$$\frac{D_S}{D_L} \times \text{gpm with larger impeller} = \text{gpm with smaller impeller}$$

$$\left(\frac{D_S}{D_L}\right)^2 \times \text{head with larger impeller} = \text{head with smaller impeller}$$

$$\left(\frac{D_S}{D_L}\right)^3 \times \text{bhp with larger impeller} = \text{bhp with smaller impeller}$$

where: D_S = Diameter of smaller impeller
D_L = Diameter of larger impeller

Note: Impellers from individual manufacturers come in standard sizes.

Note that an electric motor, even when not connected to a pump consumes power. For a typical large process plant driver, this is about 3% of the motor's rated horsepower. Thus electric savings resulting from reduction in the impeller size are somewhat less than calculated from Table 9–1.

To correct for the power lost in the electric motor, multiply the rated FLA point (full limit amperage) load of the motor by 3%. Then subtract this amperage from the current motor amp load. The difference between these two values is the electric power consumed by the pump itself.

Motor-Driven Centrifugal Compressors

The 2,000-bhp motor repeatedly tripped off on overload. This motor was driving a propylene vapor compressor. The propylene was being evolved from a propane-propylene splitter. The chief operator explained that whenever the pressure in the splitter tower rose above 105 psig, the motor would pull too many amps and trip off. To correct this situation, the chief pinched back on the block valve located in the compressor suction. The amperage load immediately fell, as did the compressor suction pressure.

This story only makes engineering sense when one knows that both the motor and compressor were constant-speed machines. For motor-driven centrifugal compressors that are oversized for a particular service, throttling the compressor suction is the most energy-efficient method to control the discharge pressure.

The reduced suction pressure satisfies the compressor's suction volume requirements with fewer pounds of flow. This reduction in weight flow reduces the compressor's horsepower use. Of course, throttling always wastes energy. In this case, though, controlling the compressor discharge pressure by throttling on the suction side only wastes 50% of the horsepower that would be lost by throttling on the discharge side.

The usual method of controlling compressor suction pressure by spill-back control is also wasteful. Again, to save driver horsepower, a suction throttle valve is suggested. Assuming the spill-back valve is partially open, proceed as follows:

1. Slowly pinch on the suction throttle valve.
2. Observe the amps on the motor drive start to fall.
3. Observe the spill-back valve start to close-off to keep the pressure at the spill-back control point from rising.
4. Continue to pinch on the suction throttle valve until the compressor surge limit is approached, or until the spill-back is shut.

Venting the discharge of an air blower to the atmosphere is the same as spill-back control and the above procedure applies for such air blowers.

Reciprocating Compressors

Reducing the gas rate through a reciprocating compressor can be a source of substantial energy savings, if the compressor cylinders are "unloaded" properly. Please see chapter 25 for information about adjustable clearance pockets and unloading a reciprocating compressor. Intelligent unloading can also reduce valve plate failures due to overheating the compression cylinders.

Modify Exchanger Pass Partitions

An exchanger tube bundle is designed to utilize the available pressure drop to provide maximum heat transfer at design flow rates. At lower throughputs, velocities through the exchanger will be reduced. Heat-transfer coefficients (U) are impaired at lower velocities because the film resistance to heat transfer is inversely proportional to mass velocity raised to the 0.7 power and fouling rates increase as velocity decreases. For many hydrocarbon services, a velocity below 3 ft/sec can mean rapid tube-side fouling.

Increasing the shell-side velocity requires decreasing the tube bundle baffle spacing. Practically speaking, this requires a completely new tube bundle. In many cases, however, the tube-side velocity can be doubled by modifying the exchanger pass partitions. Figure 9–2 illustrates one way to do this.

If either the tube-side film coefficient or fouling coefficient is the controlling resistance to heat transfer, doubling the tube-side velocity can increase the overall heat-transfer coefficient by 50%. Be careful! This modification will also increase the clean tube-side pressure drop (assuming noncompressible flow) by a factor of eight. Such an increase in pressure drop is frequently acceptable when operating at lower throughputs.

Reduce Tray Leakage

Common distillation tower valve trays are assembled from a number of sections bolted together inside the tower. The seal between the tray sections (unless gasketing strips are used) is far from leakproof. The valves themselves (i.e., the little bubbling devices on the tray decks) are also subject to leakage.

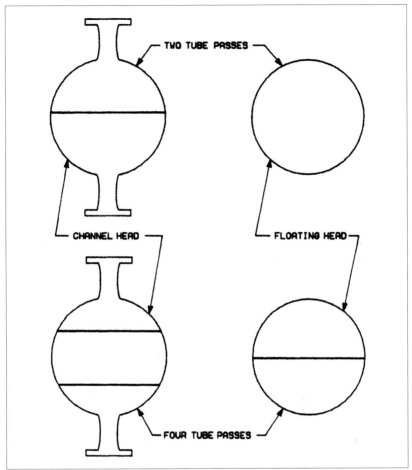

Fig. 9–2. Doubling the number of tube passes increased exchanger heat transfer.

At low vapor rates, tray pressure drop decreases and tray leakage is increased. This reduces tray fractionation efficiency. Then to achieve the desired product split, a higher reflux ratio, which wastes reboiler energy, is needed. Figure 9–3 shows a typical relationship of tray efficiency versus load.[1] When assembly of the tray sections is less than perfect, the turndown efficiency of the tray is further degraded.

Several ideas to reduce tray leakage at low rates are:

- Reassemble and gasket tray sections.
- Seal-weld tray sections (this has been done in a number of towers but obviously creates a nasty problem if the trays are ever to be removed).

- Purchase new tray decks with valves especially designed for a high turndown ratio. This suggestion is only applicable to nonfouling services because trays of this type have flat valves that easily stick to the tray deck. Note that the existing downcomers can be retained.

- Reduce tower pressure. The lower tower pressure will create higher vapor velocities and thus minimize tray leakage. This idea is discussed in detail in chapter 6, and is my preferred method.

- Make sure trays are installed level. Tray decks with low points will leak and lose tray efficiency at relatively high vapor rates. The more level the tray, the greater the tower's turndown ratio.

- Make sure weirs are installed level.

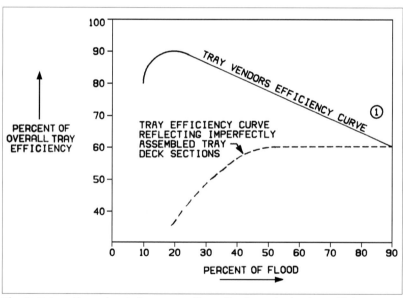

Fig. 9–3. Low throughputs increase reflux ratios by reducing tray efficiency.

Steam Balance

Many refineries generate their own electric power by passing high-pressure steam through a turbogenerator. The resulting exhaust steam is used primarily to reboil distillation towers. At reduced refinery runs, the demand for exhaust steam will usually decline faster than power requirements. Venting exhaust steam (especially in the summer) is the unpleasant consequence. Here is a technique for the process engineer to find a home for this wasted steam:

1. Make a list of all the preheaters and reboilers using high pressure steam.

2. Assuming the steam is on the tube side, check the channel head pressure of each exchanger.

3. For exchangers showing lower pressures, reduce the process side pressure and the condensate level in the channel head to a minimum. For example, as long as the overhead product can be condensed, there is nothing sacred in a particular distillation tower operating pressure.

4. If the channel head pressure is still not low enough to utilize exhaust steam instead of high-pressure steam, consider modifying the tube bundle with extended surface (fins) or nucleate boiling (Linde Hi-Flux) tubes.[2] The increase in effective heat-transfer coefficient or surface area will permit use of lower-pressure steam.

Insulation

A refinery's energy use due to ambient heat loss was 50,000 BTU/bbl at a crude rate of 30,000 B/D. At 15,000 BID, ambient heat loss was of course 100,000 BTU/bbl.

In one small refinery with a rated capacity of 4,500 BID, but which operated at 1,800 B/D, ambient heat loss accounted for 45% of the fuel consumed.

To calculate the energy lost from a process line, find the temperature drop between both ends of the line and multiply by the specific heat and flow in lb/hr. Find the temperatures with a glass thermometer inserted under a layer of insulation or with an infrared temperature gun.

Heat loss from large process vessels can best be pinpointed with an infrared picture of the unit. A less sophisticated but much more dramatic method is to observe the effect of a sudden rainstorm. On a sulfuric acid regeneration plant that was operating at 50% of design charge rate, the reactor outlet temperature dropped 100°F with every heavy rain. After reinsulation, the effect of rain was barely noticed.

During a thunderstorm, record furnace fuel and steam requirements before, during, and after the storm. Make sure to hold process conditions constant (i.e., furnace outlet and reflux temperatures). For a well-insulated unit, net energy requirements will not increase more than a few percent.

Overfractionation

Both the reflux and reboiler control valves on a distillation tower are sized for design flows. At reduced rates, these valves may be operating in a range that makes tower control difficult. Also, the dynamics of distillation control are such that overfractionation improves the stability of a tower's operation. Both of these factors encourage shift operators to overreflux towers and waste reboiler energy.

It may be necessary to reduce the size of the control valve internals to improve distillation tower control at lower feed rates. Greater attention directed toward tuning instruments will also payoff in energy savings (see chapter 28). The energy wasted by overfractionating in a distillation tower may be approximately calculated as follows:

$$H = \frac{Y_1/Y_h}{X_1/X_h}$$

$$\Delta H = \frac{\ln H_1}{\ln H_2}$$

where:

Y_1 = Mole fraction of light key component in overhead
Y_h = Mole fraction of heavy key component in overhead
X_1 = Mole fraction of light key component in bottoms
X_h = Mole fraction of heavy key component in bottoms
H = A number proportional to reboiler duty
ΔH = Percent change in reboiler duty due to changing fractionation

This calculation method assumes constant tray efficiency.[3] A plot of the preceeding equations is shown in figure 9–4.

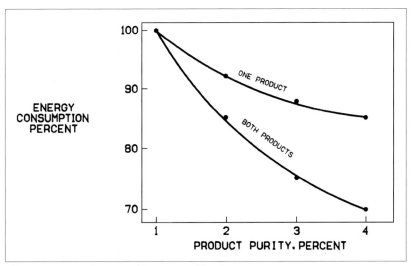

Fig. 9–4. Energy is wasted by overpurification.

Overabsorption of Light Ends

A catalytic naphtha reformer was designed to recover propane from hydrogen off-gas by contacting the gas with fresh naphtha feed. At design feed rates, the volume of light hydrocarbons absorbed—and thus reprocessed through the reformer—was 10% on feed. However, at reduced reforming rates, this internal recirculation rate (figure 9–5) rose to 30%. The increased absorption was mostly due to the lower absorber feed temperature. The excessive recirculation of light ends (mostly ethane) wasted energy.

The problem was resolved by preheating the naphtha with exhaust steam back to its original 100°F design temperature.

Vacuum Ejectors

A vacuum tower is a major component of most crude distillation units. The vacuum in these towers is usually created by a series of steam jet ejectors that typically use 100-psi steam.

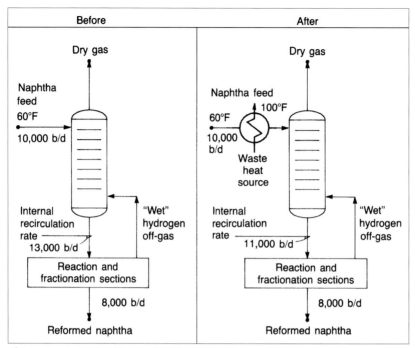

Fig. 9–5. An overly efficient absorber operation wastes energy.

To obtain the design vacuum from ejectors, the design steam pressure must be maintained. Since the ejector internals are not adjustable, the steam flow is not throttled (because then steam pressure would be lost) and therefore ejector steam use is fixed.

The ejector gas load consists of two sources: air in-leakage and uncondensed gas. The amount of air sucked into the system is independent of throughput. The uncondensed gas, which results from the thermal cracking of the vacuum tower charge in the preheat furnace, is proportional to the load on the ejector interstage condensers and thus to the feed rate.

If an ejector is not overloaded at a normal gas rate, reducing the gas load will not result in greatly improved vacuum. The ejector is simply oversized at the lower charge rate and wastes steam without obtaining any appreciable benefit in lower vacuum tower pressure (see chapter 13). To save this wasted steam, new ejector internals (e.g., steam nozzle) are needed. The internals are easy to install. As long as the gas rate does not exceed design, they will use less steam but still develop the same vacuum as the original ejectors.

Leaks

Hydrogen leaking into the plant fuel gas system from hydro-processing recycle gas loops indirectly wastes energy. Only a portion of the energy consumed in a hydrogen plant is contained in the product hydrogen exported to the refinery. Routine monitoring of the fuel gas composition will highlight this potential waste.

Leaking trapout trays above the flash zone in crude units—or above the hot vapor inlet in coker and FCCU main fractionators—cause clean gas-oil products to be reprocessed. The barrels of leakage do not vary with throughput. This source of energy waste is discussed in detail in chapter 1.

Air in-leakage into the convective section of a furnace can markedly reduce the furnace's efficiency. The cold air sucked in is unavailable for combustion but still must be heated to the stack outlet temperature. This problem is reviewed in chapter 15. Adjusting secondary air registers and atomizing steam flow to oil burners is also discussed in chapter 15.

Long-standing refinery practices, such as setting a turbine's speed, maintaining a minimum pressure on a circulating amine system, holding a temperature of a reboiler hot oil stream, or fixing the reflux drum pressure of a distillation tower, may no longer be valid at reduced throughput. Arbitrarily constraining any operating parameter will waste energy—more so as refinery runs drop. For example, maintaining a constant speed in your car, regardless of whether you are driving uphill or downhill, wastes gasoline compared to keeping a steady pressure on the gas pedal.

Troubleshooting Checklist
to Save Energy at Reduced Throughputs

Pumps and Compressors

Install small impeller
Close nozzle block hand valves
Minimize turbine speed
Reduce venting from air blowers
Throttle constant speed
  compressor suction

Fractionation Columns

Level trays
Check for leaking total trapout trays
Avoid overfractionation with
  excess reflux
Reduce overabsorption of light ends
Improve tightness of valve trays
Minimize pressure

Exchangers

Change channel head pass partitions
Retube to consume wasted
  exhaust steam
Change shell side flow from parallel
  to series
Increase onstream exchanger
  cleaning frequency

Furnaces

Minimize convective
  section draft
Reduce air in-leakage
Cut atomizing steam
  to oil burners
Close secondary air registers

Miscellaneous

Modify vacuum jet
  ejector's internals
Check for ambient heat losses
  during rainstorm
Eliminate arbitrary constraints
  of process parameters
Use reciprocating compressor
  unloader pockets

References

1. Fritz W. Glitsch & Sons, Inc., Dallas, *Ballast Tray Design Manual*,
    Bulletin 4900, p. 38.

2. R. M. Milton and C. F. Gottzmann, "High-Efficiency Hydrocarbon
    Reboilers and Condensers," AIChE 71st National Meeting,
    February 21, 1972, Dallas.

3. F. G. Shinsky, "Control Systems Can Save Energy,"
    Chemical Engineering Process, May 1978.

section **2**

Process Equipment

Man is the greatest factor in the universe;
he can do anything.   *~ Mao Tse-tung*

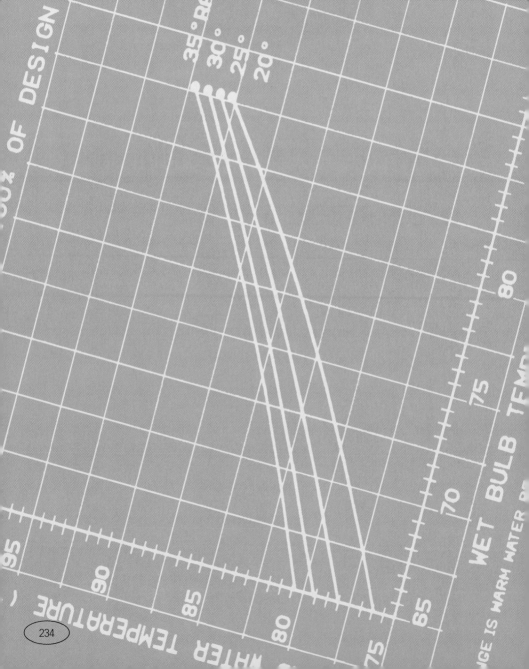

Refrigeration Systems

Refrigeration systems are widely used in a refinery. Some of the more common services are alkylation (isobutane refrigerant), ammonia production (ammonia refrigerant), and gas plant light ends absorber (propane refrigerant).

The problems that occur with refinery refrigeration systems are much the same as those we have at home with air conditioners and refrigerators, the major difference being that at home we use Freon as a refrigerant and in the refinery propane and butane are most often employed. The simplified refrigeration loop in figure 10–1 shows the major components:

- Refrigeration compressor
- Condenser
- Refrigerant accumulator
- Throttle valve
- Evaporator
- Refrigerant fluid

The evaporator is used to remove heat from a process stream. A cooler evaporator requires more work to remove one BTU of heat. The compressor supplies this work. The condenser duty equals the evaporator duty plus the work done by the compressor. The refrigerant accumulator acts as a surge drum for the liquefied refrigerant. The throttle valve is the heart of the system. Its purpose is to keep uncondensed refrigerant fluid from circulating back to the compressor suction.

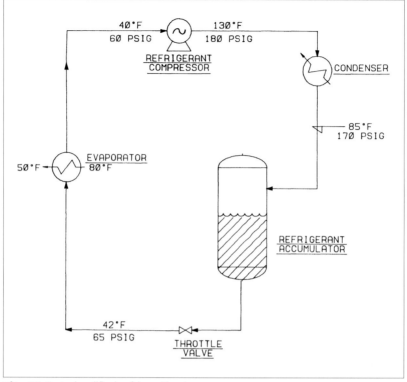

Fig. 10–1. A simplified refrigeration loop.

Is Refrigeration Efficiency Falling?

The first question that the process engineer should address is the actual amount of lost efficiency. Perhaps refrigeration efficiency has not declined at all. Higher ambient temperatures, or a colder process side temperature, could be the whole problem. The following calculation should be made.

$$\frac{(TAI - TEO)\, H_E}{TEO} = R_W$$

where:

TAI = Ambient temperature (absolute) for an air-cooled refrigeration condenser or inlet cooling water temperature (absolute) for a water-cooled condenser (*Note:* Absolute temperature = 460+°F)

TEO = Evaporator process side temperature, absolute

H_E = Evaporator duty, BTU/Hr

R_W = Refrigeration work

This calculation is a form of the second law of thermodynamics. Daily changes in refrigeration work will correlate with changes in refrigeration system efficiency. If R_W has not declined, even though refrigeration capacity has decreased, your refrigeration system is okay. If R_W has dropped, you have a problem and need to read on.

Diagnosing Refrigeration Compressor Problems

A refrigeration compressor may be speed- or horsepower-limited. A compressor with the turbine or gas engine governor or driver throttle valve wide open is horsepower-limited.

For a motor-driven compressor, the maximum amperage rating on the motor defines when the compressor is work- or horsepower-limited. Check the amp load in the motor "breaker" or "switch" room.

When a variable speed compressor is running at its maximum rated speed, it is speed-limited. The maximum rated speed is stamped on the compressor nameplate. There is no physical restraint to prevent operation of the compressor above its nameplate speed. However, compressors are equipped with overspeed trips, which automatically shut the compressor driver down when the compressor reaches a preset speed. Certainly, compressors that run well above their nameplate speed can self-destruct in a dramatic and tragic fashion.

When a compressor is reported to be speed-limited, one should take a tachometer and compare the actual speed of the machine against the nameplate speed. The machinist who set the overspeed trip may have set it too low. The operators are careful not to let the compressor trip off. They avoid this by running at speeds 100 rpm to 200 rpm below the set trip speed.

For compressors driven by an ordinary (nonvariable speed) motor, the speed of the compressor is fixed. However, there is a method you can employ to overcome capacity limitations imposed by fixed speed compressors (and compressors running at their maximum rated speed). This method is to change the composition of the circulating refrigerant. Details are discussed later in this chapter.

Short of horsepower

Many refrigeration compressors are horsepower-limited. This means the driver cannot put out enough work to drive the compressor as fast as desired. A motor-driven compressor is horsepower-limited when it trips off, because it is drawing too much current.

A motor-driven refrigeration compressor may be tripped off when the trip (or fuse) is set too low. Check this with an ammeter. The rated amperage (or horsepower) of the motor can be exceeded by trading the motor's operational life for increased horsepower. If refrigeration capacity is valuable, the amperage trip on a motor can be set higher. Anticipating reduced motor life, a replacement motor of increased horsepower should be ordered.

Cooling the motor

A motor running hot will draw more amperage than a cool motor. A refrigeration compressor motor that is tripping off at reduced refrigeration duty should be checked for proper operation of the motor's cooling fan. Sometimes, the motor's intake air vents may be partially blocked, leading to hot motor operation.

Chapter 25 presents a more detailed discussion of general troubleshooting techniques for both centrifugal and reciprocating compressors.

Steam drivers

Many large centrifugal refrigeration compressors are driven by steam turbines. A gradual loss in refrigeration duty can be due to a reduction in turbine horsepower. An approximate method for calculating small changes in compressor horsepower is:

$$\text{HP} = \text{mT}_1 \left[\left(\frac{\text{P}_\text{O}}{\text{P}_1} \right)^{\frac{K-1}{K}} - 1 \right]$$

where:

P_O = Outlet pressure (absolute) of refrigerant

P_1 = Inlet pressure (absolute) of refrigerant

m = Moles per hour of refrigerant

T_1 = Compressor suction temperature, absolute (i.e., 460+ºF)

HP = Number proportional to compressor horsepower

K = Ratio of specific heats for refrigerant; $(C_p \div C_v)$

This formula will permit one to differentiate between compressor deficiencies and losses in refrigeration from evaporator or condenser problems. A reduction in horsepower for a steam-driven centrifugal compressor is caused by one of four factors:

- Reduced driver steam pressure
- Reduced driver steam temperature
- Increased exhaust steam pressure
- Fouling of turbine internals

If steam conditions are normal, loss of turbine horsepower is likely due to fouling of the turbine wheels. This fouling may be removed by condensate washing. With the compressor running, pump clean steam condensate into the driver steam line. Enough condensate should be used to bring the driver steam to its saturation temperature. Be careful—injecting condensate in excess can damage the turbine.

If reduced driver steam pressure is to be a long-term condition, a larger nozzle block on the turbine steam inlet may be available at a relatively small price. (The nozzle block is an internal part of the turbine. When steam passes through it, the steam's pressure head is converted to velocity head.) Sometimes the turbine governor can be adjusted to open wider. Turbine hand valves, used to reduce nozzle block cross-sectional area, should be wide open when trying to maximize horsepower.

If high exhaust steam pressure is the problem, increasing the flow of driver steam will not provide much of an increase in refrigeration compressor horsepower.

Valves a problem on reciprocating compressors

Reciprocating refrigeration compressors can lose capacity due to leaking suction or discharge valves. This problem is easy to identify. The difference between the compressor suction and discharge temperature will increase as valve leakage gets worse. Refrigerant vapors are recirculating internally in the compressor cylinders. The compressor is doing the same amount of work on fewer pounds of refrigerant. Usually, the springs or plates in the cylinder valves will be found damaged. Both the suction and discharge valves must be checked.

Many reciprocating compressors come equipped with unloading pockets. This is an open area at the end of the cylinder whose length is adjustable. The compressor piston cannot compress the gas in the unloading pockets. The purpose of the pockets is to allow the operator to decrease compressor capacity without wasting horsepower. If you are looking for maximum capacity, make sure these adjustable pockets are closed.

Refrigerant Composition

In a refinery, usually a light hydrocarbon is used as a refrigerant rather than Freon. Thermodynamically speaking, propane is the best refrigerant. However, from the point of view of the amount of work required to generate a given amount of refrigeration, propane, butane, Freon, and so on are all about the same. Light hydrocarbons are used as refrigerants in refineries because of their low cost. Freon is used in homes for safety because it is nonflammable.

Changing refrigerant composition can be a simple, cost-effective method to de-bottleneck refrigeration systems and save energy. To predict the effect of changing refrigerant composition, the operating engineer will need to calculate the bubble point and dew point of hydrocarbon mixtures. The change in dew point represents the change in evaporator temperature. The change in bubble point represents the change in condenser temperature. The required calculation methods are given in the Appendix.

It is a good practice to slowly change the refrigerant composition and observe the effect on the unit. This is a more precise method than calculation.

Speed-limited

For refrigeration compressors that are speed-limited (or motor-driven compressors not drawing maximum amps), a dramatic increase in refrigerant circulation can be obtained by adding a lighter hydrocarbon to the refrigerant. Most commonly, propane is spiked into a refrigeration system using butane. This raises the compressor suction pressure. The higher suction pressure allows more pounds of refrigerant per cubic foot of suction volume.

At constant speed, the suction volume for a compressor stays the same; thus, more pounds of refrigerant will be circulated. For example, consider the case when a refrigerant stream of butane is spiked with 10% propane:

- 100% butane refrigerant
- 40°F compressor suction temperature
- 2 psig compressor suction pressure
- 100 mol/hr refrigerant circulation
- 5,800 lb/hr refrigerant circulation
- 30,000 ft³/hr compressor suction volume

Now, with 10% propane added to the refrigerant, refrigeration capacity is increased by 18%:

- 90% butane and 10% propane refrigerant
- 40°F compressor suction temperature
- 5.4 psig compressor suction pressure
- 120 mol/hr refrigerant circulation
- 6,820 lb/hr refrigerant circulation
- 30,000 ft³/hr compressor suction volume

Note that, for a given speed, the compressor suction volume is constant. The increase in refrigeration capacity is proportional to the refrigerant circulation. The evaporator temperature has remained constant.

The compressor suction pressure has increased because the refrigerant has become more volatile. This increases the density of the gas being compressed and results in a greater circulation of refrigerant. Unfortunately, the first law of thermodynamics still holds: You cannot get something for nothing. Compressor horsepower will increase proportionally to the increase in refrigeration duty.

This technique was used daily by the author in 1974 through 1976 in Texas City. on an alkylation unit that was refrigeration-compressor limited. During periods of cold weather, the compressor would change from horsepower-limited to speed-limited. Propane was then added to the refrigerant accumulator drum. The turbine governor control valve (i.e., the driver steam inlet valve) would open to maintain compressor speed. When the governor valve was fully open, the compressor was using its maximum available horsepower. The optimum propane concentration had been reached.

Horsepower-limited

When a compressor is horsepower-limited, changing the refrigerant composition will not make an appreciable difference. Superficially, it may seem that using a refrigerant with a higher molecular weight might help. This is wrong. There are many methods to overcome horsepower limitations in refrigeration systems; adjusting refrigerant composition is not one of them.

Accumulator relief valve

The person who designed your unit may have forgotten how hot it gets in summer. As ambient temperatures increase, the compressor discharge pressure goes up, typically 1 or 2 psi for each degree Fahrenheit. This is because the refrigerant condenser is running warmer.

The compressor case, condenser, and refrigerant accumulator all have a maximum operating pressure. This pressure is stamped on the nameplates. These facilities are protected from being overpressured by a relief valve (or safety) on the accumulator drum. The operators will have learned from experience to stay 5 to 10 psi below the relief valve setting. Once a relief valve opens (pops), it doesn't always close by itself. Then the unit is down.

The compressor discharge pressure depends only on two factors: the temperature in the accumulator drum and the composition of the refrigerant in the accumulator drum. The speed or operating characteristics of the compressor have nothing to do with the compressor discharge pressure.

The accumulator drum temperature equals the condenser outlet temperature. High circulating refrigerant rates increase the heat load on the condenser and raise the accumulator temperature. Hot weather has the same effect.

The operators will slow the compressor down on hot days. This controls the refrigerant accumulator pressure below the relief valve popping pressure. Slowing the compressor down also reduces refrigeration capacity.

To hold down the accumulator pressure in hot weather without losing refrigeration capacity, the operating engineer needs only to change the refrigerant composition. On one unit the capacity was substantially increased by this technique (the refrigerant was propane). A hose was hooked up from a nearby butane line to a bleeder connection on the refrigeration system. Butane was slowly added through the hose. This made the refrigerant less volatile, and the accumulator pressure started dropping. The compressor could then be speeded up safely.

Again, this technique of reducing compressor discharge pressure does not save horsepower. The compressor suction pressure will drop right along with the discharge pressure. This will hold evaporator temperature constant.

Minimum suction pressure problems

It is not safe to let compressor suction pressure fall below atmospheric pressure. If it does, the compressor will suck in air. The air will accumulate in the top of the refrigerant accumulator drum and form an explosive mixture.

When the suction pressure gets too low, the operators will slow down the compressor. This may raise the evaporator temperature above the desired level. To prevent this, one should add a more volatile component to the refrigerant. For example, add a little propane to butane refrigerant.

A word of caution

When adjusting refrigerant composition by spiking in a light hydrocarbon, remember that the spike component must be dry. A truckload of propane or butane meeting commercial liquefied petroleum gas (LPG) specs is a safe bet. To make propane more volatile, using propylene as a spike is fine. However, propylene will form hydrates (an ice-type solid) in the presence of moisture at temperatures as high as 55°F.

Importance of the Throttle Valve

The principal purpose of the refrigerant throttle valve is to prevent recirculation of uncondensed refrigerant from the compressor discharge to the compressor suction. If this happens, refrigerant capacity will fall.

One should determine if the throttle valve is allowing vapor to escape from the condenser. The way to do this is to make sure that a visible liquid level can be observed in the refrigerant accumulator drum. One of the reasons for reduced refrigeration capacity on operating units is a false liquid level. Loss of the liquid level will allow vapors to blow through the throttle valve. As refrigeration duty is supplied only by the vaporization of liquid refrigerant, allowing vapors to pass through the throttle valve robs refrigeration capacity.

To see if this is your problem, check the level in the accumulator gauge glass. First blow out the gauge glass bottom tap. Then blow out the top tap. Now drain the gauge glass down. Open the bottom tap and watch the liquid level come back up in the gauge glass. If either gauge glass tap can't be blown out, you will not be able to obtain a true liquid level reading.

Missing accumulator drum

Suppose the refrigeration system does not have an accumulator drum. Figure 10–2 shows such a system. Note that the throttle valve is now called a back-pressure control valve. Although this represents an inferior design, many refrigeration systems don't have an accumulator drum.

The trick is to set the back-pressure control valve properly. Too high a back pressure wastes compressor horsepower. Too low a back pressure allows uncondensed refrigerant to blow through the throttle valve (see figure 10–2). As indicated in the figure, 10% of the compressor capacity is wasted.

I have spent many fruitless hours trying to set such a control valve. In practice, operating personnel will not be able to put such a control valve in a proper position routinely. Even if the control valve is set correctly, a small increase in the condenser operating temperature will cause the vapor to start blowing through the throttle valve again. The only permanent solution is to install an accumulator drum between the condenser and the throttle valve. The throttle valve will then be controlled by the accumulator drum liquid level.

Evaporating Problems

In the evaporator, heat from the process is removed by vaporizing refrigerant. The evaporator may be a reactor or a heat exchanger. In applications where a process fluid is used as the refrigerant, the evaporator drum is a drum of boiling hydrocarbons.

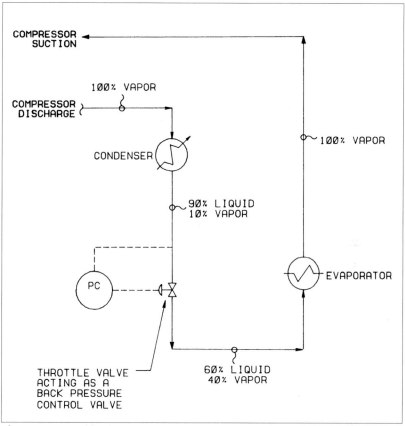

Fig. 10–2. Vapor blowing through the throttle valve wastes compressor capacity.

Reduction of the heat-transfer coefficient in an evaporator will lead directly to lost refrigeration capacity. A good indication of evaporator problems is an unusually low compressor suction temperature. For example, your home air conditioner ices up when it gets low on Freon.

You can tell how much refrigeration capacity has been lost due to evaporator heat-transfer problems as follows. Measure the refrigerant outlet temperature of the evaporator exchanger. Note the outlet temperature on the process side. Now calculate the evaporator duty. Then compute the following value:

$$\frac{H_E}{(TEO - TRO)} = UA$$

where:

H_E = Evaporator duty

TEO = Evaporator process side outlet temperature

TRO = Evaporator refrigerant side outlet temperature

UA = A number proportional to the evaporator heat-transfer coefficient

The calculated value of UA is then compared to a similar number, calculated from historical data, when the refrigeration system was operating properly. If no former data are available, compare your calculated UA against the design valve.

A decrease in UA in excess of 20%–30% is significant. On one unit a decrease in UA of 75% doubled the required refrigeration horsepower. Having determined that the problem with the refrigeration system is caused by evaporator heat-transfer deficiency, how can one pinpoint what's wrong with the evaporator? Evaporator problems come in two forms: fouling of the heat-transfer surface on the process side, and reduction of the boiling film heat-transfer coefficient on the refrigeration side.

Drown tubes in refrigerant

The refrigerant is usually clean and will not cause fouling of the evaporator heat-transfer surface. Depending on evaporator construction, reduced boiling film coefficients are caused by many factors.

A typical evaporator will consist of a shell-and-tube heat exchanger. Refrigerant will be on the shell side as shown in figure 10–3. The refrigerant liquid level should be sufficient to cover all tubes if maximum refrigeration capacity is desired. Any factor that causes tubes to be exposed in the vapor space will reduce evaporator heat transfer and, hence, refrigeration duty.

Note that in figure 10–3 the evaporator level control valve is being used as a throttle valve. If you see this valve wide open, there may not be sufficient inventory of refrigerant in the system. Check the accumulator drum level and add refrigerant to establish a normal accumulator liquid level. If there are two evaporators in parallel, pinch back on the control valve to the evaporator that you are not having trouble with.

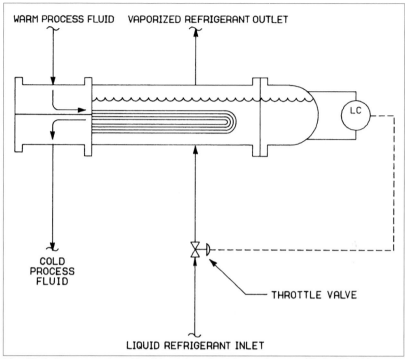

WARM PROCESS FLUID   VAPORIZED REFRIGERANT OUTLET

LC

COLD
PROCESS
FLUID

THROTTLE VALVE

LIQUID REFRIGERANT INLET

Fig. 10–3. An evaporator with all heat-transfer tubes submerged in liquid refrigerant.

Increasing plant throughput

Sometimes improving the boiling film coefficient in an evaporator will have dramatic results on a process unit. On one refrigerated polymerization unit in El Dorado, Arkansas, in 1969, plant throughput was increased by 150%.

On this unit the refrigerant was vaporized in a tube bundle immersed in the reacting liquid. To keep the reactor cold, the refrigerant was vaporized at 30 psig and 20°F. The calculated overall heat-transfer coefficient (U) was 22.

Figure 10–4 shows how the refrigerant system was changed. The reactor (evaporator) was revamped from stagnant boiling in the tubes to forced refrigerant circulation through the tubes. The heat-transfer coefficient went up to 60. The higher U permitted a higher refrigerant vaporization temperature and pressure. This raised compressor suction pressure and conserved compressor horsepower. By changing the circulation rate of the new pump, it was possible to raise or lower the reactor process side temperature.

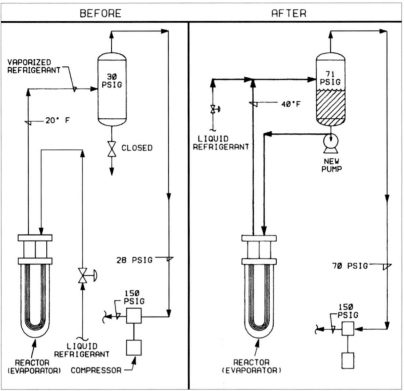

Fig. 10–4. Saving compressor horsepower by going to forced circulation in an evaporator.

The lesson is: When refrigerant is on the tube side of an evaporator, the circulation rate (pounds of liquid per pound of vapor at the outlet) needs to be kept high. On the polymerization unit described, the U reached its maximum at a circulation rate of eight to one.

Evaporator fouling

If loss of heat transfer in the evaporator cannot be accounted for by process problems on the refrigerant side, the operating engineer can assume there is fouling on the hot side. Pulling the evaporator bundle for cleaning is then in order.

On occasion, evaporator deficiencies can be due to compressor lube oil leaking into the refrigerant. The lube oil will accumulate in the bottom of the evaporator. Lube oil is not soluble in propane and can be drawn off the bottom of the evaporator.

Sometimes the lube or seal oil will not settle out until it achieves concentrations in the refrigerant high enough to interfere with efficient heat transfer. To determine if this is likely, draw off a sample of refrigerant in a quart bottle and allow the propane to weather off. This may take several hours. If you have more than a milliliter of oil left in the bottle, oil contamination of refrigerant may be diminishing refrigeration capacity. Determining the cause of excessive seal oil leaking into the refrigerant is the first step. Adding a "slip stream centrifugal separator" on the circulating refrigerant will also help. The recovered oil can then be returned to the seal oil sump after weathering.

Refrigerant Condenser Difficulties

Possible problems with condensers are:

- A shell-and-tube water-cooled condenser is fouled with dead crabs, mud, grass stalks, etc., and needs to be back-flushed.
- Cooling water flow is too low through tubes and is promoting waterside fouling.
- A bypass around the condenser is leaking vapor to the accumulator drum.
- Draining of condensed refrigerant from the condenser to the accumulator drum may be restricted due to piping hydraulics or mechanical blockage. (Once, a dead rat was found in an accumulator feed line.)
- Excessive subcooling of refrigerant liquid in the condenser reduces the heat-transfer surface available for condensing.

Methods to identify and correct these problems are discussed in chapter 12. Remember that a refrigerant accumulator drum is similar to the tower reflux drum.

Troubleshooting Checklist for Refrigeration Systems

Compressor
Overspeed trip set low
Ammeter trip (fuse) set low
Motor running hot
Low steam temperature or pressure
High exhaust steam pressure
Dirty turbine wheels
Governor not opening
Turbine hand valves closed
Leaking valves on
   reciprocating machines
Often clearance pockets on
   reciprocating compressors

Refrigerant Composition Adjustment
Speed-limited compressors
Horsepower-limited compressors
Excessive accumulator
   drum pressure
Minimum compressor suction limits

Evaporator
Low refrigerant level
Inadequate refrigerant feed
Low circulation ratio on
   refrigerant side
Fouling on process side
Lube oil accumulation in refrigerant

Throttle Valve Accumulator Drum
Accumulator drum liquid level
Effect of throttle valve too open
Effect of throttle valve
   pinched closed

Condenser
Fouling due to low water velocities
Water side needs back-flushing
Vapors leaking around condenser
Not enough air flow
Warm air circulating through
   air coils
Poor refrigerant drainage
   from condenser
Excessive subcooling
   of refrigerant
Refrigerant accumulator elevated
   above condenser
Condenser outlet nozzles
   too small

Centrifugal Pumps

One of the most frequent decisions that a process operating engineer is asked to make is whether a pump needs to be repaired. Before pulling a pump out of service and having it overhauled, an important distinction must be made. Has the pump really failed mechanically, or is a process deficiency the real problem?

Troubleshooting pumping difficulties is a job that operating engineers face every week. Count the number of pumps on your unit; the total will likely exceed 100. A major part of a unit's mechanical budget is allocated to pump maintenance. Many pumps sent to the shop are just disassembled, reassembled, and returned to the unit. The shop machinists have found nothing wrong with them. Also, many pump mechanical failures are caused by operator errors.

Hourly operators often decide when a pump is to be pulled and overhauled. Their criterion is simply that it "will not pump." It's easy to tell why a pump is inoperable once it is torn apart. The trick is to determine the trouble with the pump in place. That is the operating engineer's job.

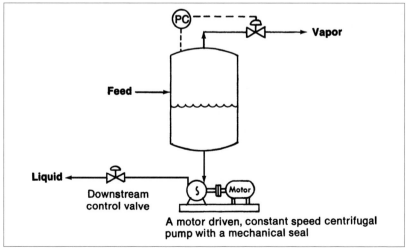

A motor driven, constant speed centrifugal pump with a mechanical seal

Fig. 11–1. Typical pump arrangement (courtesy *Oil & Gas Journal*).

Figure 11–1 shows a typical pump arrangement: A motor-driven centrifugal pump is taking suction from a vessel. This vessel is partially full of a liquid hydrocarbon at its bubble point. The flow of liquid is controlled by a control valve on the pump's discharge line.

What Can Go Wrong?

Some pumps run continuously for years. Others of similar design and service fail every few months. The difference is in the people running the equipment. The more common problems encountered with motor-driven centrifugal pumps are:

- Leaking seal (liquid squirting out around the shaft)
- Rattling noise
- Shaft not turning
- Below normal discharge pressure or flow
- Pump running roughly
- Motor tripping off or running roughly

The novice operating engineer typically has a chemical engineering background. Therefore, he can best master pump troubleshooting by inspecting disassembled pumps in the shop. The idea is to learn to relate the physical damage that has occurred to the pump's internal parts to the process unit's operating problems.

How Pumps Work

A centrifugal pump is the ultimate in simplicity. A basic representation is shown in figure 11–2. Liquid flows to the suction nozzle and into the impeller eye (center). The impeller usually rotates at 3,600 rpm. The liquid undergoes centrifugal acceleration.

Downstream of the impeller, the liquid decelerates in the volute or diffusor, and its velocity is converted to pressure according to Bernoulli's principle. The impeller wear ring minimizes backflow to the suction. The shaft, which is fixed to the impeller, rides on the bearings. The shaft is connected to the motor with a coupling.

The shaft fits into the pump case through a delicate mechanism called a seal. The seal's purpose is to keep the process liquid from leaking out around the case. In very old pumps, rings of packing (a soft gasket-like rope) are used instead of a mechanical seal.

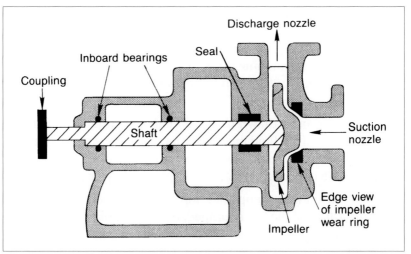

Fig. 11–2. Centrifugal pump components (courtesy *Oil & Gas Journal*).

Centrifugal pump operation should conform to the pump curve supplied by the manufacturer. A sample pump curve is shown in figure 11–3. Viscosity does not affect pump performance as long as the viscosity is below 20–40 centistokes (cs).

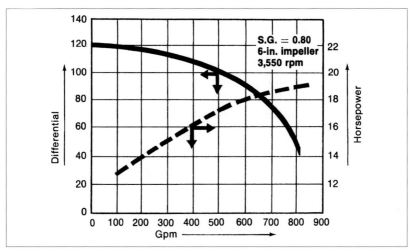

Fig. 11–3. Typical pump curve (courtesy *Oil & Gas Journal*).

A pump will develop a constant feet of differential head at a given flow rate, regardless of the specific gravity of the fluid pumped. However, the differential pressure developed by the pump is proportional to the specific gravity of the liquid. For example, a pump is able to lift water from atmospheric pressure to a tank on a hill 231 feet high at a rate of 100 gpm. The discharge pressure of the pump is 100 psig. The same pump will deliver naphtha of 0.60 sp gr to the same tank at 100 gpm, but the pump's discharge pressure will be 60 psig.

The motor amps drawn by a pump are also proportional to the specific gravity of the fluid pumped. Therefore, in the above example, the motor driver was pulling 100 amps when pumping water (1.0 sp gr); but only 60 amps when pumping 0.60 sp gr naphtha.

The concept that pumps develop a constant differential feet of head, regardless of the fluid pumped, no longer holds true when the fluid viscosity approaches 40–50 cs. At these higher viscosities, the pump efficiency rapidly declines, and the discharge pressure and flow will be greatly reduced.

Rattling Equals Cavitation

Cavitation. Everyone who works in a refinery needs to have a clear idea of its causes and consequences.

Any liquid in a partially full vessel at equilibrium with the vapor phase is at its bubble point. If this liquid is depressurized even by one psi, it will begin to boil. Liquid flowing from a vessel will lose 5 to 15 feet of head before it is picked up by the pump's impeller. If fluid enters a pump at its bubble point, it will start vaporizing inside the pump. The formation of these bubbles in the area of the impeller accounts for the noise associated with cavitation.

A cavitating pump makes a rattling sound. Operators call it slippage because the pump discharge pressure slips down. A discharge pressure rapidly rising and falling is typical of cavitation.

To avoid cavitation, a pump needs net positive suction head (NPSH).

This is the height of liquid required to ensure that the fluid is above its bubble-point pressure at the impeller eye. Assuming a pump needs 10 feet of NPSH to prevent cavitation, the minimum liquid level in the vessel the pump is taking suction from should be 12 feet above the impeller. The extra 2 feet is needed for piping friction head loss.

Why pumps cavitate on start-up

The correct way to start up a centrifugal pump is with the discharge valve barely cracked open. If the operator starts the pump with the discharge valve wide open, or if he suddenly opens discharge after initiating flow, the pump will cavitate. Why does this happen?

When a pump is idle, the velocity in the suction line is zero. When we start the pump, velocity must increase. We must accelerate the velocity of the fluid in the suction line to several ft/sec. The energy to provide this acceleration cannot come from the pump; it must come from the liquid itself. The only source of energy available to the liquid in the suction line is gravity (i.e., the static head of liquid in the suction line). The available NPSH to the pump is partially converted to increasing the kinetic energy in the suction line while the operator opens the discharge valve.

This means that a pressure gauge placed on the suction of a pump would show a relatively large loss in pressure while the pump was being lined out. Once the flow has been increased to its steady-state value, the suction pressure would come partly back up. Hence, if the operator opens the discharge valve too quickly, he can cause an excessive loss in suction pressure and the pump may cavitate during start-up. This is the reason many pumps experience seal failure due to cavitation when they are first put online.

Origins of cavitation

The larger the volume of vapor generated in the pump suction, the worse the cavitation. Therefore, the result of insufficient NPSH is most apparent in light hydrocarbon and vacuum services.

Many pumps are sent to the shop due to vibration. Frequently, there is nothing wrong with the pump. The problem may be low liquid level in the vessel the pump is taking suction on. The level controller on the pump discharge line may have malfunctioned. To check this, blow out the taps on the vessel gauge glass and verify that there is a good level in the vessel.

A suction-line restriction will also cause a pump to cavitate. To verify this, run the pump with its discharge valve pinched back just enough to suppress cavitation. Then measure the pressures at the upstream vessel and the pump suction (use the same gauge). Convert to feet of liquid by multiplying the ΔP reading by 2.31/sp gr of liquid. Then subtract the vertical distance between the two gauges from the pump suction reading. The resulting normalized head at the pump suction should only be 1 to 2 feet less than the head of liquid in the vessel. If the pressure difference is quite a bit more, there is probably a plugged suction line.

A slight warming of the fluid at the pump suction promotes flashing. In refrigeration service, deterioration of insulation on a suction line has caused cavitation. Conversely, cavitation in a pump moving 500°F oil has been stopped by spraying water on its uninsulated suction line.

Pumps require more NPSH when their discharge flow is increased. A pump may be running fine at 2,000 B/D. Increase the flow to 2,100 B/D and the flow starts to wobble. Put the flow back to a rate of 1,000 B/D and the pump lines out. The simplest way to correct this difficulty is to raise the level in the upstream vessel a few feet before turning up the flow.

Increasing the pressure in a partially full vessel does not stop cavitation. The contents of the drum are still at their bubble point. A sudden decrease in vessel operating pressure will usually initiate pump slippage. This is only temporary; the liquid will soon reach its bubble point again.

Sometimes a bad seal will allow air to be drawn into the suction of a pump. This also promotes cavitation and only happens with a low (but not necessarily a negative) suction pressure.

Starting troublesome pumps

Since pumps always require extra NPSH on start-up, pumps with a marginal suction head tend to slip on start-up. To overcome this problem, proceed as follows:

1. Suddenly increase the pressure of the vessel providing suction pressure.
2. Start the pump and open the discharge valve within five minutes of the above step.

Before the bubble point liquid reaches the pump, the flow will have been established.

Consequences of cavitation

A pump running with insufficient NPSH vibrates. Eventually, its seal will be damaged. In propane service, running with too low a liquid level for a few hours will often injure the mechanical seal sufficiently to require taking the pump out of service.

Passing vapor through the impeller causes rapid changes in the density of the fluid pumped. This uneven operation forces the impeller and shaft to shake, and the vibration is transmitted to the seal.

A mechanical seal consists of a ring of soft carbon and a ring of hard metal pressed together. Their smooth, polished surfaces rotate past each other. When either surface is chipped or marred, the seal leaks. You can imagine the effect of severe vibration on such a delicate mechanism. Continued operation of a cavitating pump will damage its bearing and eventually the impeller wear ring and shaft.

Oversized Pumps Surge

You may sometimes hear a large centrifugal pump making a low-key, asthmatic, surging sound. Such pumps are subject to frequent impeller, seal, and bearing failures. The problem is that the pump is too big. Some pumps, operating at less than 60% of design, are subject to excessive internal recirculation. For large, high-head machines, internal damage can be severe. If you hear a pump making a surging sound, determine if it is operating on its curve. If the flow is less than 60% of design, install a smaller impeller with a smaller width. This will save maintenance and energy dollars.

For many smaller pumps, it is not clear how far back on the pump curve the pump can be operated before it is damaged due to low flow. When in doubt, ask the maintenance department to check the pump for vibration before you reduce flow. Then, with the vibration monitor in service, reduce the pumping rate to the desired level. If the indicated vibration amplitude does not increase noticeably, one may safely assume that throttling on the pump discharge will not damage the pump.

When Not to Pull a Pump

On coming out to the unit in the morning, the operating engineer will frequently be faced with the following report from the midnight shift: "P-17 won't pump its normal volume; it needs to be overhauled." Before sending a pump into the shop, check the following:

- **Block in the discharge.** Calculate the head (feet of liquid) put up by the pump at zero flow. This head is:

$$\frac{2.31}{\text{sp gr of liquid}} \times (\text{discharge pressure} - \text{suction pressure})$$

  Compare this value to the indicated head on the pump curve. For instance, on figure 11–3, the shut-in (zero flow) head is 120 feet. If the observed shut-in head is close to the curve value, this may indicate that there is nothing wrong with the pump.

- **Low NPSH.** Some pumps do not vibrate noticeably when they are marginally short of NPSH—they just lose capacity. Raise the liquid level in the upstream vessel a few feet. If the pump then picks up nicely, you have a partially plugged suction line or too low a liquid level in the upstream vessel. See the section on "Origins of Cavitation."

- **Downstream restriction.** At the maximum obtainable flow, check the feet of head (as described previously) on the pump curve. If the pump is putting up the head indicated on the curve, do not pull the pump. Ask the instrument mechanic to check the control valve on the discharge line. Perhaps it is not opening all the way or the downstream pressure requirement has increased due to a lunch bucket lodged in an exchanger inlet.

- **Low suction pressure.** The hydrocarbon in the vessel upstream of the pump may run colder than normal. This lowers the pressure (to hold the liquid at its bubble point) in the vessel. The pump then has to put up more head and will consequently transfer less volume. This problem has nothing to do with inadequate NPSH.

- **Pump won't turn.** A good operator will turn a pump by hand before switching on the motor. If the shaft does not turn easily, wash out the pump case with hot water. For vertical pumps in light hydrocarbon service, caustic may have solidified inside the pump case. Try uncoupling the motor from the pump. Maybe the motor bearings are ruined, but the pump is okay. If the pump still won't turn smoothly by hand, send it to the shop. If you run it, you can wreck it.

- **Erratic Flow.** The usual cause of erratic flow is cavitation due to poor suction conditions. However, if the discharge pressure is erratically high (i.e., varying from normal to high), the problem is downstream of the pump. A heat exchanger leaking light liquid hydrocarbons into the pump discharge can cause an erratic pump discharge pressure. Also, a loose seat in a control valve downstream of a pump may vibrate and also cause an erratically high discharge pressure. Cavitation is indicated by an erratically low pump discharge pressure.

Internal Recirculation

The purpose of pump wear rings is to reduce to a minimum internal recirculation of the process fluid. Figure 11–4 shows a cross section of an impeller and wear ring. The wear ring fits tightly against the pump case and clears the edge of the impeller by about 0.02 inch. The space between the wear ring and the impeller is called the *diametrical clearance*. This clearance will vary with the impeller diameter and operating temperature. At design conditions about 5%

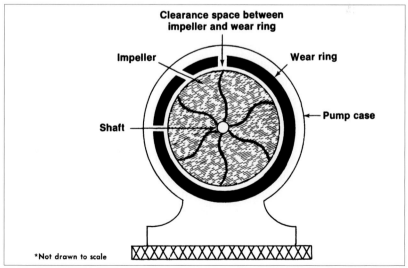

Fig. 11–4. The function of a wear ring (courtesy *Oil & Gas Journal*).

of the flow from the impeller short-circuits back to the pump suction through this clearance.

If the wear ring erodes, the clearance could double to 0.04 inch. Internal circulation then increases to 10%.[1] The recirculated liquid is always warmer than liquid flowing to the pump. Therefore, as the clearance increases, the pump runs hotter. For a really badly eroded wear ring, you can feel a slight temperature increase by putting your hand on the discharge line. Also, hotter operation will cause a pump to cavitate.

Inability to operate the pump on its design curve, coupled with an accelerating rise in fluid temperature across the pump, means the impeller wear ring clearance has widened. Take a look at the wear rings when they are removed in the shop. Sometimes a change in process conditions (temperature, composition, particulates) on the unit will cause a rapid erosion of the rings. Possibly a change of wear-ring metallurgy could help.

For example, in a case involving hot sulfuric-acid pumping service, Carpentor 20 wear rings were being used. Every two weeks the pump regularly lost 50% of its capacity. Illium was substituted for Carpentor 20, and the pump's visits to the shop ceased.

If nothing else, excess wear ring clearances waste horsepower. Tripling the design clearance increases motor power requirements 10%.

Worn-out impeller

If the impeller has suffered substantial erosion, pumping capacity will decrease. Not infrequently, the impeller comes loose from the shaft, and pumping capacity falls to essentially zero. Check the driver motor with an ammeter. Low amperage indicates the impeller is no longer fixed to the shaft.

Blowing seal

A mechanical seal's principal components are two smooth rings: carbon and a hard metal alloy. One ring rotates on the shaft while the other is stationary and connected to the pump case. The two faces are forced together with a spring, called the bellows. When a seal leaks, it is usually because these faces become separated or are no longer smoothly rotating against each other.

Gland or seal oil is sometimes used to lubricate the faces. More often, the liquid being pumped is used as the lubricant. This lubricant must be clean. If a process change has introduced particulates into the liquid being pumped, the seal on a self-lubricated pump can be damaged. Installation of a small cyclone separator on the seal flush line (i.e., the small line from the pump discharge to the seal) may help. Better yet, obtain a source of clean seal flush liquid.

In propane and butane service, water in the process fluid can turn to ice in the seal. These ice crystals get between the faces and cause the seal to leak. Keep the seal warm with steam to melt these crystals.

The pressure of the gland oil or seal flush fluid to the seal should be roughly 10–30 psi higher than the pump suction pressure. This is a fine point many operators miss. Too high a pressure can damage the seal.

Careful examination of the carbon face will often indicate the cause of seal failure. The wear tracks left in carbon faces are clues. A complete discussion of this subject is given by McNally.[2]

Safety Note

Seal leaks are dangerous. Because small pump seal leaks are common and are slowly increasing in volume, we sometimes let them go. This is dangerous. Some very heavy hydrocarbons autoignite at temperatures below 420°F. Light hydrocarbons will not flash without a source of ignition, but experience has shown that an ignition source eventually appears. Clouds of propane, butane, and pentanes that flash are a major cause of refinery catastrophes.

Upgrading Seals for Reduced Seal Flush Flow

For an older type mechanical seal, good rules of thumb for seal flush usage are:

- Multiply the shaft diameter in inches by three gpm to obtain seal flush usage per seal.
- Multiply the gpm per seal by the number of seals per pump.
- Multiply the gpm per pump by the number of pumps.
- For pumps on standby, multiply the above result by two-thirds, because a pump not running consumes less seal flush.

Here's an actual example from a fluid coker at the Syncrude Tar Sands plant in Alberta, Canada. They have eight black oil pumps (pumping recovered bitumen from the tar sands). The total metered flushing oil flow, which is clean coker gas oil, averages 85 barrels per hour.

This means 85 barrels an hour of recovered product is recycled back to the fluid coker reactor where I estimate 15% is converted to low or zero value coke and gas. That's a net loss of:

$$15\% \times 85 \, \text{B/H} \times 24 \times 330 \, \text{days} = 100,000 \, \text{bbl/yr.}$$

As I write this, crude is selling for $80 (U.S.) per barrel. That means Syncrude is wasting $8 million dollars a year by using seal flush. What can be done?

- About 35% of the above loss could be eliminated today by blocking in the seal flush to idle pumps. One needs just to wait about 30 minutes to flush out the seal chamber. Of course, the operator must remember to turn the seal flush back on, perhaps ten minutes before he restarts the pump.

- A modern pump seal uses no seal flush (see *Dry Gas Seals Handbook,* by John Stahley, Pennwell Publications, 2004.) Such seals are kept pressurized by nitrogen, which separate the seal faces. This works because the seal face clearances are only 20%–30% of conventional seal face clearances. The amount of nitrogen entering the pump's suction is too small to cause cavitation. Of course installation and maintenance of dry seals, because of the smaller clearances, is more difficult then conventional seals.

- Purchase a premium seal rather than a low cost seal. Eagle-Burgmann produces seals that require less flushing oil, but they can be very expensive. Actually, their seals may be very cheap if one takes into account the cost of seal flush lost in a black oil resid pump.

- Avoid excessive seal flush pressure. The manufacturer of the seal defines the required stuffing box (i.e., seal flush chamber) pressure, as a function of the pump's suction pressure. Too much seal flush pressure forces the seal faces too far apart and may damage the seal faces and will obviously increase the flow of the seal flush oil into the process fluid.

Balanced Seal Faces

A modern, externally flushed, mechanical seal circulates a barrier fluid like diesel oil, from a pressurized pot. The pot pressure is maintained with nitrogen. The pot may have to be refilled once every few days. The circulation is created by a rotating propeller type device in the seal chamber.

The important idea about balanced seal faces is that the seal flush pressure can be quite variable, without compromising the seal integrity. The seal oil pot pressure must still be above the pump suction pressure, but by a variable amount (plus 20 to 70 psi, on a recent job I worked on).

Rough Running

A noisy or vibrating pump requires immediate attention. You can feel the vibration by touching your fingernails to the bearing housing. There are four common reasons for a noisy pump.

- **Cavitation.** Discussed previously.
- **Driver troubles.** Uncouple the motor from the pump. If the driver runs roughly when uncoupled, send it to the shop. The pump is okay.
- **Bad bearings.** The pump bearings housing will often be hot to the touch if the bearings are bad. There is no sense running a pump with damaged bearings; this will just wreck the rest of the pump. When the pump is disassembled, ask the shop machinist which components of the pump required replacement. If the bearings were the only problem, chances are the damage was initiated by inadequate pump lubrication, that is, low oil level in the bearing housing. You had better check on the lubrication schedule (if any) that the unit operators are following.
- **Impeller balance.** Uneven erosion-corrosion of the impeller will unbalance its rotation and cause vibration. In normal refinery service, corrosion on one side of an impeller occurs when the pump is idle.

Check Spare Pumps

Most pumps in a refinery are spared. Spare pumps should be run once a week, or once a month, to be sure that the pump will actually be operable when needed in an emergency. Another important reason is that machinery tends to deteriorate when idle. Specifically, the impeller in a centrifugal pump may corrode unevenly when it has not turned for a long time.

Pumps that remain idle for an extended period will also tend to develop flattened bearings. While the bearings will resume a round shape while the pump is running, the pump will run roughly for some period. During this rough running interval, the pump's mechanical seal may be damaged.

A thorough pump-switching program is a great aid in troubleshooting. It allows early identification of problems. This gives you a chance to use technical judgment in analyzing the deficiency. The engineer should try to be present on the shift when spare pumps are switched. Witnessing the initial run-in of a faulty pump can give you a head start in troubleshooting.

Avoiding Motor Failures

The ordinary 3,600-rpm electric motors used to drive refinery pumps are quite reliable. If an uncoupled motor is running roughly, the bearings need replacement. This is often due to insufficient lubrication.

Some large motors have their lubricating oil cooled by cooling tower water. If the bearings on these motors go out frequently, check the lube oil for water. The small lube-oil coolers may be leaking due to corrosive cooling tower water. On one unit this problem was solved by retubing the coolers with 316 stainless steel.

Some pump motors may trip off or burn out at high pumping rates. Good design practice is to oversize motors to prevent this. However, if you have a pump that trips off as the discharge valve is opened, consider the following process variables:

- Increased density of the fluid pumped. A 20% density increase will call for a 20% or more increase in motor amperage.
- A large (several orders of magnitude) increase in viscosity will require considerably more horsepower.

Specifically, when the viscosity approaches roughly 40 to 60 cs (depending on the size of the pump), a rapid increase in driver horsepower will be required. For most refinery applications, viscosity is in the range of 0.5 to 3.0 cs.

- Flow beyond the design point. Figure 11–3 shows how motor horsepower required goes up as gpm increases.
- Has the size of the original pump impeller been increased? A larger impeller will cause the pump's motor to draw more amperage.
- Excessive internal pump bypassing (worn wear rings) or external pump bypassing (through a discharge to suction recycle line) also takes more horsepower. A useful troubleshooting tool for pump motor problems is a permanently mounted ammeter.

Expanding Pump Capacity

The cheapest way to de-bottleneck a process operation that is limited by a pump is to reduce downstream pressure drop. A detailed hydraulic survey is the key. Do not forget to check the pressure drop across the downstream control valve (see chapter 9).

A larger suction line sometimes eliminates NPSH problems. Also, many pumps do not have the maximum-size impeller that can fit in the existing pump case. Changing impellers is an inexpensive method to gain capacity if the motor can supply the extra horsepower (see chapter 10 for de-bottlenecking motors).

A conservative method to ascertain if an existing motor can accommodate a larger pump impeller is as follows:

1. Measure the amp load in the motor breaker room (AI).
2. Calculate the new maximum motor amperage, as follows:

$$(AI) \cdot (DN/DO)^3$$

where:

DN = Diameter of new impeller

DO = Diameter of old impeller

Do not assume that the existing motor will perform at its original efficiency. Motor windings deteriorate with age and will pull more amps as they get older.

Reduction of Line Delta Pressure Drop

Another method of expanding pump capacity is to reduce downstream pressure drop losses. Perhaps there is some unwarranted pressure loss in the downstream piping. The pressure drop between a pump discharge and the downstream equipment or control valve should be measured. Next this observed pressure loss should be compared to the calculated pressure drop. Assuming nonviscous flow (i.e., less than ten centistokes), hydrocarbon or aqueous fluids and ordinary carbon steel pipe, the following simple equation can be used:

$$\Delta P = [0.003 \cdot (DF) \cdot (V^2)] \div ID$$

where:

ΔP = Piping head loss per one hundred equivalent feet in PSI

DF = Fluid density in pounds per cubic feet

V = Velocity in feet per second

ID = Inner diameter of pipe in inches

Increasing the diameter of downstream process piping is an effective method of increasing centrifugal pump capacity. For shorter piping runs, increasing the nozzle sizes for vessel and heat exchangers is also needed to achieve a significant increase in pumping capacity.

Troubleshooting Checklist
for Centrifugal Pump Problems

Leaking Seal
Cavitation-induced vibration
Ice crystals
Dirt in seal flush
Seal oil pressure too high

Rough-running Pump
Infrequent operation
Bearing lubricant
Water in lube oil
Motor bearings bad
Cavitation
Impeller imbalance

Pump oversized
Pump motors
Impeller too large
Lubrication schedule
Increased density of process liquid
High viscosity
Wear ring/impeller
  clearance excessive

Cavitation
Discharge pressure fluctuating
Pump vibrating
Low liquid level
Suction line plugged
Temperature rise in suction line
Flow exceeds design capacity
Badly worn wear rings
Seal sucking air

Capacity Decrease
Check shut-in pressure
Insufficient NPSH
High downstream pressure drop
Low suction pressure
Excessive internal pump
  case erosion
Excessive wear-ring erosion
Impeller erosion

References

1. J. Lightle and J. Hohman, "Keep Pumps Operating Efficiently—Calculate the Cost of Your Excess Clearances," *Hydrocarbon Processing,* September 1979.

2. L. McNally, "Increase Pump Life," *Hydrocarbon Processing,* January 1979.

Distillation Towers

Operators often report that major process units are limited by the capacity of a particular distillation tower. The process engineer assigned to that unit should take such reports as a challenge, investigate the situation, and determine precisely what aspect of the distillation operation is really limiting.

Experience has shown that many reported process limits are really trivial in nature. For example, an overhead condenser may not have been backflushed in years, or the tower trays are plugged and need to be waterwashed. These minor problems can be quickly rectified at little cost. In general, misassembly, dirt accumulation, and piping errors are more commonly the cause of the capacity limitations rather than undersized equipment.

Light Hydrocarbon Distillation

Whether one is troubleshooting a distillation tower in a natural gas liquids recovery facility or in a petroleum refinery, the components of the problem causing inadequate fractionation are the same:

- Damaged tower internals
- Low reflux rates
- Erratic control
- Flooding

Control Problems

There is a very straightforward method to determine if a control problem is leading to the production of off-spec products. This method is based on the following premise: *If you can't run it on manual, you'll never run it on auto.* Certainly, for the relatively simple operations of concern here (i.e., deethanizers, debutanizers, propane-butane splitters, and deisopentanizers), liquefied petroleum gas (LPG) and gasoline fractionation specifications are achievable with manual control. If you cannot successfully operate a distillation column on hand control for a few hours, check to see if the control valves in the field are properly responding to the control center valve position signal.

Next, increase the reflux rate and the tower bottoms heat input (reboiler duty). An unexpectedly large improvement in the separation between the light and heavy key components indicates poor vapor-liquid contacting due to dumping liquid through tray decks or vapor channeling in packed beds. On the other hand, if separation efficiency does not noticeably increase, or even decreases as reflux is raised, flooding is indicated.[1]

Flooding

There are two commonly accepted terms to describe flooding in a trayed distillation tower:

- Liquid flood
- Jet flood

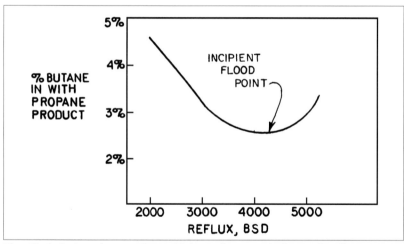

Fig. 12–1. Reduced separation at higher reflux rates indicates flooding.

Figure 12–1 shows the effect on tower pressure drop as the reflux rate and reboiler duty are increased. The liquid flood point is characterized by a sudden increase in measured ΔP. At this point, the capacity of the downcomers to drain liquid pouring over the outlet weir is insufficient. The height of liquid in the downcomers shown in figure 12–2 increases until the top of the weir is reached; at this point liquid begins to stack up on the tray deck and drainage from the downcomer on the tray above is reduced. Thus, all trays above a tray that is flooding will also begin to flood. Fully developed downcomer flood will always greatly reduce fractionation efficiency.

Downcomers can flood at even relatively low liquid rates if the bottom edge of the downcomer is not submerged in the liquid on the tray deck. Unsealing the downcomer in this manner permits upflowing vapor to interfere with liquid flow in the downcomer, and thus reduces downcomer liquid handling capacity. Downcomers will also flood at low liquid rates due to high pressure drop across the tray decks. After all, the height of liquid in the downcomer must be sufficient to overcome the pressure differential between trays. Thus, fouling deposits on tray decks, which reduce the open area available for vapor flow, can precipitate downcomer backup and liquid flooding.

The downcomers do not have to be overflowing for excessive tray loads to reduce tower efficiency. Figure 12–1 shows data collected on one NGL propane-butane splitter. The cause of increased butane content of the tower overhead at the higher reflux rates was excessive entrainment of liquid between trays. Factors, such as increased reflux rates, which raise the liquid level on the tray decks, promote entrainment which, when excessive, is called "jet flood."[2]

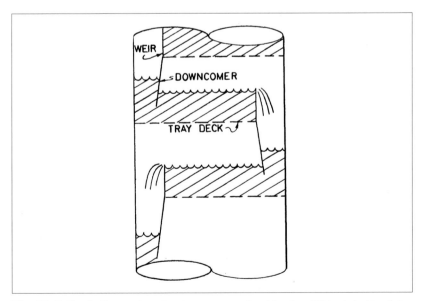

Fig. 12–2. The bottom edge of a downcomer should extend ½ inch below the top of the weir.

Jet flood

Of course, flooding induced by downcomer backup will also raise the liquid level on the tray decks, while liquid entrainment increases downcomer loading. Thus the terms *jet* and *liquid flood* are somewhat a matter of semantics. Flooding—as far as operational personnel are concerned—is indicated by the following:

- The temperature spread between the reboiler outlet and the tower overhead is reduced as the feed rate is raised.
- The concentration of the heavy key component increases as the reflux rate is increased.
- With the reboiler heat duty fixed, increasing the reflux rate does not result in a proportionate increase in the bottoms product rate.

Tray loading which corresponds to the type of incipient tray flooding described above is characterized by the following:

$$\frac{\Delta P}{(sp\ gr)(T_N)(TS)} \geq 2\ 2\%\ to\ 25\% \qquad \text{(Equation 1)}$$

where

ΔP = Pressure drop, inches of water

sp gr = Specific gravity of the liquid on the tray at the appropriate temperature

T_N = Number of trays

TS = Tray spacing, inches

Measuring tray pressure drop

To obtain an accurate pressure drop measurement across the trays in a low-pressure fractionator (less than 15 psig) a single pressure gauge can be used. However, for most services involving propane, butane, and natural gas condensates, a ΔP gauge such as a Magneholic must be employed. Magneholic gauges are available for pressures up to 500 psig. For troubleshooting purposes, it is best to locate the Magneholic gauge as shown in figure 12–3, rather than rely on inert gas purging of the pressure taps to prevent liquid condensation in the pressure drops which distort the true measured tray ΔP. Alternately, the sensing legs may be filled with glycol.

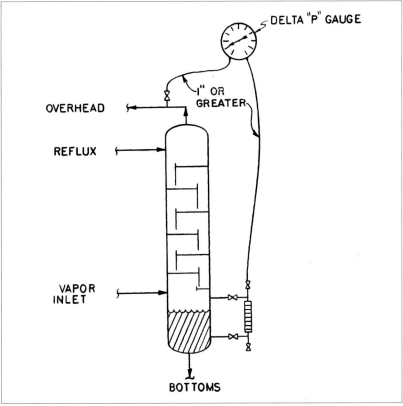

Fig. 12–3. Installation of a ΔP gauge with self-draining is vital to detect tray flooding.

Confusing incidents

Process plant troubleshooting is characterized by a descent into realism. As a proof of this axiom, consider that measuring a low tower pressure drop does not prove the tower is not flooding, and that measuring a high tower pressure drop does not establish that the trays (or tower packing) are overloaded. For example, the operators at one NGL plant noted unusually high butane content in their depropanizer overhead product. To rectify the problem, they increased the reflux rate by 20%. This only seemed to degrade the fractionation further. The operators checked the rectification section pressure drop, expecting to find the trays flooded. To their surprise, the measured pressure drop per tray was quite low. Did this mean the tower was not flooding?

When I inspected the tower, I found an absolutely positive indication of floodings; a vent line on the tower overhead line emitted liquid when cracked open. This observation, coupled with the low measured pressure drop per tray, indicated that only the top tray was flooding. When we opened the tower for inspection, we found the deck of the top tray encrusted with corrosion products. These deposits had caused a high pressure drop on only the top tray and had caused the downcomer on only the top tray to back up.

High liquid level induces flooding

Figure 12–4 illustrates the operation of a debutanizer equipped with a kettle reboiler. The liquid level in the bottom of the tower is determined by the pressure difference between the reboiler and the tower. (As the reboiler vapors vent back to the tower, the reboiler must be at the higher pressure.) The plant operators observed that when they increased the reboiler heat duty beyond a certain point, the tower's ΔP gauge reading would dramatically increase, indicating flooding. They therefore erroneously concluded that the trays were overloaded.

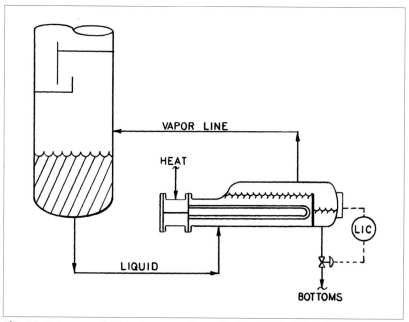

Fig. 12–4. Increased ΔP in the vapor line will raise the liquid level in the bottom of the tower to the vapor return nozzle, at which point the bottom tray will flood.

However, the tower flooding was being induced by a high tower bottoms liquid level. As the reboiler duty was increased, the reboiler pressure also increased. This raised the level in the tower bottoms. When this level reached the reboiler vapor return nozzle, the tower flooded due to entrainment of liquid from the bottom of the tower up to the bottom tray. The operators had noticed that the liquid level rose when the reboiler duty was increased, but they thought that the bottoms liquid level would not cause tray flooding until this level reached the bottom tray. This point deserves emphasis:

> *When the tower bottom's liquid level rises to the vapor inlet or the kettle reboiler return nozzle, the bottom tray begins flooding. The flooding spreads up the tower until liquid is carried overhead. However, even at this point, the indicated liquid level in the tower is still at the level of the vapor inlet nozzle.*

A careful pressure drop survey indicated that there was an inexplicably high pressure drop in the liquid inlet line. When the tower was opened for inspection, the carcass of a dead rat was discovered lodged in the reboiler liquid inlet nozzle. Inadvertently, this rat caused the tower to flood as the reboiler rate was increased. If the rat had expired in the vapor outlet line, the effect would have been the same.

Vertical temperature survey

One of the most powerful tools available to identify tray flooding is a radiation scan. A vertical survey of a column using a radioactive source measures hydrocarbon density inside the column at various elevations. Such a survey will reveal both foam and liquid levels in downcomers and tray decks. Unfortunately, this is a rather expensive procedure—$25,000 per tower survey being a typical cost.

The objective of a radiation scan is to pinpoint the particular tray that is initiating flooding. This tray acts as a pinch point in the column. That is, the trays above the tray which is flooding will also flood (but to a somewhat lesser extent). The trays below the pinch point tray will not be particularly affected.

A "poor man's" method of locating the "pinch point" tray is the vertical temperature survey. This is accomplished as follows:

1. Using the tower vessel sketch as a guide, locate the position of the downcomers.
2. Cut a 1-inch diameter hole in the tower's insulation in line with the center of each downcomer.
3. Using a surface pyrometer (find a probe with a flat, flexible head), carefully and accurately measure the external

temperature of each downcomer. An infrared temperature gun will also do nicely (see chapter 29).

Figure 12–5 shows the results of one such survey. If a tower is functioning properly, the tray temperature will always decrease at higher tray elevations. Even when tray efficiency is low, there will still be a temperature gradient of constant direction. However, the data shown in figure 12–5 indicate a temperature inversion in the stripping section of the column. There are two possible explanations for this inversion:

- Tray flooding
- Upset tray decks

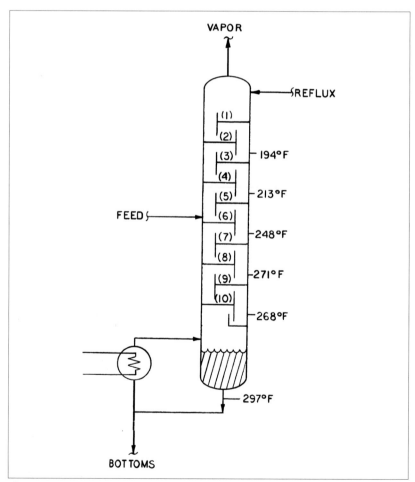

Fig. 12–5. Temperature inversion is a definite indication of flooding.

If the observed pressure drop across the trays is small, the cause of the temperature inversion is upset trays. On the other hand, if an increased reflux rate enhances a temperature inversion between trays, flooding may be predicted with confidence.

Two-phase bottom level problem

When does the level indicated in a gauge glass not correspond to the liquid level in a vessel? Assuming the level taps are unplugged, is it possible for the liquid level in the bottom of a distillation tower to be higher than the level in the gauge glass?

Figure 12–6 illustrates a rather common occurrence in a natural gas amine H_2S scrubber. A light hydrocarbon with a 0.7 sp gr is floating atop amine with a 1.0 sp gr. Due to the location of the level taps, the liquid in the gauge glass is only amine. If we think of the gauge glass and the bottom of the tower as the two legs of a two-phase manometer, it's apparent why the liquid (i.e. hydrocarbon) level inside the tower is above the level in the gauge glass.

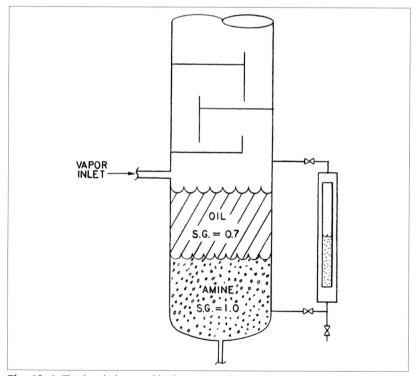

Fig. 12–6. The level observed in the gauge glass may not correspond to the liquid level in the tower.

This phenomenon is a common cause of flooding in gas scrubbers. Due to an erroneous level indication, the liquid hydrocarbon is permitted to rise to the vapor inlet. The liquid is then entrained by the upflowing gas onto the bottom tray. The mixing of the liquid hydrocarbon with the aqueous amine solution promotes foaming and thus flooding of the bottom tray. The foaming/flooding spreads up the tower until gross quantities of amine are carried overhead with the sweet natural gas. Once the accumulated liquid hydrocarbon in the bottom of the tower is flushed overhead by the flooding trays, the tower's bottom level settles out below the vapor inlet nozzle and the flooding and excess rates of amine carryover stop.

To the operators, the cause of the flooding appears inexplicable. It has apparently started and stopped by itself. However, it is possible to locate the level in the bottom of an amine natural gas scrubber by touch. The liquid outlet normally is 20°F to 40°F warmer than the vapor inlet. This temperature difference, when noted on the outside of the tower's shell, corresponds to the true bottom's liquid level.

To prevent this situation from developing, hydrocarbon skimming taps should be provided in the bottom of natural gas amine H_2S scrubbing towers.

Foaming

Equation 1 states that a tray will become less efficient due to incipient jet flood when the pressure drop per tray, expressed in inches of liquid, equals 22% to 25% of the tray spacing. The inches of liquid term assumes the liquid is deaerated. Of course, the liquid in the downcomers and on the tray decks is closer to a froth than to a flat liquid. The more highly aerated the liquid is (i.e. the more foamlike it becomes), the greater will be the depth of liquid corresponding to a measured external pressure drop. So liquids which foam in distillation columns (such as dirty amine and ethane-rich fractionators) reach their incipient jet flood point at pressure drops below the 20% indicated in equation 1.

Ordinarily, this foam cannot be observed in a column's bottom single-sight glass. Even if the bottom of the tower is retaining a foam level above the vapor inlet (or reboiler return nozzle), the fluid in the sight glass is a flat, deaerated liquid. To observe a high foam level in the bottom gauge glass proceed as follows:

1. Block in the bottom gauge glass level tap.

2. Open the top tap completely.

3. Crack open the gauge glass drain valve.

Assuming the foam is above the top tap, the foam will slowly start running down the gauge glass and can be observed readily.

Expanding tray capacity

The capacities of existing towers can best be expanded by changing from trays to structured-type packing. Although this is an expensive modification, increases in tower capacities of up to 35% can be achieved. Unfortunately, one study showed that the internals for one 8-foot ID isobutane-normal butane splitter would cost $45,000 for trays and $290,000 for structured packing.

A less ambitious plan to expand distillation capacity is illustrated in figure 12–7. Combining the ideas shown in this sketch with some other common tray features can easily increase capacity by 25% in many instances:

1. Convert from single- to dual-pass (or even four-pass) trays. Avoid using the three-pass trays, as poor vapor-liquid distribution may result. Keep to a minimum flow-path length of 18 inches.

2. Use "anti-jump" baffles on the center downcomer of two-pass trays at weir loadings above 8 GPM per inch.

3. Maximize the number of caps or sieve holes on the tray decks. Often, a minor change in a downcomer dimension or the width of a tray ring can accommodate an additional row of valve caps or many more sieve holes. Do not exceed 13% hole area as a percent of bubble area.

4. Use a valve-cap retainer assembly that does not restrict vapor flow. This, and the following items, all add slightly to the cost of a tray.

5. Install tray decks fitted with venturi openings. This will reduce tray ΔP.

6. Slope the downcomers so that the outlet area is 65% of the top downcomer area.

7. Use swept-back downcomers. This will reduce the height of liquid over the weir.

8. Install recessed sumps under the downcomer outlets. In effect, this modification increases the open area under the

downcomer and this reduces downcomer pressure drop. Clean services only.

9. Reduce the height of the outlet weir on each tray.

10. Use "push-type" grid caps instead of valve trays.

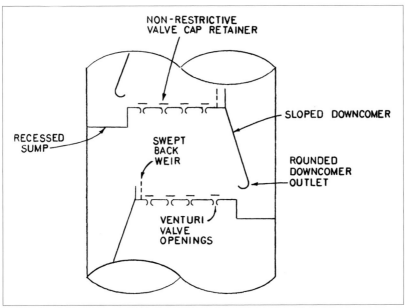

Fig. 12–7. Methods to enhance tray capacity without sacrificing fractionation efficiency.

Many trays come with adjustable weirs, and if a tower is flooding due to liquid backing up out of the downcomers (which is indicated by a step increase in tray ΔP, as the reflux rate is increased), reducing the weir height may significantly enhance tower capacity. Ordinarily, the top edge of the weir is ½ inch above the bottom edge of the downcomer. This dimension keeps the downcomer sealed in liquid so as to prevent vapor blowing up through the downcomer. However, this seal height may be changed to zero (for small-diameter towers where the trays have been carefully and accurately leveled). A zero seal depends on the hydraulic height of liquid over the weir to submerge the downcomer outlet. Reducing the weir height by ½ inch will drop the liquid level in the downcomer by 1 inch. Decreases in weir heights should be undertaken with the knowledge that, on occasion, rather unpleasant reductions in tray efficiency have been observed after such decreases.

Damaged trays

Most often, tower packing supports are dislodged, or distillation trays upset, when a tower is operated with an excessive bottom liquid level. Forcing heat into a tower when the liquid level is several feet above the bottom tray deck often results in dislodging the bottom few trays. Occasionally, trays are misassembled during a turnaround. The result of either of these misadventures is diminished fractionation efficiency.

While x-ray pictures of tower internals easily detect most types of tray damage, this can be a cumbersome and expensive troubleshooting procedure. A simpler way to obtain almost the same information is by a pressure survey. Tower pressure drops of less than 1 inch of water per tray typically indicate tray damage, assuming a tray spacing of 24 inches.

Liquid-filled towers

Whether the overhead vapor line contains liquid can be a good indicator of tower flooding. For example, a propane-isobutane splitter had a high C_4 content in the overhead vapor line, and high C_3 content in the bottoms. Both reflux and feed were running at rates that normally resulted in good fractionation.

On examination of the unit, it was discovered that ice had formed on the surface of the line going to the flare from the splitter's overhead relief valve. The bypass valve around the relief valve was leaking slightly.

Normally, vapor expanding from about 200 psig through the bypass would cool only about 20°F. But there was a temperature drop of more than 80°F, indicating that liquid propane was expanding through the bypass and that liquid propane was present in the overhead vapor line, and that at least the top tray was flooding.

When tower flooding is suspected, the overhead vapor line should be checked for liquid.

Reflux changes

Normally, an increase in reflux rate should improve fractionation. If certain changes in overall tower operations do not occur as a result of increasing the reflux rate, tower flooding can be suspected.

Two rules of thumb can be used to determine the effects of an incremental change in reflux: For every 1,000 B/SD increase in reflux flow rate, the reboiler steam rate should increase about 1,200 lb/hr; alternately, an increase in reflux of 1,000 B/SD should increase the bottoms flow rate about 650 B/SD. If neither of these effects occurs when reflux rate is changed, suspect tower flooding.

For example, on a C_3–C_4 splitter on an FCCU, reflux was increased to 16,000 B/SD from the normal 9,000 B/SD in an effort to minimize the butane content in the splitter overhead. The steam rate to the reboiler did not increase as the reflux rate was increased. (The steam flow was automatically controlled by the bottoms temperature.)

The extra 7,000 B/SD of reflux did not flow down through the tower because the tray decks were already flooding at the 9,000 B/SD reflux rate. The extra reflux overflowed the tower and recycled back to the reflux drum via the overhead condensers.

The flooding problem was attributed to a plugged level tap on the level control system that prevented the level control from sensing the bottoms level. The incorrect level reading allowed liquid to rise in the tower until the trays flooded.

Level control

Improperly operating level controls can cause tower flooding. In the example just discussed, the level controller is the type that uses a separate float chamber, with displacers to sense the liquid level in the chamber (figure 12–8). These types of controllers can sense incorrect level.

In another situation, a level controller had been an indication that the tower bottoms level was at 85% of the level controller span. In fact, the controller had showed a constant 85% for several days.

To verify the controller indication, the liquid level was observed in the gauge glass on the tower, after carefully blowing out both gauge glass taps separately. Blowing the taps is essential to obtain a true level reading in the gauge glass.

After this procedure, the gauge glass was full. Because the instrument mechanic had been confident in the level controller reading, he suggested that the level was in the range of Area A (figure 12–8). He also remarked that the float chamber had been recently insulated.

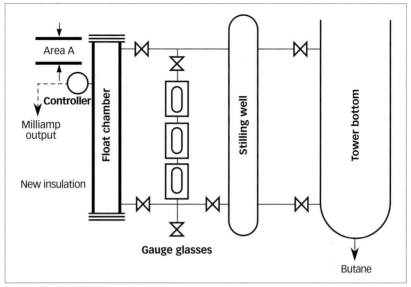

Fig. 12–8. Insulated float chamber gives incorrect level (courtesy *Oil & Gas Journal*).

The insulation caused the level controller to sense incorrectly the true level in the tower bottoms. The insulated chamber actually contained liquid about 60°F hotter than it did before it was insulated. The higher liquid temperature reduced the specific gravity of liquid in the chamber from about 0.51 sp gr to 0.45 sp gr, or about 15%.

Level is determined by these types of controllers by multiplying the milliamp output from the controller (proportional to the height of liquid in the chamber) by the specific gravity of the liquid in the tower. Before the instrument was insulated, the specific gravity in the chamber equaled the specific gravity of the tower liquid, and the instrument was calibrated accordingly. However, the lower specific gravity in the chamber after it was insulated meant that the instrument would indicate 85% level when the chamber was actually full. Therefore, the tower bottoms level could reach 100% or higher without a reaction from the level controller.

To rectify the problem, the bottoms level was brought down by increasing the bottoms product flow rate. The level controller was recalibrated in its insulated condition and fractionation efficiency returned to normal.

Reboiler Problems

Troubleshooting malfunctioning reboilers can be the toughest part of determining capacity limitations on a distillation tower.

Assume a typical reboiler configuration as shown in figure 12–9. The process fluid is on the shell side of a horizontal thermosiphon exchanger with condensing steam on the tube side providing the source of heat. The common problem with thermosiphon reboilers is reduced circulation. This condition is diagnosed by determining the temperature rise across the reboiler. Insert a glass thermometer under the insulation on the inlet and outlet process piping of the reboiler.

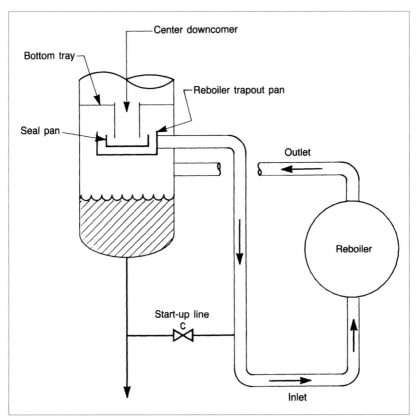

Fig. 12–9. Typical reboiler configuration.

The temperature difference obtained should be compared to the design ΔT (temperature difference). If the measured ΔT is much greater (more than 50%) than the design ΔT, circulation through the reboiler is insufficient. The two usual causes of decreased reboiler circulation are plugged shells and leaking trapout pans.

The process engineer can perform a series of experiments to determine whether the reboiler bundle needs to be cleaned to remove plugging or whether the distillation tower must be opened to repair a malfunctioning trapout pan. Following are details in diagnosing and correcting insufficient reboiler circulation.

Trapout pans

As a first step in testing for leaking trapout pans, consult figure 12–9. If the tower has a reboiler start-up line, open it. Increase the tower bottom level to maximum. If the reboiler outlet temperature drops rapidly, this is a positive indication of a malfunctioning reboiler trapout pan.

If the tower is not equipped with a reboiler start-up line, try increasing the flow of liquid to the trapout pan. This is done by increasing the tower feed or the reflux rate. If the trapout pan is leaking, the increased liquid flow to the pan will cause the ΔT between the inlet and outlet to decrease.

Plugged reboilers

If neither of the above methods helps increase reboiler circulation, the shell-side flow areas are likely restricted. This is cross-checked by slowly closing the reboiler inlet block valve and observing the change in shell-side outlet temperature. Closing the block valve 60%–70% without causing an increase in shell-side outlet temperature is proof that there is substantial shell-side plugging.

Repair of trapout pans

The exact cause of malfunctioning trapout pans can easily be determined once the tower is out of service. Hose water into the center downcomer, disconnect the tower from the reboiler inlet, and observe the flow of water.

Steam-side problems

Steam reboilers are subject to a wide variety of serious capacity problems. The most common is steam condensate backup into the channel head. This is checked by manually draining the steam condensate to a nearby sewer and observing if the reboiler duty increases or the steam inlet control valve closes.

In one instance the effect of steam condensate backup was clearly observed during a unit turnaround. As shown in figure 12–10, the rust layer on the inside of the channel head of a reboiler showed clearly where the condensate level normally ran. This was a positive indication that 20% of the heat-transfer surface area of the reboiler was waterlogged and, therefore, useless. Steam condensate backup can be due to one of the following:

- Malfunctioning steam trap.
- Excessive pressure in the steam condensate collection header.
- Installation of a control valve directly on the condensate drain line. This is usually only a problem when the reboiler steam supply is 30 psig or less.

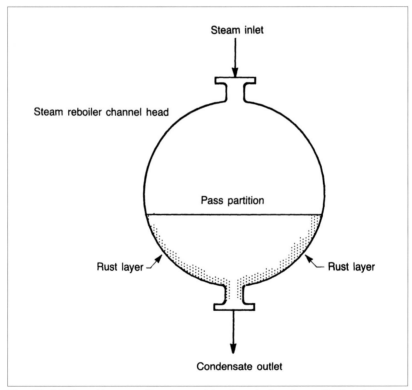

Fig. 12–10. A rust layer on the steam side of a reboiler shows the level at which steam condensate usually runs.

I have often maintained that I could retire and live in luxury on the proceeds from correcting this latter common design error. The problem is easily eliminated by the addition of an intervening condensate level control drum between the reboiler and the control valve, provided that downstream piping is sized to accommodate the flashed stream condensate.

Blown condensate seal

The opposite problem to steam condensate backup is blowout of uncondensed steam through the reboiler and out the condensate drain line. This phenomenon causes a loss in heat transfer entirely out of proportion to what might be expected. Literally half of a reboiler's duty can be lost by an apparently small amount of steam blowing out the condensate drain line.

To determine if this problem exists on a reboiler, establish a definite condensate seal by restricting the condensate effluent line. If the tower bottom's temperature increases, the reboiler has a blown condensate seal. This wastes steam and reboiler capacity. The correction is the same for condensate backup: installation of a steam condensate seal drum yun on level control.

Accumulations of noncondensibles in the channel head will also retard the condensation rate of steam and lead to a loss in reboiler duty. Typically, CO_2 contained in the steam supply collects below the bottom channel head-pass partition baffle. If left to accumulate, the CO_2 will dissolve in the steam condensate and form corrosive carbonic acid.

The channel head cover should have a ¾-inch plug located below the bottom pass partition baffle which is intended for venting noncondensibles. *Note:* Venting from the top of the channel head is useless as only steam from the header is vented.

Reflux Problems

A tower overhead system consists of a condenser, reflux drum, and reflux pump. The condenser, in particular, is subject to a variety of problems.

Leaking vapor bypass control valve

A typical installation of a hot vapor bypass controller is shown in figure 12–11. At one refinery a large debutanizer had been limited for many years by overhead condensing capacity during the summer. A new condenser was purchased to relieve the problem. Prior to its installation, an engineer observed that the hot vapor bypass line was warm to the touch, even though the bypass control valve showed it closed. The control valve was manually blocked in; condenser capacity instantaneously increased by 50%. The new condenser was not installed. It still lies in the spare equipment yard—a monument to inadequate field observation.

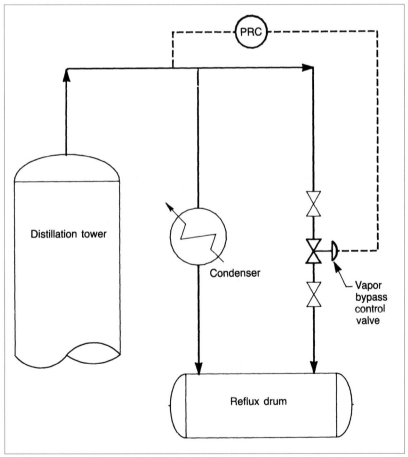

Fig. 12–11. A leaking vapor bypass control valve effectively reduces condenser capacity.

The purpose of the hot vapor bypass controller is to pump heat into the reflux drum. Obviously, if one is limited by condensing capacity, introduction of extra heat to the reflux drum aggravates the limitation. Usually, the rubber-type seat in the hot vapor bypass butterfly control valve dries out with age and needs to be renewed. My experience is to eliminate the hot vapor bypass control scheme entirely and convert the tower to flooded condenser-type pressure control.

Back-flushing condensers

Absolutely the first things to look for on water-cooled condensers are back-flush connections. Such condensers must be back-flushed periodically to obtain peak performance. This is especially true if the water outlet temperature is higher than design. Plastic bags are typical of the type of material that blocks the tube inlets. Mud, grass, crabs, and paper cups are also common.

The best type of back-flush connection is a nozzle welded onto the inlet portion of the channel head cover. The nozzle should be the same size as the cooling water inlet line. For best results, proceed as follows:

1. Block in the cooling water inlet.
2. Blow high-pressure air into the cooling water outlet line.
3. Fully open the back-flush line.
4. Continue to back-flush for 20 seconds or until the water clears up.

Incidentally, back-flush connections are easily hot-tapped while onstream. Small back-flush connections are useless (see chapter 16).

Water velocity

The rate of fouling deposits inside the cooling water tubes can increase exponentially as water velocity decreases. Water velocity should be at least 3 ft/sec. To determine water velocity, first approximate the condensing duty from the observed reflux and distillate rates. Then measure the water side inlet and outlet temperatures and calculate the volume of cooling water. If the calculated velocity is less than 3 ft/sec, decrease the number of tube side passes by modifying the pass partitions. As the water side ΔP is constant, this will increase water flow.

Air-cooled condensers

The usual problem with air-cooled condensers is insufficient air flow. Calculate the overhead condenser duty from process side data. Next, measure the air inlet and outlet temperatures and calculate the volume of air being moved.

Compare this number against the design value and then measure the amperage being drawn by the fan motors and compare this number against the design data sheet. Check the air-pressure drop across the tubes with a water manometer and compare with the design value.

This data will provide a guide on whether the fan pitch should be increased to move more air. Watch the motor amperage rise as the pitch is increased. If the increased fan pitch does not provide sufficient cooling, you may need to replace the fan (if pitch limits) or the motor (if amperage limits) or both.

Air recirculation

Locating air coolers too close together or improper plenum design can cause the air discharging from one fan to be pulled into the fan of another condenser. This recirculation can be verified by comparing the average air inlet temperature to the ambient temperature.

Condensed liquid drainage

Significant heat-transfer rates in condensing service are only obtained in the portion of the condenser tubes not submerged in liquid. For a shell-and-tube condenser with cooling water on the tube side, tubes submerged in condensate are being used to subcool the condensate. These tubes contribute nothing to condensing the tower overhead vapors. Figure 12–12 shows how the condensate level might look inside such a heat exchanger. About 30% of effective heat-transfer surface is wasted, as shown in this illustration.

One way to find the condensate level in a condenser is to wait for a warm, humid day. If the cooling water is cold enough, atmospheric moisture will condense on the outside of the shell. The wet shell area will roughly correspond to the level of condensate inside the shell. A simpler method is to feel along the outside of the shell. You will be able to locate a 2–4-inch wide band on the shell's surface with a noticeable temperature gradient. Mark this line, which is the level of subcooled liquid, and calculate (from the tube layout drawing) the percent of tubes submerged in condensate.

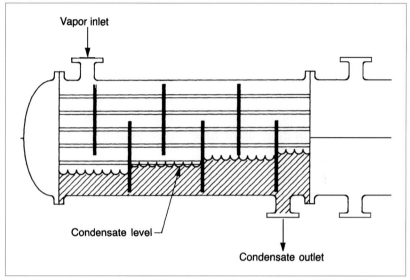

Fig. 12–12. Condensate level in an overhead condenser. About 30% of the effective heat-transfer surface is wasted.

Reasons for condensate subcooling

Condensate subcooling occurs when the distillation tower overhead product is all liquid. This means the contents of the reflux drum must be at or below its bubble point. The condenser operation will automatically adjust itself to satisfy this condition.

For example, assume a depropanizer reflux drum is located 12 feet above a condenser. Since the specific gravity of propane is 0.52, the reflux drum pressure is 3 psi lower than the condenser outlet pressure. Because the liquid entering the reflux drum is at its bubble point, the liquid leaving the condenser is, of necessity, subcooled by 3 psi. A 3-psi pressure differential for propane is about 1.5°F on a vapor pressure chart.

Now, assume a lunch bucket becomes stuck in the depropanizer reflux drum inlet line. This introduces an additional pressure drop between the condenser outlet and the reflux drum inlet of 6 psi. Now 4.5°F of subcooling is required.

On the same tower, a hot-vapor bypass around the condenser starts leaking. This puts a small amount of vapor into the reflux drum inlet. On mixing with the condenser outlet, the vapor condenses and increases the reflux drum temperature 3°F. For the contents of the reflux drum to remain at its bubble point, the condenser outlet temperature must drop an additional 3°F. Now, a total of 7.5°F subcooling is required. This is a lot of subcooling.

Reducing liquid subcooling

Devoting 10% of a condenser's surface area to subcooling is about right when the reflux drum is elevated above the condenser. If one finds 30% or more of the tubes in subcooling duty, something is wrong.

Check the pressure drop from the condenser outlet to the reflux drum inlet. Subtract the calculated static pressure loss (i.e., the height of liquid) from the measurement. If the result is more than one-half psi, consider enlarging the condenser outlet line.

If the contents of the reflux drum are warmer than the condenser outlet, then vapor is bypassing the condenser.

Vapor-bound condensers

In one debutanizer, the ability to condense the overhead product was lost. Investigation showed that a nitrogen-purged instrument had been recently installed. The nitrogen had accumulated in the top of the condenser shell, and gas blanketed the water-cooled tubes. A vent on the top of the condenser shell was opened, and condenser capacity was restored. Note that a small amount of really noncondensible gas can effectively vapor bind a shell-and-tube condenser.

Distillation Tray Types

The industry standard for a modern tray design is the grid tray, sometimes called a fixed valve assembly. The tray should be constructed from 10 or 12 gauge, 410 S.S. or 316 S.S.

- Do not use carbon steel trays or trays made from metal thinner than 14- or 16-gauge material.

- Do not use movable valve caps. They serve no purpose and can stick to the tray floor. They do not increase tray turndown rates as advertised.
- Sieve trays are almost as good as grid trays. They have somewhat reduced turndown capacity, and perhaps 5% to 10% less vapor-handling capacity then a grid tray.
- Bubble cap trays have excellent turndown capacity and should be used where vapor rates are very variable or where small liquid leakage rates through the tray deck cannot be tolerated. Bubble cap trays do have about 15% less capacity than grid trays and for that reason are not in widespread use.
- Trays with complex, proprietary features are best avoided as the installations are difficult and maintenance is problematical. Their incremental capacity over a conventional, well-designed grid tray is 5% to 10%, at best.

Troubleshooting Checklist for Distillation Towers

Tray Capacity
Downcomer backup
Jet flood
Plugged trays
Increased downcomer width
Decreased liquid height on tray
Venturi-type valve trays
Check incipient flood factor

Reboilers
Reduced thermosiphon circulation
Leaking trapout pan
Plugged reboiler
Steam condensate drainage
Blown condensate seal

Reflux Problems
Leaking vapor bypass control valve
Back-flush condensers
Increased fan pitch on air coolers
Air recirculation
Condenser liquid drainage
Subcooling reflux
Vapor-bound condensers

References

1. Lieberman, N.P. *Troubleshooting Process Plant Control.* Wiley, 2008.

2. Lieberman, N.P. and E.T. *A Working Guide to Process Equipment, 3rd Ed.* Mcraw-Hill, 2008.

Vacuum Towers

Anyone who has ever seen crude oil distilled in the lab under atmospheric pressure will appreciate the importance of a vacuum tower. At about 680°F–700°F, the residual liquid will start producing a yellowish vapor. This is an indication of the thermal cracking that degrades the quality of virgin distillates and gas oils.

To distill most of the gas oil out of the crude while still avoiding excessive temperatures, a vacuum tower is used. The crude unit's primary tower is intended to fractionate between naphtha, kerosene, and furnace oil. The vacuum tower only has one function: to produce a clean, high-boiling gas oil suitable for cracking-plant or lube-oils refining feed.

A vacuum tower's flash zone typically operates at 1–2 psia and 720°F to 780°F. The tower is designed to tolerate a small degree of thermal cracking. A sketch of a typical vacuum tower is shown in figure 13–1. Some of the more common troubles associated with operating a vacuum tower are:

- High flash-zone pressure
- Black gas oil
- Excessive production of trim gas oil
- High-gravity resid
- Ejector deficiencies
- Bottoms-pump NPSH problems
- Low gas-oil draw temperatures
- Transfer-line failures

Loss of Bottoms-Pump Suction Pressure

Providing net positive suction head (NPSH) for any centrifugal pump can be a tricky business. For vacuum tower bottoms pumps, the difficulties are magnified. A few pounds of vapor that another pump might pass will completely gas up a centrifugal pump in vacuum service. This is because the low absolute suction pressure expands a small amount of vapor into a very large volume. Moreover, the possibilities of introducing vapor into the suction of a vacuum tower bottoms pump are more numerous than in other services.

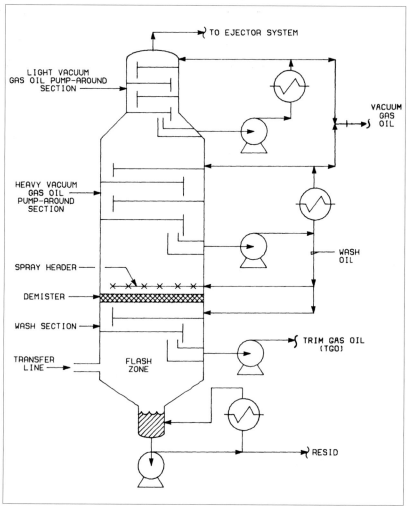

Fig. 13–1. A typical crude unit vacuum tower.

Especially on start-up, initiating and maintaining good suction conditions for the bottoms pump is one of the most difficult aspects of vacuum tower operation. A few of the more noteworthy problems are discussed below and summarized in figure 13–2.

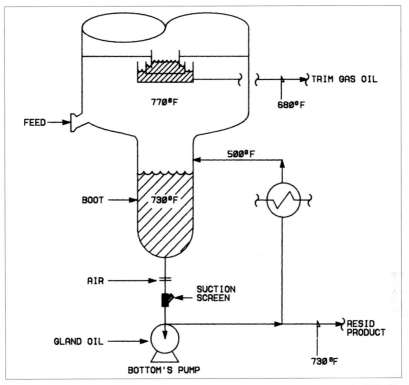

Fig. 13–2. Providing suction head for a vacuum tower bottoms pump is difficult.

Insufficient quench

Most vacuum towers are provided with a means to reduce the bottom boot temperature 20°F–50°F. This is done with a circulating quench as shown in Figure 13–2. The purpose of the quench is to reduce thermal cracking of the bottoms product and to suppress vaporization at the suction of the bottoms pump. If the bottoms pump is truly losing suction because of insufficient NPSH, increasing the circulating quench rate or reducing the quench return temperature will help.

TGO pan overflows

Trim gas oil (TGO) is a black oil stream withdrawn immediately above the vacuum-tower flash zone. It consists of 20%–50% resid and 50%–80% gas oil. This relatively light material may cause the bottoms pump to cavitate when it overflows its trapout pan at a nonuniform rate. Reduce the TGO pan level to see if this helps the NPSH problem on the bottoms pump.

Gland oil

The purpose of gland oil is to keep black oil away from the pump seals. A gland oil pressure of 10 psig is usually sufficient to keep black oil out of the seals. Naturally, a small amount of gland oil will leak through the seals into the vacuum bottoms stream. This is of no consequence.

If gland oil leakage becomes excessive, either because of a defective seal or excessively high gland oil pressure, the vacuum bottoms pump will lose suction. The gland oil, which consists of a relatively light hydrocarbon, flashes on contact with the hot resid product. The evolved vapor gases up the pump.

Try pinching back on the gland oil pressure to reduce cavitation. If this helps, but only at a very low gland oil pressure, the pump's seal is bad. One refiner substituted a heavy vacuum gas oil for a lighter gland oil to eliminate this problem.

Suction screen

Thermal cracking will eventually produce coke in the vacuum tower. The coke will wash down into the pump suction and plug the suction screen. A simple pressure survey will identify this problem.

Measure the pressure at the pump's suction and at a point in the boot above the liquid level. The difference in pressure, expressed in feet of liquid (assume a 0.75–0.80 sp gr) should equal the height of liquid in the boot above the pump suction. If the pressure difference is quite a bit less than this delta height, the pump's suction screen is plugged and should be cleaned.

Air leak

A rather small leak in a flange on the pump's suction piping will cause loss of NPSH. Any air sucked in will reduce the average density of the resid in the boot and suction line. The reduction in density cuts the head of liquid and usually precipitates cavitation.

On one unit, the level float in the boot would jump an inch or two when the bottoms pump started to lose NPSH. An air leak in a suction line flange was later found.

Air can also be drawn into the suction of the pump through a bad seal. Try increasing the seal oil pressure to see if this suppresses cavitation. It is also possible to pour seal oil from a can over the seal to test for a leaking seal. However, take precautions because this could cause a seal oil fire.

High Flash-Zone Pressure

The reduction of resid in a vacuum tower is a function of the flash-zone temperature and pressure. A rise in this pressure increases production of resid at the expense of the more valuable gas-oil product.

The key tool in troubleshooting flash-zone pressure problems is a vacuum-tower pressure survey. The time to initiate this survey is just after start-up when the trays, demister, and ejector system are clean and in good condition. Pressures are best measured with a portable mercury-filled vacuum manometer. Using a vacuum pressure gauge will reduce the accuracy of observed pressure drops. Relying on permanently installed gauges for pressure drop data will not give reliable results.

Figure 13–3 summarizes two vacuum-tower pressure surveys: one just after unit start-up and the other a year into the run. The data clearly show that the demister is partially plugged with coke.

Pressure drop data should be normalized by correcting for flow rates and pressure as follows:

$$\Delta P_1 = \Delta P_N \left(\frac{M_1 \times V_1}{M_B \times V_B} \right)$$

where:

M_1 = Mass flow through the tower

V_1 = Superficial velocity through the tower

M_B = Mass flow of the base data to which ΔP, is to be compared

V_B = Superficial velocity through tower from base data

ΔP_1 = Measured pressure drop

ΔP_N = Normalized pressure drop

Any restriction to vapor flow above the flash zone must increase the flash-zone pressure. An increase in ΔP across the wash trays below the demister or across the demister itself is almost certainly due to coke buildup. A generous wash oil flow (see figure 13–1) will inhibit coke formation—but, of course, this increases the production of undesirable trim gas oil. Once the coke is formed, only a shutdown will correct the situation.

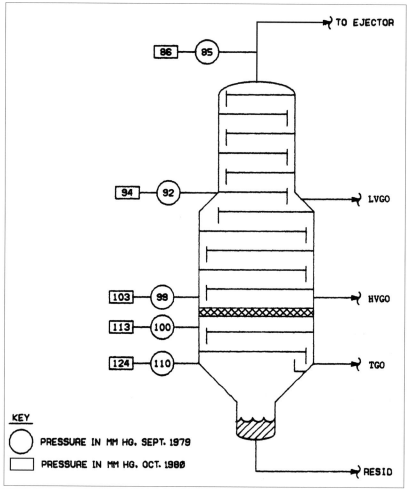

Fig. 13–3. Pressure survey is key to troubleshooting high flash-zone pressure.

Large increases in ΔP across the top or bottom pumparound trays are an indication of flooding. A reduction in liquid pumparound rate might alleviate the problem.

A high vacuum-tower top pressure is a result of air leaks, excessive production of hydrocarbon gases due to thermal cracking, or a host of ejector deficiencies.

Thermal cracking

Increasing the flash-zone temperature will reduce the gas oil left in the resid. Unfortunately, thermal cracking rates double for every increase of 20°F–25°F. As shown in figure 13–4, the production of noncondensible gas (a product of thermal cracking) rises with higher transfer-line temperature. This can overload the ejector system. Also, the rate of coke buildup on the demister pad is a function of flash-zone temperature. For these reasons simply raising the furnace outlet temperature to overcome the effect of high flash-zone pressure is not always a good idea.

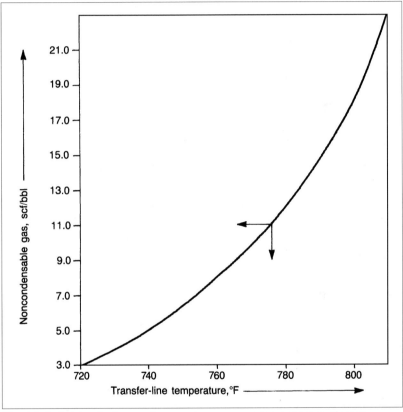

Fig. 13–4. Raising flash-zone temperature can overload vacuum ejector.

On some units, the vacuum is observed to deteriorate as the tower bottoms level increases. This is due to thermal cracking in the bottoms. If the tower is equipped with a circulating bottoms quench, reduce the bottoms temperature by 20°F. This will reduce the volume of cracked gas to the jets.

Ejector Problems

Reduced vacuum at the top of a tower may be due to a number of problems:

- Noncondensible hydrocarbons
- Air leaks
- Worn ejector internals
- Degradation of ejector steam pressure or temperature
- Fouled or overloaded condensers
- Plugged seal legs

Air leaks

A rather sudden loss in vacuum or failure to reestablish normal tower top pressure after a unit turnaround is most likely a consequence of an air leak. Air or oxygen compounds (CO, CO_2) in the ejector tail gas are the usual indication. CO may be due also to the cracking of naphthenic acids.

Figure 13–5 shows the components of a typical steam jet. The pressure head of the steam is entirely converted to velocity in the steam nozzle. Tail gas from the vacuum tower flows into the vacuum thus created in the mixing chamber. The velocity of the steam is partially converted back to pressure in the diffuser.

If CO or CO_2 is present, the air leak is probably in the hot part of the vacuum tower or the transfer line. Look for a leak in the ejector or vacuum-tower overhead piping if oxygen is present in the tail gas. On several units, leaks in the seal leg piping have proven troublesome. A sketch of a single-stage ejector system (figure 13–6) shows the location of the seal leg piping. Use aerosol shave cream to test for vacuum leaks. When all else fails, pressure the tower to 3–5 psig. The leak can then be located by the noise it will emit.

Motive steam quality

The ejector steam undergoes an isoentropic expansion, which converts much of its enthalpy and pressure into velocity. It is this high velocity (up to 600 mph) that pulls a vacuum. A reduction in steam temperature and pressure or an increase in the steam's moisture content reduces the motive steam's ability to pull a vacuum.

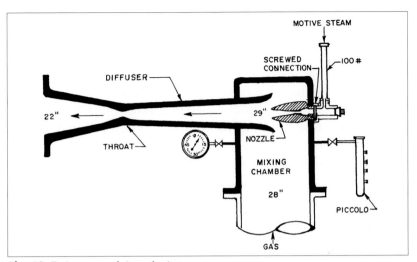

Fig. 13–5. A vacuum jet or ejector.

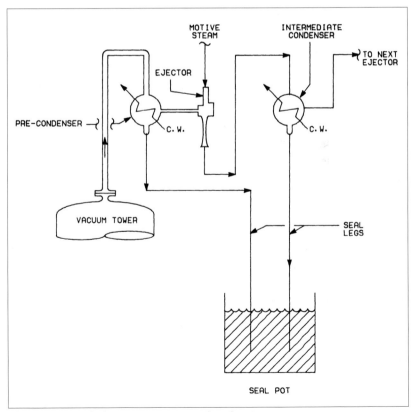

Fig. 13–6. A typical vacuum tower overhead system.

A steam jet emitting a "surging" sound is overloaded, most often due to moisture in the steam supply, or is suffering from high exhaust pressure.

Motive-steam pressure that is too high will also hurt the vacuum. This is because the steam nozzle will pass more steam than the diffuser throat was designed to handle. The diffuser chokes on the extra steam and the vacuum is adversely affected. Or, the downstream condenser may be overloaded by the increased steam flow and cause back pressure against the vacuum ejector discharge.

As the steam nozzle erodes with age, the optimum vacuum that can be obtained will call for a reduced motive-steam pressure. The steam flow (F) to the jet varies as follows:

$$F \sim (D^2) \cdot (P_1)$$

where:

D = Nozzle ID

P_1 = Motive steam, absolute pressure

Because the diffuser is designed to handle a fixed amount of steam, as the nozzle erodes, P_1 must be reduced.

Condensers

The precondenser shown in figure 13–6 removes the bulk of the mass of steam and hydrocarbons evolved from the vacuum tower. The purpose of the downstream intermediate condensers is to condense the motive steam used in the ejectors. Fouling of the condenser tubes and high cooling water temperature will increase the pounds per hour that the ejectors must handle.

Regardless of ejector capacity, if the vacuum tower uses stripping steam, the pressure at the top of the tower cannot be lower than the vapor pressure of water at the precondenser vapor outlet temperature.

Plugged seal legs

As shown in figure 13–6, the seal legs function to drain oil and water from the shell side of the condenser. The inability to maintain condenser drainage will severely reduce ejector efficiency by backing liquid up in the condenser shell. With experience, you can find the liquid level inside the condenser shell by feeling for the temperature gradient by hand. Condenser tubes submerged in liquid are effectively out of service.

A sudden rise in vacuum tower top pressure, which occurs simultaneously with the onset of cold weather, can be due to freeze-ups of the seal legs. Plugging with waxy deposits is also possible.

One positive method to determine if the seal legs are backed up is to check the seal leg temperature versus the vapor temperature leaving the condenser. If the seal legs are running even a few degrees cooler than the vapor, liquid is backed up in the condenser.

Another test for seal leg backup into the precondenser requires raising the pressure in the seal drum by 0.5 psi. If liquid had been backed up in the condenser, the vacuum tower pressure will go up by 1 in. Hg. If the condenser was running properly drained, the vacuum tower pressure will remain unaffected.

Ejector internals

The very high velocities that the ejector is exposed to subject the nozzle and diffuser throat to excessive wear. Low-quality steam will accelerate this erosion. A gradual loss in vacuum may be due to enlargement of the ejector clearances. It is a good practice to caliper these clearances when the system is out of service. A fuller description of ejector operation, steam requirements, and pressure capabilities can be found in Robert Frumerman's article "Steam Jet Ejectors," appearing in the June 1956 issue of *Chemical Engineering*.

To determine if a jet has suffered mechanical deterioration, proceed as follows:

1. Take the jet off-line and vent it to the atmosphere. Insert a blanking plate in the mixing chamber inlet.
2. Install a vacuum pressure gauge as shown earlier in figure 13–5. Using the design steam pressure, measure the vacuum developed.
3. Compare the observed vacuum with the predicted vacuum at zero air flow as shown on the manufacturer's jet performance curve.
4. If the vacuum is below the predicted curve value, the jet is mechanically deficient. However, if this is not the case, continue to the next step.
5. Install a "piccolo" as shown in figure 13–5. The piccolo is nothing more than a pipe with calibrated holes. It is also obtained from the jet manufacturer. Remove one plug at a

time. From a chart that comes with the piccolo, read the air flow to the jet based on the pressure in the mixing chamber and the hole size. Using the piccolo, you can generate a jet performance curve and compare it to the original performance curve.

Almost always, it is erosion of the jet's nozzle that accounts for loss of vacuum due to deterioration of the jet. To prove this, simply throttle the motive-steam inlet (with the jet vented to atmosphere) and the vacuum will improve. Sometimes, steam leaks around the steam nozzle into the mixing chamber. Sealing these leaks will also enhance vacuum.

Black Gas Oil

The heavy vacuum gas oil (HVGO) produced in most refineries is the principal component charged to an FCCU. The catalyst at the FCCU is adversely affected by the nickel, vanadium, and sodium that become concentrated in the vacuum-tower bottoms resid. When resid is entrained into the HVGO, the metals in FCCU charge will dramatically increase.

The operating engineer should check the color of the HVGO periodically. A darkish yellow color is normal; black gas oil indicates resid (and metals) are being entrained into the gas oil. Some of the more common causes of resid entrainment are:

- **A section of the demister pad or wash oil grid has become dislodged.** A pressure drop survey showing a substantial decrease in ΔP across the demister is evidence of such a failure.

- **The demister pad or wash oil grid is partially coked.** Quite often, the demister spray header (see figure 13–1) is designed for too large a wash oil flow. At low flow, the wash oil does not distribute to the ends of the spray header, and the peripheral area of the demister dries up and cokes. The sections of the demister that are still open are exposed to velocities high enough to promote resid entrainment. A high ΔP across the demister shows that coke plugging is the problem. An examination of the demister before it is removed from the tower will indicate which areas are not being wetted with wash oil from the spray header. This information should then be used to revamp the spray header.

- **A decrease in flash-zone pressure** increases velocity and promotes entrainment. Throttle the steam to the ejector system (to temporarily raise flash zone pressure) and see if the HVGO color clears up.
- **Inadequate flow of wash oil** will allow entrained resid to reach the HVGO drawoff tray. Naturally, increasing the wash oil flow puts HVGO into black trim gas oil.
- **A high tower bottoms level** will cause liquid to flood the bottom stripping section. This backs up resid into the flash zone. When the frothy liquid level rises to the flash zone inlet nozzle, entrainment of bottoms will carry up into the wash oil section, which will also flood.

Excessive Production of Trim Gas Oil

Trim gas oil (TGO) or slop gas oil is an intermediate cut made in a vacuum tower between HVGO and resid. The TGO consists of 20%–50% entrained resid and 50%–80% wash oil (HVGO).

In some refineries, TGO is dumped directly into resid; in others, the TGO is recycled to extinction through the vacuum furnace. Either way, TGO is an undesirable byproduct of vacuum-tower operations.

Excessive production of TGO is often a consequence of efforts to clean up dirty HVGO by overrefluxing with wash oil. If TGO production has increased after a unit turnaround, the operating engineer may be sure that the HVGO trapout pan is leaking. The evidence for this is a lower than normal TGO drawoff temperature and increased TGO gravity (°API). A low HVGO draw temperature is further proof that the HVGO is leaking into the TGO section. Leaks from the HVGO pan may be initiated by cleaning the tower trays during a turnaround or by bumping the vacuum tower during start-up—typically with a shot of water. Sometimes the leaks will coke up. Once, workmen left pieces of a trapout pan out of the tower after a turnaround. The resulting leak did not coke up (see chapter 11).

Low Pumparound Draw Temperatures

Why do most vacuum towers have two pumparound (PAR) streams and produce both a light vacuum gas oil (LVGO) and HVGO streams? After all, the LVGO and HVGO are usually combined after they are drawn off the vacuum tower (refer to figure 13–1). Try letting the LVGO drawoff pan overflow for 15 minutes, and the answer will become apparent. Both the HVGO and LVGO drawoff temperatures will fall as both products become lighter.

A leak in the LVGO pan will have the same effect as letting the pan overflow; draw temperatures and then crude preheat will decline. Tight trapout pans in vacuum towers almost always save energy.

Light Resid

Gas oil left in resid will often have a value equivalent to fuel oil. Gas oil recovered for FCCU feed will have a value equivalent to crude. It follows, then, that every barrel possible ought to be boiled out of the vacuum tower bottoms. The best troubleshooting tool for judging overall vacuum-tower performance is to vacuum distill off in the lab the 1,000°F gas oil in the resid. Based on plant data, the following rules of thumb may then be applied:

- 5%–10% 1,000°F gas oil on resid indicates excellent performance.
- 30%–35% 1,000°F gas oil on resid indicates poor performance.

Some of the more common reasons for increased losses of gas oil to resid are:

- **Leaking TGO drawoff pan.** Declining TGO production highlights this failure.
- **Low flash-zone temperature.** Have the instrument mechanic check the furnace outlet thermocouple.
- **Low tower top temperature.** The optimum tower top temperature for a vacuum tower equipped with a precondenser is usually not the minimum temperature. As the tower top temperature is raised, heavy naphtha boiling-range materials are flashed overhead into the precondenser. Acting as an absorption oil, they absorb a portion of the light hydrocarbons that would otherwise overload the jets. However, getting the vacuum tower top too hot can overload the precondensers. By field trials, find the tower top temperature (usually 230°F to 280°F) that minimizes flash-zone pressure.[1]

- **High flash-zone pressure.** Perform a ΔP survey.
- **Inadequate velocity steam.** Steam is injected into the vacuum furnace coils to promote vaporization by reducing the hydrocarbon partial pressure. Too much steam will increase residual entrainment into the gas oil and will ruin the FCCU feed.
- **Stripping deficiencies.** Stripping the resid with exhaust steam is a cost-effective method to recover gas oil. A properly operated stripper may easily remove half of the gas oil that is left in the resid after it drops out of the flash zone. Using 0.2 pound of steam per gallon of bottoms is a typical steam rate.

Effective steam stripping is indicated by a 30°F–50°F temperature drop between the flash zone and the tower bottoms. A lower ΔT means that the steam is not effectively contacting the resid. This can be due to upset or corroded trays or tray flooding. Often, poor level control in the bottom of the tower will permit resid to back up and flood the stripping trays.

Steam to Healer Passes

I was all set to land a nice contract to revamp a vacuum column in a West Coast refinery. The revamp of the tower was intended to recover an incremental 3,000 B/SD of gas oil from the vacuum tower bottoms.

The process conditions shown in figure 13–7 were the design-basis operating parameters. Note that the 715°F flash-zone temperature and the 25-in. Hg flash-zone pressure (128 mm Hg) are indicative of an operation that results in excessive gas oil left in the vacuum tower bottoms. This downgrades virgin gas oil from FCCU feedstock to delayed coker feed at a penalty of $5/bbl. A properly designed and operated vacuum column that employs steam in the heater coils operates at 27-in. Hg flashzone pressure and 760°F flash-zone temperature.

When a tabulation of operating data was examined, there was no indication of the amount of steam going to the heater coils. The unit operating superintendent said that he was not told to use steam in the heater passes.

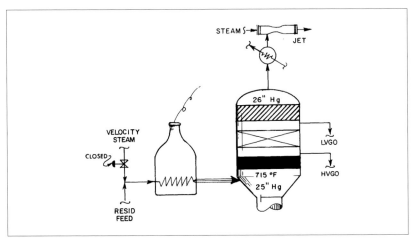

Fig. 13–7. Velocity improves flash-zone pressure.

The superintendent asked about the reason for steam in the heater coils. I told him the steam is used to suppress temperature peaking in the heater. In vacuum heaters (and to a lesser extent, delayed coke heaters), the maximum coil temperature does not occur at the heater outlet because most of the absorbed heat goes into heat of vaporization rather than into sensible heat. Most of the vaporization occurs in the last 5 or 10 tubes of the heater.

Figure 13–8 illustrates the effect. Even though the absorbed heat is similar in each tube, the temperature of the vacuum tower charge drops due to the rapid vaporization of the resid as it approaches the heater outlet.

When steam is added to the heater passes, earlier and more uniform vaporization of the resid is promoted. The steam therefore suppresses the peak temperature reached in the heater passes.

The production of cracked gas in the heater is largely a function of the peak temperature developed inside the heater coils. When the peak temperature is suppressed, the load of cracked gas to the vacuum tower overhead steam ejectors is reduced. The ejectors can, therefore, pull a deeper vacuum, lowering the tower flash-zone pressure and increasing gas oil recovery.

The amount of steam can be determined by adding enough steam or condensate so that the pressure drop through the heater increases by 50% to 60%. Over a number of days, several adjustments were made to the tower operating parameters.

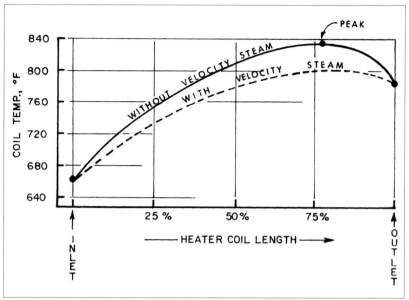

Fig. 13–8. Velocity steam suppresses peak temperatures.

As a first step, steam was supplied to the heater passes on the unit and the flash-zone pressure dropped from 25 in. Hg to 27 in. Hg. Gas oil recovery increased by 1,800 B/SD.

The heater outlet temperature was then raised by 10°F. The superintendent questioned this move because in the past, raising the heater temperature overloaded the ejectors with cracked gas. But because of steaming, the heater passes were running cooler even though the heater outlet temperature was higher.

The increase in heater outlet temperature raised the flash-zone temperature to 723°F from the previous 715°F. Pressure rose a bit to 26.9 in. Hg., but the gas oil production rose another 800 B/SD.

Then, steam pressure to the ejectors was adjusted to achieve the highest possible vacuum. Cooling water pump turbine speeds were increased to push the flow of cooling water to the overhead condensers. Also, condensers were backflushed (see chapter 16).

After these moves, pressure decreased and the heater temperature was increased another 5°F. Finally, the vacuum tower bottoms stripping steam rate, heater outlet temperature, and heater pass flow were optimized to obtain maximum gas oil production. These last items usually take some trial and error.

The final result on this tower was a 3,200 B/SD gain in FCCU feed at the expense of coker feed, about 200 B/SD more than the revamp was supposed to achieve. At $5/bbl, that amounted to a $16,000/day yield improvement on the tower.

It turned out that after the operating adjustments on the tower were made, there was no need for revamping the tower. Unfortunately, I never did get a contract for the revamp.

Projects to Improve Gas-Oil Recovery

In a crude-short world, projects that recover gas oil (which is readily converted to gasoline) from resid will almost always have a good payout. Residual fuel prices must, in the long run, compete against coal, while gasoline prices are a direct function of crude oil costs. A few projects that have seen successful field application are as follows:

- **Improve light gas-oil recovery in the crude unit's primary tower.** This will unload the vacuum tower and furnace.
- **Reduce the number of wash trays in the vacuum tower.** In one installation, two bubble-cap trays replaced four tunnel-cap trays. The flash-zone pressure was therefore reduced. The net effect was beneficial in that the lower pressure drop was more useful than the loss in fractionation represented by the two trays removed.
- **Replace pumparound trays with packing.** Structured packing is a good bet for this application. The payoff is reduced ΔP and flash-zone pressure. For example, a West coast vaccum tower was revamped from bubble cap trays to packing as follows:
  - LVGO PAR—7 feet of structured packing.
  - LVGO-HVGO Fractionation—4 feet of pall rings.
  - HVGO PAR—5 feet of structured packing.
  - Wash Oil—2 feet of structured packing on top of 3 feet of heavy-duty grid. The pressure drop across the tower was reduced from 50 mm Hg to 15 mm Hg.
- **Improve ejector performance.** Providing a source of colder cooling water for the interstage condensers is one

method of achieving this objective. Lower top pressure will translate into lower flash-zone pressure.

- **Increase the number or efficiency of resid stripping trays.** Increasing the superheat of the exhaust stripping steam will also multiply the amount of gas oil stripped out of the vacuum-tower bottoms resid product.

- **Seal-weld the HVGO and TGO drawoff trays to eliminate leakage into the resid.** Welding a dam around the periphery of the tray between the outermost rows of caps and the tray ring is an effective method to cut tray-ring leakage. The height of the dam must be greater than the height of the outlet weir. Alternately, a seal ring can be specified.

Transfer-Line Failures

It seems as if the vacuum-tower transfer line is a weak point in many crude units. It is possible, because of an incorrectly sized transfer line, to approach sonic velocity in these lines. Such superhigh velocities have led to rapid erosion and failure of the transfer line. If a unit's transfer line is experiencing an accelerated rate of failures, the operating engineer should consider several questions. Has the flash-zone pressure been substantially reduced? Has the furnace charge rate (including velocity steam) been increased? Is the vacuum-tower feed lighter than it used to be?

A positive response to these questions means a higher transfer-line velocity—and enhanced rates of erosion. Operating data indicate such erosion can become a problem at velocities exceeding several hundred ft/sec. An increase in the naphthenic acid content of crude will also accelerate transfer-line corrosion (see chapter 22).

Furnace-tube failures

The furnace outlet (or transfer-line) temperature is not usually the highest temperature that the vacuum-tower charge reaches. The high pressure drop in the transfer line (caused by high velocities) suppresses vaporization in the furnace's tubes. In turn, this raises the oil temperature in the tubes so that the required duty may be provided even at the reduced percent vaporized. This is called temperature peaking. The high peak temperatures cause furnace-tube failures.

The process operating engineer, faced with a series of transfer-line and furnace-tube failures in vacuum service, should consider enlarging the transfer-line diameter and the size of the last few tubes in the furnace.

The Tale of Weak Steam

Sometimes I get paid quite a lot for very little. Here's such an example.

Quite recently I had a project to improve the vacuum on a column at a refinery in Corpus Christi, Texas. I climbed to the top of the tower with two operators. I could hear the three parallel first stage primary jets surging before I could see them. The vacuum was really bad.

"The jets are surging." I observed the obvious.

"Yes, Mr. Lieberman they do that a lot," said Tommy, the younger operator.

"What do you mean 'a lot.' Does that mean that sometimes they don't surge?" I asked.

"Well Mr. Lieberman," answered J.J., the older operator. "Not when we have strong steam. Then the jets don't surge. We get two types of steam here. Weak steam and strong steam. It's the weak steam that causes the jets to surge and hurts vacuum."

"You mean the weak steam is at a lower pressure than the strong steam," I tried to clarify.

"No, No!" Tommy shouted above the jets. "It's all the same 400 psig steam. We get the weak 400-pound steam from our No. 5 boiler. We get the strong 400-pound steam from the Union Carbide plant across Navigation Boulevard. It's all the same steam. It's just that the Carbide steam is strong, and our steam is weak. All the steam we get is at the same temperature and pressure. It's like me and J.J. He's weak, and I'm strong, even though we're both married," Tommy laughed.

The strong steam was dry, saturated steam. The weak steam had a very high moisture content. I recommended to my client that they re-install the inline steam water separator that had been removed many years previously. And for this they paid me $5,000 (less the $20 for pizza I bought for Tommy and J.J.).

Troubleshooting Checklist for Vacuum Towers

Bottoms Pump NPSH
Insufficient quench
Overflow of TGO pan
Too much gland oil pressure
Leaking seal
Suction screen plugged
Air leak in suction piping

High Flash-Zone Pressure
Tower pressure survey
Normalized pressure survey
Coke buildup on demister
Flooding of PAR trays
Ejector deficient
Excess thermal cracking

Ejector Problems
Air leaks (CO, CO_2, or O_2)
Noncondensible hydrocarbons
Worn ejector internals
Low motive steam pressure
  of temperature
Wet motive steam
Fouled condenser

Transfer-Line Failures
Approach to sonic velocity
Erosion due to high velocity
Reduced flash-zone pressure
Lighter feed
More velocity steam
Increased charge rate
Furnace tube peaking temperature
Naphthenic acid in crude

Black Gas Oil
Indicates metals in FCCU feed
Demister section upset
Demister pad partially coked
Too low a flash-zone pressure
Wash oil flow too low

Excessive Production of Trim Gas Oil
Symptom of wash oil
  section problems
Leaking HVGO drawoff tray
Low HVGO temperature—HVGO
  drawoff tray leaking
Low HVGO and LVGO—LVGO
  drawoff tray leaking

Light Resid
Inadequate velocity steam
High flash-zone pressure
Low flash-zone temperature
Leaking TGO drawoff pan

Low Pumparound Draw Temperatures
Stripping steam too low
Stripping trays upset
Adjust steam to jets

Projects to Improve Gas Oil Recovery
Reduce light gas oil in vacuum
  tower feed
Reduce number of wash trays
Replace PAR trays with packing
Colder cooling water for ejector
  condensers
Seal-weld HVGO and TGO
  drawoff trays
Increase the number of
  bottom-stripping trays
More superheat of exhaust
  stripping steam
Add velocity steam

Reference

1. Lieberman, N.P. *Process Design for Reliable Operations, 3rd Ed.* Lieberman Publications.

Treating Liquid Hydrocarbons

Many operating problems are blamed on bad feed—a catalytic cracking unit loses conversion due to sodium in its charge; an alkylation unit has an acid runaway because of high mercaptans; a polymerization unit's catalyst consumption doubles due to feed contamination with ammonia. From the process operating engineer's point of view, there is a common thread to these problems. They are all emergencies requiring immediate correction.

Too often, feed cleanup facilities are located at the tail end of a process unit whose treated product is charged to a second unit. This means that the feed washing system is often neglected by the operators. After all, contaminants left in their products won't affect the operation of their unit.

Operators at the downstream receiving unit often blame all of their problems on poor feed. The upstream operators are labeled as careless; the downstream operators are complainers. The operating engineer is the one in the middle who has to solve the puzzle.

Ailments

Improper cleanup of hydrocarbon streams can be caused by a variety of ills: poor mixing, inadequate settling and circulation rates, incorrect solution concentrations, and faulty pH control. On one unit, for example, channeling of packing was suspected in the contacting vessel. Upon opening the vessel for inspection, it was found to be completely empty. The packing had disappeared.

How does one know if a salt drum or a calcium chloride dryer has anything left in it? Is caustic carryover from a settler normal, or is it a sign of excessive mix-valve pressure drop?

Components

Figure 14–1 shows a typical feed cleanup system. The system's components are a settling drum, circulation pump, mix valve, interface level control valve, and wash liquid. The wash liquid, usually caustic or water, is contacted with a hydrocarbon stream. The mix valve provides the turbulence required for good contacting. The settling drum separates the two phases. The circulating pump returns the wash liquid to the mix valve. Finally, the interface level control valve sets the relative volumes of the wash liquid and hydrocarbon in the settler.

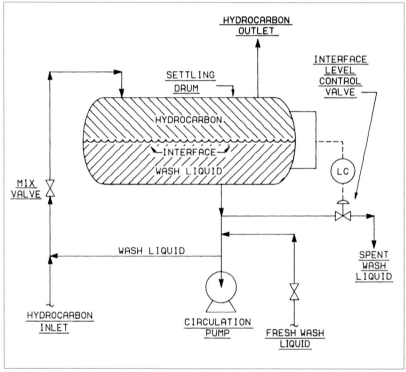

Fig. 14–1. Components of a simple feed-wash system.

A malfunction of any of these components will result in bad feed. The trouble may be due to mechanical problems (a worn circulating pump impeller) or to process errors (the mix valve set incorrectly.) By making a series of field observations, the operating engineer can diagnose the difficulty.

Carryover Problems

Caustic or water will often be found in the hydrocarbon leaving a settler. This problem is called carryover. The entrained liquid will show up downstream. At the first place that the flow rate of the settler effluent slows to a few feet per minute, caustic or water will drop out.

The causes of carryover can be related to high settler interface level, excessive mix-valve pressure drop, inadequate settler residence time, dirt, settler internal mechanical damage, or all of the above.

Missing interface level

As a general rule, the simplest solution is probably the right one. A large and unusual carryover is likely caused by improper control of an interface level. The way to check this is to find the interface level in the level gauge glass. Observing an interface in a gauge glass is always hard. This is why the shift operators have lost track of the level and let the drum carry over. The operating engineer should personally find the actual interface level.

Determining the interface level in a water/light hydrocarbon settler can be extremely difficult. If both phases are clean, they will also be transparent. First, blow the gauge glass assembly taps out after draining down the gauge glass, allow it to slowly refill (see chapter 12). The interface can then be seen by its upward motion. For many settlers the aqueous phase will occupy the lower third of the vessel.

If the interface level is unusually high, your problem is solved. The interface level controller is broken or set incorrectly. Perhaps the level control valve has jammed in a closed position. Maybe a capricious sprite has blocked it in.

Mix-valve pressure drop

After checking the interface level, consider the mix valve as a possible cause of the carryover problem. To get good contacting between the hydrocarbon and wash liquid, a pressure drop of 10 psi to 25 psi is about right. Use the same pressure gauge to check pressure up and downstream of the mix valve. Try reducing the pressure drop to 10 psi and see if the carryover problem is diminished. Remember that setting it too low can decrease hydrocarbon cleanup efficiency.

Experiment in settling time

Inadequate settler residence time can cause a persistent carryover. Required settler residence time may be calculated with Stoke's Law. However, for those of us who have hours, not years, to solve problems, the following procedure is recommended:

1. Obtain a glass cylinder about 12 inches long.

2. From a bleeder connection downstream of the mix valve, draw a sample into the cylinder. For propane or butane, the cylinder will have to be packed in ice. Dry ice will be needed for propane.

3. Observe the time for the two phases to separate. The height of the sample divided by this time is called the settling velocity. Typically, for well-mixed water-gasoline systems, this velocity is ½ ft/min.

4. Calculate the required settling time:

$$T_R = \frac{H}{V}$$

where:

T_R = Required settling time

H = Height between the interface level and the hydrocarbon outlet nozzle

V = Settling velocity as measured previously

5. Calculate the residence time of the hydrocarbon in the drum. Include only that part of the settler above the interface level.

Compare the required settling time (T_R) against the actual residence time in the vessel. If the actual residence time is a fraction of the required settling time, this is a good explanation for persistent carryover. *Caution:* Settling velocities are a function of system properties. Settling velocities from 2 ft/min to 1 in./hr have been measured.

Turn down circulation pump

For a typical refinery hydrocarbon cleanup application, only a small amount of fresh wash liquid is needed. To obtain good contact through the mix valve, the wash liquid is recirculated by a pump. Usually 1 vol of the wash liquid is circulated for each 10 vol of hydrocarbon. If the circulation rate is well beyond this guide, reduce it by throttling the pump discharge and see if carry-over is diminished.

Carryover can be caused by dirty wash liquid or particulates in the hydrocarbon feed. Caustic strengths in excess of 15%–20% can contribute to carryovers. Extraneous chemicals that reduce surface tension (surfactants) and thus promote foaming are obvious culprits.

Correcting Inadequate Treating

Having spent many humid Texas nights struggling with the operation of extracting mercaptans from butylenes with caustic, I will discuss this common refinery example of a feed cleanup system. Figure 14–2 illustrates the process.

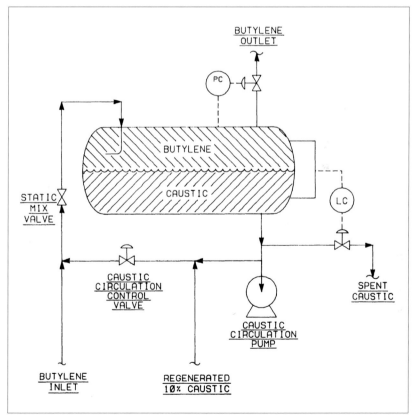

Fig. 14–2. Extracting mercaptans from butylene with caustic.

Mercaptans are acidic; they are more soluble in caustic than in butylene. However, ethyl mercaptan, which is usually associated with butylene produced from an FCCU, has a small equilibrium driving force for caustic extraction. Also, caustic and light hydrocarbons do not mix readily. Therefore, the first job in analyzing this type of treating problem is to decide if the process is mass-transfer limited or equilibrium limited.

Mass-transfer limits

The caustic circulating pump and mix valve work together to provide intimate contact between caustic and butylene. A large reduction in caustic circulation would increase the residual mercaptan content of the butylene. Circulating 1 vol of caustic to 10 vol of butylene was our target. The flow was checked with a rotometer.

One evening, the pump impeller came loose, and the rotometer was broken. We defined the problem by blocking the pump discharge and comparing the observed shut-in pressure against the design shut-in pressure. The low discharge pressure indicated internal pump mechanical damage (see chapter 11).

Mix valve pressure drop should be about 10 psi for good mixing. On our caustic wash facility, we had a static mixer. This meant no adjustment could be made. Usually, one should increase mix-valve pressure drop when inadequate treating is reported.

In another case we could not circulate caustic because of excessive pressure drop across the static mix valve. To conserve energy, we had started diluting our 50% caustic with river water. Formerly, we had used clean steam condensate. Calcium salts precipitating inside the mix valve plugged it. Acid washing with dilute HCI relieved the restriction. The environmental engineer who had convinced us to try river water was ejected from the unit.

On the rare cold days we had in Texas, high mix-valve pressure drop would reappear. Caustic would stick to the cold metal walls of the mix valve. Steam tracing the mix valve corrected this difficulty.

If substantial increases in caustic circulation and mix-valve pressure drop do not improve treating, the operating engineer must assume he is faced with an extraction equilibrium problem.

Equilibrium limitations

Most caustic used in mercaptan extraction is regenerated to remove the mercaptans. In any feed cleanup system, the quality of the fresh wash liquid should always be checked when a treating problem is encountered.

In our extraction facility, regenerated caustic was supplied to us by another unit. During one period when the mercaptan level of treated butylenes was excessive, the mercaptan content of our regenerated caustic was discovered to be higher than our spent caustic. Insufficient addition of 10% caustic was the major problem. The pressure control valve on the effluent butylene had to be pinched back on hot days to keep butylene from vaporizing in the settler. The higher pressure backed regenerated 10% caustic out of the caustic circulating stream. This was apparent when the interface level control valve was found closed. Any time this happened, four hours later we would get a lab report showing high mercaptans in treated butylenes.

A potential case of ulcers was avoided by relocating the 10% caustic addition point. Figure 14–2 shows this point just upstream of the caustic circulation control valve. The system was revamped to add 10% caustic into the suction of the circulating pump. The pump suction pressure was 50 psi lower than the pump discharge pressure.

When analyzing a problem, consider the refinery as a whole. For example, on our extraction unit, the mercaptan content of our treated butylene once jumped from 20 ppm to 400 ppm. At the same time, the cracking unit, which produced the butylene, experienced an upset on its depropanizer.

A sample of untreated butylene was checked for H_2S with lead acetate paper. (A roll of lead acetate paper is a good item to keep in your desk.) The lead acetate paper turned black when exposed to vapors from the butylene sample. This indicated that there was H_2S in the butylene feed. The H_2S reacted with the caustic and formed NaHS. This is an irreversible reaction. Therefore, our supply of regenerated 10% caustic quickly went bad. Ordinarily, H_2S was distilled overhead in the cracking unit depropanizer. When this tower was upset, the H_2S came out the bottom of the depropanizer with the butylene.

Vertical Wash Towers

For difficult cleanup jobs, a vertical tower is used instead of a mix valve and settler. It will provide several theoretical equilibrium stages. The mix valve provides one stage. Figure 14–3 shows a typical arrangement: propane scrubbing to remove H_2S with amine. Carryover of amine in the treated propane and high residual H_2S are the two usual complaints.

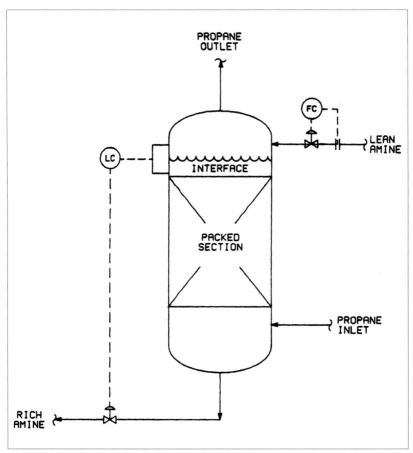

Fig. 14–3. Packed tower used for washing liquid hydrocarbon.

In figure 14–3, the amine level is held above the top of the packed bed. This means the amine is the continuous phase and propane is the dispersed phase. Propane is bubbling up through a column of relatively stagnant amine. A correctly designed packed wash tower has the stream with the smaller flow as the continuous phase.

When troubleshooting a carryover problem for a wash tower, first calculate the vertical velocity of the discontinuous phase, propane in this case. Divide the volume of propane in cubic feet per minute by the tower cross-sectional area. For velocities of less than 1.0 ft/min, entrainment should be negligible. For 1.2–1.6 ft/min, entrainment may be marginal. For more than 1.8 ft/min, excessive entrainment can be expected.

Regardless of its vertical velocity, the propane requires a minimum settling time to remove entrained amine. About five minutes above the amine feed nozzle is right.

Find that rag

Next, check the interface level. You may find that the level glass keeps filling up with a dirty-looking emulsion, no matter how many times it has been blown down. If so, you have found the cause of the carryover. A layer of emulsified material consisting of heavy hydrocarbons, dirt, and amine has accumulated at the interface. Because its density is less than amine and more than propane, the material is trapped. This layer is called a rag and will eventually fill the entire top of the tower. The rag can be drained off through the level taps.

Amine with a high concentration of absorbed H_2S is more difficult to separate from FCCU C_3s and C_4s than leaner amine. This means that reducing amine circulation may promote amine carryover. I don't know the reason for this. However, many times amine carryover has been stopped by increasing amine circulation rates.

The first indication that the two phases are starting to separate more easily is a drop in the interface level and a loss in amine flow from the bottom of the column. Unfortunately, many operators misinterpret this initial response as an increase in amine carryover and cut back on the lean amine circulation rate.

Poor treating

A loss in treating efficiency in a packed tower is often due to poor-quality wash liquid. In the propane-amine system, the amine is not being regenerated correctly (see chapter 4). If this is not the case, there are two other possibilities.

The packing may have disintegrated and is no longer in the tower, or the support grid has broken and the packing has passed out the bottom of the tower. On one unit the packing was found downstream in a caustic storage tank. Check the pressure drop across the packing to uncover this problem. A lower tower pressure drop (less than 0.1–0.2 inch of water per foot of packing) is a good tip-off. The other possibility is that an internal distributor plate has become damaged. This will cause channeling of the liquid phases and result in poor contacting.

Water Washing

A caustic wash is typically followed by a water wash. This will remove entrained caustic from the treated hydrocarbon. Also, water washing is commonly used to remove salt from crude oil.

Clean steam condensate is the best source of wash make-up water. Condensate collected from fractionator reflux drums is sometimes acceptable. River, lake, or well water with high total dissolved solids (TDS) is bad. These solids can precipitate out on the mix valve and cause excessive pressure drop. In one refinery, using untreated river water in a crude-unit desalter for a few months reduced desalting efficiency from 90% to 65% salt removal.

Remember that the treated hydrocarbon effluent will probably be vaporized downstream. When this happens, the small amount of entrained water will also flash. The dissolved solids will precipitate out in heat exchangers. For example, a 5,000-B/D butane stream contains 0.1 wt% entrained water. It has been washed with river water containing 1,000 ppm TDS. When vaporized in a steam heater, 400 lb/yr of solids will accumulate in the heat exchanger.

Ammonia

Ammonia is the hidden enemy of propylene polymerization processes. Propylene streams produced in cracking units usually contain ammonia. This ammonia will pass through a caustic wash untouched. When it arrives at the propylene polymerization unit, the ammonia neutralizes the acid-base catalyst.

One might expect that ammonia, being infinitely soluble in water, would be entirely removed by a water wash downstream of a caustic wash. This is not so. When troubleshooting a report of high nitrogen from a downstream unit, check the pH of the water wash drum. If the water wash follows a caustic wash, chances are you will see a pH of 10 to 12. This is caused by entrained caustic. Caustic, being a stronger base than ammonia, prevents the ammonia from completely dissolving in the water.

Do not increase the water make-up rate in the hope of improving ammonia removal. To reduce the water wash pH from 12 to 11 would require an increase in water use of 1,000%.

The way to remove ammonia from propylene is to control the circulating wash water at 5.5–6.5 pH. This is done by adding sulfuric acid to the circulating water stream. The required chemical injection pump, analyzer, and instrumentation are standard equipment for cooling tower pH control.

Drying

After water washing, a liquid hydrocarbon stream is often dried. A salt dryer will remove all entrained water, but only a small part of the dissolved water. For really bone-dry streams, calcium chloride is used. Walnut-sized KOH is also good. The materials are loaded as solids into a vertical vessel. When the vessel is empty, it is opened up and refilled. Simple. But how does one know when the vessel is empty?

With practice, some refinery labs can consistently measure the water content of dried light hydrocarbons. By periodic sampling, it may be possible to determine when a dryer is spent. A simpler procedure is to check the amount of brine being drawn off the bottom of the vessel. When the brine drained decreases markedly, the desiccant has mostly dissolved.

If a $CaCl_2$ drying vessel develops a large pressure drop and prematurely loses drying capacity, take a look inside the vessel. If it is still partly full and the $CaCl_2$ has a hard surface crust, caustic carryover from an upstream treater is the problem. This can also occur in KOH dryers.

Haze in Jet Fuel

A large refinery in southern England had a jet fuel haze problem. Haze is caused by a collodial suspension of water in hydrocarbon. The configuration of the plant is shown in Figure 14–4.

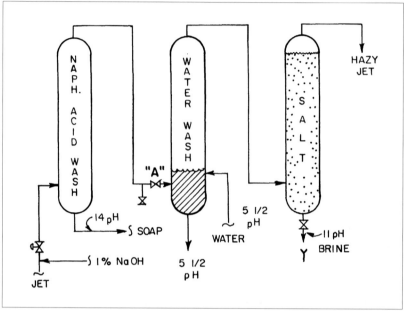

Fig. 14–4. Sodium naphthenates can render a salt dryer ineffective.

Naphthenic acid in the jet fuel is removed by reacting with caustic (1% NaOH) and is converted to sodium naphthenates. This is required to meet the acid number (neutralization number) specification for jet fuel. Entrained sodium naphthenates (soaps) are extracted from the jet fuel with water. The soap is alkaline and will increase the pH of the wash water.

Entrained as well as some dissolved water in the jet fuel is finally removed in the salt dryer. Enough of the dissolved water is removed so that a water haze does not form in the jet fuel when it cools in storage. In this English refinery, however, the salt dryer was being rendered ineffective, and hazy product was accumulating in tankage.

The first step in this assignment was to list the factors that could result in poor performance of the salt dryer:

- Low level of salt in dryer
- Channeling of jet-fuel flow through the water wash drum
- Coating of salt with alkaline soaps

The hourly operators said that when they opened a salt dryer to renew its charge, they found 90% of the salt still in the drum. The dryer was then topped up and returned to service.

At first, the jet fuel would be haze-free. After only four to six weeks, however, the haze problem would return. I asked the operators if the salt in the drum appeared to be channeled. They replied that other than having a slightly darkened color, the salt in the dryer looked about the same as when it was initially loaded in the drum.

These comments eliminated the first two components of the problem listed above. The possibility that water carryover into the dryer was excessive was eliminated by the fact that very little brine was routinely drained from the salt drum. Any water entrained into the dryer would dissolve the salt and appear as brine.

The remaining component of the problem was coating the salt with soaps. If sodium naphthenates were not efficiently removed in the water wash drum, the naphthenates would coat the salt and render it ineffective. Alkaline soaps were being carried over into the salt dryer—as proven by a check of the pH of the brine drained from the salt drum.

The 11 pH of the brine shown in figure 14–4 proved that soap carryover was the cause of the poor dryer performance. The factors that could contribute to soap carryover from the water wash vessel were:

- Inadequate water make-up rate
- Excessive soap content of jet fuel to the water wash
- Poor jet fuel-water contacting
- Low water level in the water wash vessel

The higher the water level in the wash drum, the better the soap removal. I checked the level and found that the operators had previously raised it to the upper level indicated by the seven tricock nozzles. I next checked the pH of both the water makeup and the water drain from the water wash vessel. Both pHs were 5.5. As long as the pH of the drained water is well below the pH of the brine drained from the salt drum, the water makeup rate must be more than sufficient. As a matter of fact, I later found that my client was using 1,000% more steam condensate makeup water than required, at a penalty of $20,000/year.

The observed water wash pHs (see figure 14–4) proved that neither inadequate water makeup nor excessive soap carryover from the naphthenic acid wash drum, was promoting soap carryover to the salt dryer. This left poor jet fuel-water contacting as the culprit. A sparger in the bottom of the water wash drum was supposed to disperse the oil phase in the aqueous phase. Evidently, the sparger was ineffective in bringing the oil and water into intimate contact. This was proved by the lack of any pH increase in the water phase.

To mix the jet fuel and water more effectively, I took a hose and connected the discharge of the water makeup pump to the ¾-inch bleeder upstream of valve A shown in Figure 14–4. I then throttled valve A to set up a 10–15 psi pressure drop. Although this valve was an ordinary gate valve, it could still be used as a "poor man's" mix valve with sufficient pressure drop. Finally, I reduced the total water makeup rate until a noticeable increase in the pH of the wash water was observed. Why load up the refinery's effluent water treatment system to no purpose?

Over a period of three days, the pH of the brine drained from the salt dryer slowly declined. Evidently, the entrained low pH water from the water wash vessel was washing off the alkaline soaps coating the salt bed. After a week, my client reported that the haze problem in jet fuel had disappeared. Further, the salt dryer could be operated for many months before it had to be opened and recharged with salt.

Electrostatic Precipitator

Many feed-washing systems are built with an electrostatic precipitator instead of a settling drum. The electrical field in the precipitator supplements gravity to speed settling. The higher the voltage on the precipitator, the less carryover there will be. When the precipitator starts drawing more amps at constant voltage, something is awry. The precipitator, like a motor, has a maximum amperage rating. When this rating is exceeded, the precipitator will trip off.

If the amperage is slowly increasing over a period of days, the precipitator is fouling. Probably the internal insulators are accumulating deposits. A thorough water wash of the precipitator may correct the problem.

If the precipitator amperage starts jumping violently (arcing) and the system trips off and can't be reset, a dead short has occurred. Something may have broken loose inside. More likely, the interior walls of the precipitator have developed a layer of rust scale that is falling off. The rust is shorting out the electrical grids. Either way, the precipitator will have to be reopened for inspection and repair.

Troubleshooting Checklist
for Treating Liquid Hydrocarbons

Vertical Wash Towers

Excessive vertical velocity
  causes carry-over
Formation of a rag at interface
Flooding of packing or trays
Poor treating due to loss of packing
Poor treating resulting from
  distributor plate failure

Water Washing Problems

Dissolved solids
High pH or NH_3 removal
Desalter fouling

Inadequate Treating

Inadequate circulation rate
  of wash liquid
Mechanical damage
  to circulation pump
Low mix-valve pressure drop
Mix valve plugged
Quality of fresh make-up wash liquid
Insufficient supply of make-up
  wash liquid
Upset changes composition
  of hydrocarbon feed
Coating salt with soap

Carryover or Entrainment

High settler interface level
Excessive mix-valve pressure drop
Settler residence time too short
High wash liquid circulation rates
Emulsion caused by dirt (rags)
Surfactants
Excessive caustic strength
Mechanical failure
  of settler internals

Drying Liquid Hydrocarbons

Determine when drying
  solid consumed
Crusting over drying solid

Electrical Precipitator

High amperage due
  to insulator fouling
Rust scale causing arcing
Dead short due
  to mechanical failure

Process Heaters

Energy represents two-thirds of the cost of running a modern refinery. Of this, the largest component is furnace fuel. The process operating engineer, assigned to follow a large crude or reforming unit, can have more influence on the nation's energy resources than many of our congressional leaders.

Saving fuel without shortening the operational life of a heater is a challenge that the process engineer must meet. Acquaintance with the principles of fired heater operation is a prerequisite for success.

This chapter outlines the basic principles of furnace operation and troubleshooting. For simplicity, the natural-draft, gas-fired, horizontal heater shown in figure 15–1 will be used as a basis for discussion. Problems dealing with draft control, excess air, hot tubes, expanding heater capacity, and oil burning will be detailed. By making the suggested field observations, significant fuel savings can be realized on most refinery process heaters.

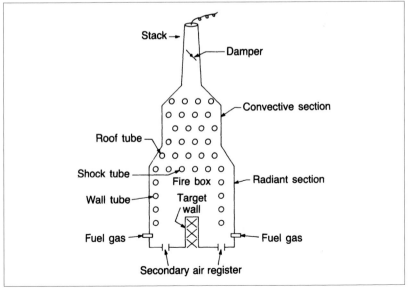

Fig. 15–1. A typical process heater.

Draft

The radiant section (i.e., the combustion chamber) of a natural draft furnace is under a vacuum. The amount of this vacuum, called furnace draft, is measured in inches of water. A high draft—measured near the burners—would be 1 inch of water. A more typical value would be 0.4 inch.

The draft is a result of the density difference between hot flue gas in the heater stack and the cooler ambient air. Therefore, the hotter and taller the stack, the greater the draft. Draft is measured with a water-filled inclined manometer.

Reduced draft

Loss of draft will make it impossible to provide sufficient air to combust all of the heater fuel entirely. The oxygen concentration in the flue gas will approach zero; carbon monoxide in the flue gas will increase significantly. If the furnace is oil-fired, the stack will emit black smoke. Some of the more common causes of reduced draft are:

- **Fouled convective section.** For oil-fired heaters with finned convective tubes, fouling may reduce draft. A gradual loss of draft probably means the convective section needs to be water washed. Deposits from the burnt fuel oil have built up around the tubes thick enough to restrict the passage of flue gas. The increased pressure drop through the convective section reduces the vacuum in the firebox. Permanently installed soot blowers using steam are one answer to this problem.

- **Leaking ducts and skin.** The firebox and the convective section and ducts are all under a vacuum. Holes in the convective or radiant section skin will permit cold air to be drawn into the flue gas. This air reduces the draft in two ways: by cooling the flue gas in the stack and by raising the pressure drop due to the increased volume of flue gas. In effect, combustion air is bypassing the radiant section.

- **Closed stack damper.** At a reduced firing rate, the damper may have been partially closed to control excessive draft. The damper can become stuck—or perhaps someone forgot to open it—when the firing rate was increased. Occasionally, the shaft of the damper will come loose. The operator opening the damper will see the shaft moving, but the damper itself remains closed. An x-ray will tell if this is so.

Combustion Air Supply

The air supply for a gas-fired, natural-draft heater consists of two portions: primary air and secondary air. (In some newer burners, tertiary air is use to control nitrogen oxide emissions.) The primary air is educted or sucked into the burner through a venturi by the rapidly flowing fuel gas. The air is well mixed with the gas prior to combustion. Thus the name *premix burner*. A Bunsen burner is an example of a premix burner.

The air required to complete combustion is called secondary air. The air flows into the firebox through adjustable ports called secondary air registers. A typical premix fuel gas burner is illustrated in figure 15–2.

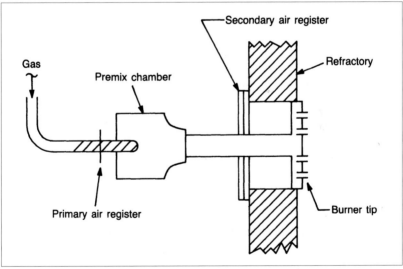

Fig. 15–2. Fuel gas premix burner.

Trimming burner operation

One should use as much primary air as possible and minimize the use of secondary air because primary air mixes better with fuel gas than secondary air. A short, compact flame results. Such a flame evenly heats the furnace tubes and thus minimizes the chances of flame impingement on the tubes. In turn, less excess air is required to cool off the firebox and so compact flames save energy.

Opening the primary air register too far will lift the flame off the burner. One should continue opening the primary registers until this happens or sufficient excess air is obtained.

If the whole firebox is full of yellow flames, slowly and uniformly open the secondary air register until flame impingement is eliminated. If only one or two burners are generating long, yellow flames, pinch back on the fuel gas to these burners.

Excess air benefits

Why do operators often run with too much air and thus waste energy? For one thing the flame temperature is reduced. Also, more excess air produces more flue gas. This increases heat pickup in the convective section. Finally, flame length is shortened and flame impingement is reduced. These factors make it easier to fire the heater without overheating the tubes. Unfortunately, the price must be paid in lost furnace efficiency and wasted energy.

Optimizing Heater Draft

If excess air can be adequately controlled with the secondary air registers, what is the purpose of the stack damper? The optimum positioning of both the secondary air register and the stack damper are related. Two objectives must be satisfied if an optimum operation is to be approached: Excess air in the flue gas to the convective section should be minimized, and there should be a very small negative pressure at the convective section inlet.

The stack damper and the secondary air registers must be used as a team to satisfy these requirements. If the stack damper is mostly closed and the secondary air dampers are mostly opened, a positive pressure can develop at the convective section inlet (figure 15–3).

A positive pressure forces hot flue gas to leak outward. This can overheat the steel structure, brickwork, roof arch supports, and so on of the heater and consequently shorten heater life. Also, loss of hot gas reduces the convective section heat recovery. "Blue smoke leaking out of roofs and flue ducts and whitish or yellowish deposits

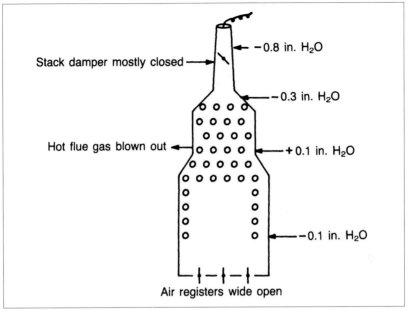

Stack damper mostly closed — ← −0.8 in. H₂O

−0.3 in. H₂O

Hot flue gas blown out ← +0.1 in. H₂O

−0.1 in. H₂O

Air registers wide open

Fig. 15–3. Positive pressure damages a heater.

around openings at the top of furnaces indicate positive pressure."[1]
To correct this situation, one should:

1. Slowly open the stack damper until the draft gauge indicates a small negative pressure (minus 0.1 in. H_2O) at the convective section inlet.

2. Pinch back on the secondary air registers until the desired excess air is obtained. Make sure that it is the firebox effluent flue gas, not the convective effluent flue gas, that is being sampled.

3. The stack damper and air registers may have to be adjusted in several small increments to meet the desired targets.

Excessive draft

Flue gas samples drawn from the stack are not necessarily representative of the oxygen available for combustion in the firebox. For natural draft and induced draft heaters, leaks in the convective section exterior walls permit air to bypass the combustion zone. The oxygen measured in the stack is the sum of unused oxygen from the firebox plus oxygen drawn into the convective section. If the heater air registers are being adjusted based on the oxygen level measured in the stack, it is probable that partially oxygenated hydrocarbons are entering the convective section.

Mixed with fresh air flowing through convective section leaks, the flue gases may reignite. This phenomenon, called afterburn, results in damage to the convective section finned tubing. Afterburn is promoted by insufficient oxygen in the firebox, excessive draft, and leaks in the convective section exterior walls.

If the stack damper is wide open and the secondary air registers are pinched back, too low a pressure can develop in the convective section. Figure 15–4 illustrates this condition. Depending on the furnace skin integrity, a great deal of cold air can be drawn into the heater through the holes in the metal sides of the furnace or through leaks in the roof tiles.

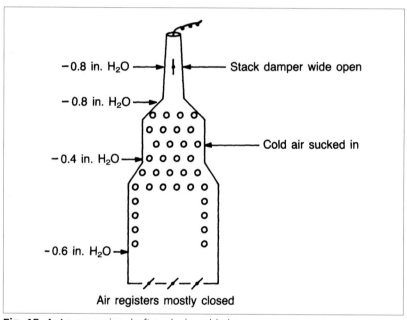

Fig. 15–4. An excessive draft sucks in cold air.

The following equation may be used to approximate the percentage of furnace fuel wasted in heating air that has been inadvertently sucked into the convective section:

$$H_W = \frac{7.4(O_S - O_{FB})(T_S - T_A)}{3,800°F}$$

where:

H_W = Percent energy wasted due to air in-leakage

7.4 = An approximate coefficient to convert oxygen measurements to percent in flue gas flow; use 18%–45% excess air only

O_S = Percent oxygen in flue gas measured at the stack

O_{FB} = Percent oxygen in flue gas measured at the outlet of the firebox

T_S = Stack temperature, °F

T_A = Ambient air temperature

3,800°F = Approximate adiabatic flame temperature (assumes no radiation) for a typical refinery fuel gas

On one 50,000-B/D crude unit, steps were taken to eliminate air leakage through the furnace skin; fuel use subsequently dropped by 6%.[2] To correct excessive draft, one should slowly pinch back on the stack damper until the draft gauge indicates a small negative pressure and open the secondary air registers until the excess air target is reached.

Plugged draft gauges

One of the problems that the operations engineer needs to avoid is a false draft reading because of plugged tubing. Simply push the flexible tubing connected to the draft gauge down. If the indicated pressure increases (i.e., draft decreases), the tubing between the gauge and heater is almost certainly plugged.

Insufficient Air

Table 15-1 shows the relationship between percentage of O_2 in the flue gas and excess air.

A gas-fired heater firebox will have a hazy appearance when short of air. A heater making a regular thumping sound is also likely to be air deficient. Sometimes an increase in fuel gas flow will result in a decrease in process-side temperature. This is a sign that the firebox does not have enough air.

Table 15–1. Excess air versus oxygen in flue gas.

% Oxygen	% Excess Air
0	0
2	10
3	16
4	22
5	29
6	37
7	46
8	57
9	69
10	83

On one unit, the process-side temperature reset the fuel gas flow via closed-loop control. Once, a sudden increase in the process fluid flow automatically increased the fuel gas flow. Since the heater had just been trimmed to use almost no excess air, the firebox went air deficient.

As the percentage of combustibles in the flue gas increased, the firebox cooled off and the process-side temperature dropped. This called for even more fuel gas, which lowered the temperatures further. Eventually, flames were observed at the exit of the heater stack. This is an example of a positive feedback control loop.

Optimizing Excess Air

Operating a process heater simply to achieve a minimum excess oxygen target in the flue gas can waste a great deal of energy. The proper way to adjust O_2 to a heater is to target for the point of absolute combustion, as shown in figure 15–5. The point of absolute combustion is defined as that air rate that maximizes heat recovery to the process. That is, either a decrease or an increase in the combustion air supply will reduce heat absorbed in the heater.

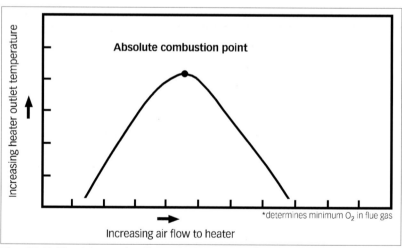

Fig. 15–5. Absolute combustion point (courtesy *Oil & Gas Journal*).

The primary objectives in the operation of a fired heater are to avoid excessive heat density in the firebox, maintain a small negative pressure below the bottom row of convective tubes, and obtain optimum combustion of the fuel in the firebox with minimum oxygen in the flue gas. Regardless of the flue-gas oxygen content, however, the point of absolute combustion is the combustion air rate that maximizes process-side heat absorption for a given amount of fuel.

The oxygen target required to achieve the point of absolute combustion may be 2%, or it may be 6%, depending on the operating characteristics of the heater. Often, energy is wasted because operators and engineers fail to understand:

- Excess oxygen of 2% to 3% in the flue gas is not a good target for all heaters.
- Insufficient oxygen will not necessarily cause the evolution of black smoke from the stack when gas is being fired.
- Excess air requirements are increased for many heaters when an air preheater is commissioned.

As shown earlier in figure 15–1, as the stack damper or air register is closed, the flow of air into the firebox through the burners is reduced. If both the process-side flow and the fuel-gas rate are held constant, the following sequence of events occurs.

1. The heater outlet temperature begins to increase as the excess air is reduced. That is, more heat goes into the process fluid and less heat goes up the stack.
2. The heater outlet temperature declines as air flow is diminished past the absolute combustion point. That is, products of the partial combustion of hydrocarbons (aldehydes, ketones, CO, etc.) are formed, and the heating value of the fuel gas is effectively reduced.

When the air register is closed to suppress air flow below what is required for absolute combustion, black smoke will not immediately issue from the stack. Depending on the hydrogen-to-carbon ratio of the fuel, varying amounts of the potential heating value of the fuel may go up the stack as products of partial combustion before the flue gas turns black.

For instance, in excess of 10% of the heating value of methane can be lost in this manner, while the stack gases remain clear. However, if heavy fuel oil is burned, the flue gas will quickly turn black as the combustion air rate falls below the absolute combustion point. For an extreme contrast, the combustion of hydrogen will never turn a stack black, regardless of combustion air insufficiency.

Flue gas oxygen

One of the more common factors causing high oxygen levels in flue gas at the absolute combustion point is poor air-fuel mixing. To promote air-fuel mixing, the following parameters should act as a guide:

1. Maximize primary air (on the premix burners) and minimize secondary air. Using too much primary air will lift the flame off the burner tip.

2. Close the openings to pilot lights, sight ports, and other holes around burners. Combustion air only mixes properly with fuel gas when it flows through the burner air registers.

3. Shut down burners when the heater is firing at reduced rates. Burners work more efficiently when operating close to their design capacity. Don't forget to shut the air registers on the idle burners.

4. Try to minimize poor lateral air distribution in the firebox by adjusting air registers on individual burners. Having a low air flow in one portion of a heater will lead to higher overall oxygen requirements.

Partially plugged burner tips also promote the use of excess oxygen. To suppress flame impingement on heater tubes, operating personnel will typically open the burner air registers. In petroleum refineries, the usual cause of plugged burner tips is sulfur deposits. Hydrogen sulfide (always present in fuel gas) combines with oxygen accidentally introduced into the fuel-gas system to form a sulfur precipitate.

A solution to burner tip plugging problems is to preclude oxygen from entering the fuel-gas system.

Flame appearance

Regardless of oxygen measurements, the appearance of the firebox must always be taken into account. A combustion zone that looks bright and clear has excess oxygen. Long, licking, yellow, smokey flames and a hazy firebox indicate low oxygen. Yellow flames in themselves, however, are fine. A gas burner may be perfectly adjusted and still have a short, bright, yellow flame.

The yellow flame is caused by thermal cracking of the fuel. Anyone who has fired a residual oil burner knows that a yellow burner flame can be normal, depending on the fuel composition.

Fin tube damage

On one particular unit, the fins on the convective section tubes had been reduced to a brittle metal scale. In this condition the fins retarded rather than enhanced heat transfer. The cause of this fin damage was secondary combustion in the convective section. This is an extremely common but frequently unrecognized problem.

Natural-draft heaters cannot be entirely leakproof. A portion of the air found in the stack flue gas will have been drawn in through holes in the convective section and roof tiles. On the heater discussed above, these leaks were substantial. The flue gas was sampled for oxygen at the base of the stack. Using these samples of convective section outlet flue gas as a guide, the operating engineer had been pinching the secondary air registers to reduce excess air.

In reality, the heater box was short of air. The unburnt hydrocarbons completed combustion upon mixing with the air drawn into the convective section. The liberation of radiant heat in the midst of the convective tube bank oxidized the steel alloy fins on the tubes. R. D. Reed has discussed this subject in detail.[3]

Sealing skin leaks

This fin tube failure could have been prevented by sampling the flue gas just below the convective section inlet. The oxygen concentration at this point represents the true amount of excess air available for combustion in the firebox.

A roll of heavy-duty aluminum tape and an interested operator looking for convective section leaks can save many fuel dollars. Welding loose sheet metal, mudding up header boxes, and using new, high-temperature sealants are probably the most cost-effective activities in a refinery. Leaks through roof tiles are also a major source of extraneous air. These tiles should be inspected during every turnaround.

Convection and Radiation

Duties between the convective and radiant sections of the furnace must be balanced to maintain efficiency. The radiant tubes in a firebox are kept cool by the flowing process fluid. On the other hand, convective section finned tubes, especially the fins, will heat up to a temperature approaching the flue-gas temperature.

Because the convective section's fins are constructed from high-chrome steel, they can withstand typical maximum flue-gas temperatures of 1,200°F. Should afterburn occur in the convective tube bank, however, the fins may oxidize when exposed to the 2,000°F that results from the localized combustion caused by afterburn.

One way to minimize the possibilities of afterburn is to pinch back on the stack damper and further open the burner air registers (figure 15–6). The resulting higher pressure (low draft) reduces the rate of air into the convective section tube banks.

Of course, if the stack damper is closed too far, a positive pressure may develop below the bottom row of convective tubes, forcing hot gas out against the structural members of the heater, possibly reducing the structural integrity of the heater.

The following example shows how imbalance between the radiant and convective sections affects the operation of the furnace.

Several years ago, a large refinery crude heater was retrofitted with an air preheater. When the revamped heater was returned to service, the following changes were observed: Consumption of fuel gas decreased by 10%; at the same heat absorbed as before the revamp, the firebox temperatures were higher, and the crude-oil side convective section tube outlet temperature had decreased.

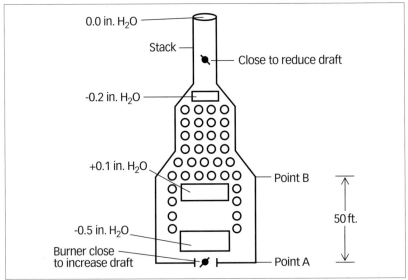

Fig. 15–6. Draft control with dampers (courtesy *Oil & Gas Journal*).

Analysis of the post-air preheater data revealed that the volume of flue gas had diminished in proportion to the reduced fuel consumption. The reduced flue-gas flow had reduced the heat-absorbed duty in the convective section. To achieve a higher radiant section duty, but really as a consequence of the preheated combustion air, the burner flame temperatures had increased.

Basically stated, preheating air by 300°F raises the burner flame temperature by 300°F. The hotter flames radiate more heat to the firebox tubes per pound of fuel consumed, and the reduced fuel consumption also reduces the flow of flue gas to the convective section.

To reduce the firebox temperatures to the pre-revamp level, the operators increased the excess oxygen from 3% to 6%. This adjustment reduced the burner flame temperature, increased the volume of flue gas, and, consequently, increased the convective section heat absorbed.

The increased flow of combustion air was not required for combustion. It was intended to transfer heat absorption from the radiant to the convective section, with the objective of cooling the firebox.

The conclusion from the preceding example is that excess air is often used to prevent overheating of the combustion zone, and the oxygen requirements to reach the absolute combustion point are, in this case, irrelevant.

Draft measurements

During calculations of pressure drops through a heater, corrections to draft gauge readings must be made for elevation of the measuring point. This is true regardless of the elevation of the draft gauge relative to the point at which the pressure measurement is taken. If elevation is not taken into consideration, some misleading results can occur.

Figure 15–6 shows a heater with a 0.1-inch water-positive pressure below the first row of convective tubes. The differential pressure between the burner and the tubes appears to be a pressure increase of 0.6 inches of water. Obviously, this cannot be correct.

A draft measurement compares the pressure inside a heater to the pressure outside the heater at the same elevation. In order to yield the actual pressure differential in the heater, the draft readings must be corrected for elevation.

The difference in elevation between the burner and the convective tube is 50 feet. The weight of the 50-foot column of air corresponds to 0.7-inch of water. Therefore, the differential pressure corrected for the 50-foot elevation difference is actually: 0.1 in. – (–0.5 in.) – 0.7 in. = minus 0.1 in. of water.

Adjusting the pressure below the first row of convective tubes to maintain a small negative pressure (0.05 in. to 0.10 in. of draft) should be achieved without changing the oxygen level in the combustion zone. This can be done by checking the pressure drop across the stack damper, checking the pressure drop across the burner air registers, adjusting the stack damper (closing if the draft is too large), and making an equal but opposite adjustment to the pressure drop across the burner air registers.

Leak prevention

To suppress afterburn and minimize energy losses caused by leakage of cold ambient air into the convective section, holes in the convection section exterior should be patched.

To determine the location of leaks during a turnaround, close the stack damper slightly and ignite colored smoke bombs in the firebox. If a forced draft fan is available, turn it on. The colored smoke will escape from the leaks.

Air preheaters

A typical air preheater will reduce the fuel required to liberate a given amount of heat by 10%. The debit for this improvement in thermal efficiency is a hotter flame temperature and the possibility of over-heating the radiant section.

This effect can be offset by increasing the amount of excess air. As shown in figure 15–7, after an air preheater has reduced stack temperatures to 400°F, increasing excess air results in a little incremental energy loss.

Three types of air preheaters are typically used: direct heat exchange, heat exchange via an intermediate circulating oil, and heat exchange by a massive heat-transfer wheel with metal baskets (Lungstrom type).

A common problem with the wheel-type exchanger is air leaking past the mechanical seal between the air and flue-gas sections of the preheater (figure 15–8). This leakage can be identified by increased oxygen content in the flue gas, low flue-gas outlet temperature, and greater temperature loss in the flue gas than the rise in air temperature.

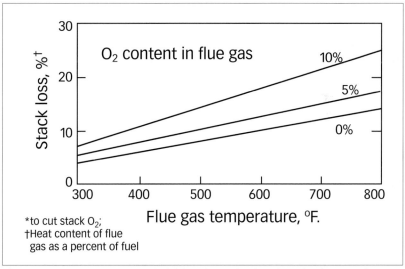

Fig. 15–7. Low temperature diminishes incentive for air preheater retrofit.

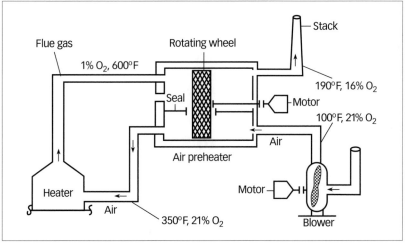

Fig. 15–8. Air preheater with leaking seal (courtesy *Oil & Gas Journal*).

Seal leaks can reduce the thermal efficiency of an air preheater. Also, air that leaks past the seal bypasses the combustion zone and may reduce excess air below that required to achieve the point of absolute combustion.

Destructive afterburn in the preheater can then result. The leaking seal also causes the forced-air blower to operate out of its normal operating characteristics, requiring higher driver horsepower.

All air preheater heat exchangers are subject to corrosive attack caused by the condensation of sulfur trioxide. Refineries must burn fuel gas with less than 160 ppm of sulfur. Regardless of the calculated dew point of the sulfur trioxide, operating experience has shown that a minimum temperature of 300°F–350°F in the outlet flue gas is required to minimize corrosion and fouling of the heat exchanger. Uneven cooling of the flue gas results in the need to keep the flue gas 50°F–100°F hotter than its calculated dew point.

If the heater is not equipped with an air-side bypass, the operator can increase the preheater outlet flue-gas temperature by increasing the excess air in the combustion zone.

Preheater vibration

Air preheaters, or any type of waste-heat recovery devices that are designed for horizontal flow of fuel gas across vertical tubes, are subject to vibration produced by the velocity of the gas across the tube banks. The velocity produces a vortex-shedding wave pattern that may correspond to the natural harmonic frequency of the tube bank.

At a particular gas velocity, excessive vibration of the tubes can result. Operators have been forced to bypass hot flue gas around the heat recovery device to control vibration. Redesign of the internal baffling by insertion of dummy baffles can stop the vibration by changing the natural frequency of the tube bank (see chapter 23).

Other Ideas to Save Energy

A few common methods, other than a careful control of air, to improve the energy efficiency of a heater are:

- Clean the outside of the radiant tubes onstream. This is best done by blasting the tubes with nutshells or with a thin jet of high-pressure water. The trick is to hit the tubes—and not the easily damaged refractory.

- Steam-air decoke the tubes. This very effective method of reducing firing requires a two- or three-day shutdown to complete.
- Water wash onstream the convective tube banks with a fire hose. Soot blowers also successfully combat convective section deposits.
- Conduct an infrared thermograph survey of the heater while onstream. This will locate areas of bad refractor that can be repaired during the next unit turnaround.
- Use blanket tile insulation in back of furnace tubes.
- Minimize the use of atomizing steam in oil-fired heaters. Typically, 1 pound of steam is used for each gallon of fuel oil.
- Most energy conservation schemes entail reducing stack gas temperature. Corrosion problems place a limit on how far this temperature may be reduced. Part of the sulfur contained in refinery fuels will be converted to SO_3. Depending on the unevenness of flue gas cooling and the amount of sulfur in the fuel, sulfuric acid will condense out of the flue gas between 300°F and 400°F. Excessive corrosion in the stack or in the convective section is likely due to the precipitation of sulfuric acid.
- Optimize the use of firebox thermocouples. An experienced operator can balance furnace firing and thus minimize excess air by visually observing the flame pattern. A substitute for an experienced eye is a large number of firebox temperature indicators (TIs). The TI tells an operator how hot it is in a particular part of the firebox. An optimized firebox should look very slightly hazy because the fuel is groping for the last part of oxygen in the firebox. A reasonable number of TIs for a large firebox (100 MM BTU/hr heat release) is eight.

If a computer is available on the unit, a valuable aid to assist operators in trimming the furnace is a computer printout of firebox temperatures. Such a pictorial display is shown in figure 15–9. The operator takes the printout, which consists of a single sheet of paper, to the heater and uses it as a guide in adjusting individual burners. In the example shown in figure 15–9, the operator might decrease fuel to the southeast burners and pinch down on the secondary air registers in the northwest corner.

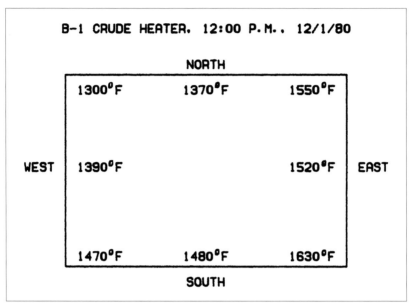

Fig. 15–9. A computer printout of firebox temperatures.

Measuring heater temperatures

Heater temperatures measured with a static thermocouple will read low. This is because of the reradiation of heat from the thermocouple. The higher the temperature, the greater the error. For example, the flue gas temperature to the convective section, measured by a static thermocouple, may read 100°F–200°F lower than the real temperature. A velocity thermocouple will correct this problem. The velocity thermocouple device channels hot flue gas past an ordinary thermocouple. This largely eliminates the error due to reradiation.

Spotting Hot Tubes

Heater tubes will glow different colors, depending on their exterior temperature. A freshly cleaned heater fired at a moderate rate will have black-appearing tubes. In contrast, the hangers (i.e., the brackets supporting the tubes) will be glowing red. The process fluid keeps the tubes, but not the hangers, cool. A moderately hot temperature is indicated by a dark cherry-red color. As this color brightens, the tube is becoming hotter.

A hot spot on a tube is usually caused by a partial loss of flow through one pass of a multipass heater. In most refinery services, a sudden reduction in flow will cause tubes to overheat. The low velocity combined with high tube-wall metal temperature results in localized coke lay-down. The layer of coke insulates the tube wall from the cooling effects of the process fluid. The insulated portion of the tube overheats. Typically, the hot spot looks like a silver streak or a silver dollar.

In oil-fired heaters, ash accumulates on the radiant tubes. These ash deposits, which are largely vanadium, may glow red, while the tube wall metal is relatively cool. This can be confusing to the novice observer.

Cooling overheated tubes

A tube with a silver streak will begin to bulge at its hottest point. Eventually, the tube wall will thin out over the bulge and split open. Figure 15–10 illustrates this problem. Process-side hydrocarbons will spill into the firebox and ignite. Depending on the size of the leak and how quickly the operators shut the unit down, the heater may burn down. Proper shut-down requires use of the box snuffing steam.

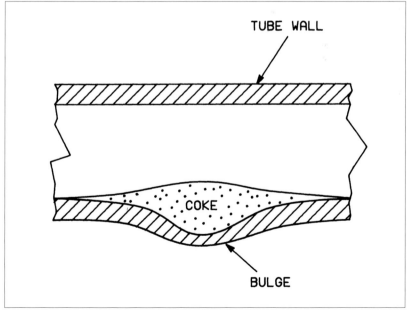

TUBE WALL

COKE

BULGE

Fig. 15–10. Coke lay-down causes tube thinning.

To cool off small hot spots, operators can direct a jet of steam onto the affected area. Reducing burner firing in the dangerous area is also helpful. Maximizing the flow of fluid through the coked-up tubes is usually the best way to prevent failure from overheating. For orientation, many crude furnaces are designed to operate at 100–150 lb/ft²/sec. tube side mass velocity.

Coke deposition

Radiant tubes don't fail because they burn up. Coke lay-down inside the tubes insulates the tubes and impedes heat transfer to the process fluid. This causes the tube walls to operate at higher temperatures. The pressure inside the tubes results in high-temperature creep of the tube material, leading to bulges at hot spots. This eventually causes the tube to burst at the bulges.

Low flow of process fluid is a principal cause of coke lay-down. Reduced tube-side mass velocity results in low fluid shear at the tube wall. For many services, a mass velocity of 200 lb/ft²/sec will minimize coke lay-down and hot spots.

Flame impingement is another cause of hot spots. Increasing combustion air, especially premix air, will shorten flames. Keeping burners clean is an easy way to help control flame impingement. Almost all burners can be cleaned without shutting down the heater. In refineries, a frequent cause of tube coking is associated with a complete but temporary loss of feed. By flushing the tubes with steam, operators minimize the formation of coke deposits when the feed interruption is first recognized.

During the interval before feed flow is restored, however, the furnace tubes will continue to absorb heat from the refractory walls after the burners are shut off. If the firebox temperature had been running at 1,700°F, it is probable that the tubes will heat up to about 1,500°F shortly after the flow of process fluid is lost.

If the unit operators then rapidly restore flow through the tubes, the initial flow will be exposed to the 1,500°F tube walls. A small amount of oil will form coke that then deposits on the tube wall.

If the operators have restarted the flow without determining the cause of the original flow loss, the sequence can repeat, causing further coke deposition. To prevent coke buildup in this manner, the tubes should be cooled to about 800°F–900°F before flow is reintroduced.

Oil Burning

In general, firing oil in a process heater is more difficult than burning gas. A simplified sketch of an oil burner is shown in figure 15–11. Steam is used to atomize the oil prior to combustion.

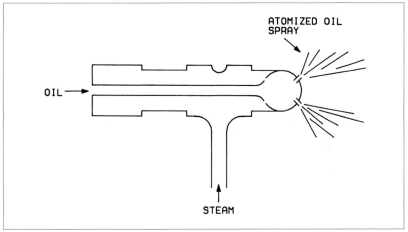

Fig. 15–11. Oil burner using atomizing steam.

As in gas-fired heaters, the objective of an operating engineer should be to ensure complete combustion of the oil fuel with minimum excess air. A few of the more common troubles peculiar to oil burning that cause excess air to be increased are:

- Oil pressure is too high, indicated by smoke puffs at the stack.
- Oil temperature is too low. The more viscous oil does not atomize properly in the furnace, and incomplete combustion is promoted. Try raising the oil temperature and see if the air dampers can then be pinched back.
- Plugged burner tips. Again, oil atomization is adversely affected.
- Enlarged holes in the burner tips. The oil will drool out of tips onto the refractory.
- Loss of insulation on the atomizing steam lines. Water in the steam supply to the oil burner interferes with atomization process.

A detailed discussion on troubleshooting oil combustion systems is given in Fletcher's article "Maintain Combustion Systems" in the January 1979 issue of *Hydrocarbon Processing*.[4]

Heater Huffing and Puffing

A client recently called about a problem with an FCCU CO boiler that was making a huffing sound. Moreover, the problem was becoming gradually worse, with the convective tube bank's sheet metal sides moving an inch or more.

The erratic pressure in the boiler was caused by lack of adequate combustion air in the firebox. My client stated that he had 5% O_2 in the stack. On my advice, the refinery sealed off the leaks in the convective section skin. The O_2 in the stack dropped; steam production went up; and the huffing and puffing stopped.

Expanding Heater Capacity

Process heaters are limited by combustion air, heat release, draft, or heat absorption. The process operating engineer, seeking to expand furnace capacity, must first determine which of these problems is limiting.

The rate of heat release is the measure of burner capacity. If the fuel gas control valve to the heater is wide open, the burner capacity limits furnace duty. Gas burners can easily be expanded by enlarging the holes in the burner tip. Alternately, piping changes will sometimes increase the gas pressure to the burners. However, never exceed design burner tip pressure.

In one refinery unit, the hydrogen content of refinery fuel was much higher than the furnace design specified. Since the volumetric heating valve of hydrogen is comparatively low, the furnace was heat-release limited. Enlarging the burner tip holes rectified the problem.

Draft-limited

Draft and air limits often coincide. With both the stack damper and the secondary air registers fully open, not enough air can be drawn into the firebox to optimize combustion. If a positive pressure develops and the air registers are pinched, the heater is draft limited.

Reducing air in-leakage in the convective section may correct this problem. Perhaps the convective tube bank is partially plugged on the flue-gas side. A pressure survey with a draft gauge is the first step in identifying the restriction that limits draft. Occasionally, the size of the secondary air registers can be increased.

Heat-absorption limited

Overheating tubes is a commonly encountered limit. One way to increase duty without overheating tubes is to add convective tubes. This improves furnace efficiency and reduces firing. Of course, this also reduces draft. Another way is to replace raw-gas burners with premix burners. Raw-gas burners rely exclusively on secondary air and therefore produce a less dense flame than premix burners. A denser flame makes better use of the available firebox volume. Raw-gas burners are used because they are quieter, produce less NOX, and are not subject to flashback when burning hydrogen-rich steams. A third method is to replace large oil-fired wall burners with smaller gas-fired floor burners.

If none of these three suggestions help, increase tube-side mass velocity. The overall tube heat-transfer coefficient is largely a function of the process fluid mass velocity. On one unit, changing from four parallel passes to two parallel passes increased the rated furnace capacity by 20%. Maintaining a high mass velocity ($lb/ft^2/sec$) is the best way to keep tubes cool. Rating furnace capacity on the basis of heat flux ($BTU/hr/ft^2$) without consideration of tube mass velocity is a serious mistake.

Burner school textbook

The John Zinc Burner Company has conducted a furnace operations school for many years. The textbook for this school details many of the operating problems that have been briefly discussed in this chapter.[5]

Troubleshooting Checklist for Process Heaters

Insufficient Draft	**Controlling Air Supply**
Draft gauge plugged	Maximize primary air
CO_2 in flue gas	Flame lifts off burner
Fouled convective section	Open secondary dampers
Leaks in furnace skin	to control yellow flame
Stuck stack damper	Cool box with excess air
Damage to furnace structure	Seal convective section doors
Smoke leaks out of convective section	Burner flashback

Energy-Saving Ideas
Onstream blasting of radiant tubes
Steam-air decoking tubes
Onstream washing convective tubes
Infrared thermograph survey
SO_3 dew-point limit
Minimize atomizing steam
Install soot blowers
Fix convective section air leaks

Excessive Draft
Pinch stack damper
Wastes energy
Sucks cold air into convective section
Secondary combustion in
    convective tube bank

Insufficient Air
Firebox looks hazy
Heater makes a thumping sound
Temperature drops with increase
    in fuel
Fin-tubing damage
Sample firebox effluent, not stack gas

Oil Burning
Oil pressure too high
Oil temperature too low
Plugged burner tips
Enlarged holes
Wet atomizing steam

Hot Tubes
Cool tubes are dark to cherry red
Silver streaks are hot spots
Low flow causes coke laydown
Steam jet cools hot spot
Maximize flow through hot coil
Use more air

Expanding Heater
Capacity
Drill out burner tips
Draft limits
Pressure survey on flue gas side
Add convective tubes
Replace raw gas burners
Increase tube-side mass velocity
Install floor tubes

References

1. R. Fletcher, "Maintain Combustion Systems," *Hydrocarbon Processing*, January 1979.

2. R. D. Reed, "Furnace Operational Factors Save More Heat Energy," *Oil & Gas Journal*, Dec. 31, 1979.

3. Ibid.

4. Fletcher, 1979.

5. R. D. Reed, *Furnace Operations* (Houston: Gulf Publishing Company, 1973).

Water Coolers

A shell-and-tube heat exchanger used to cool hydrocarbon streams is a common piece of heat-transfer equipment. The troubleshooter will find that the deposits that foul the water side are the principal reason for reduced cooling efficiency. Deterioration of a cooling water tower or insufficient water circulation will also reduce heat-transfer capacity. When investigating a cooler deficiency, you will want to consider the following factors:

- Biological fouling
- Hardness deposits
- Plugged tubes
- Cycles of concentration
- Cooling tower wet bulb temperature
- Hydrocarbon leaks

Plugged Tubes

Maximizing the water flow through a cooler using recirculated cooling tower water is the best way to keep it clean and efficient. For exchangers with water on the tube side, major obstacles to maintaining high water velocities are plugged tube inlets or accumulated deposits inside of the floating head.

Figure 16–1 shows a typical refinery water cooler. The cold water flows into the bottom of the channel head and then passes through the lower half of the tube bundle. The water makes a 180° turn in the floating head and flows back through the upper half of the tube bundle. The warm water exits from the exchanger through a nozzle located at the top of the channel head.

Back-flushing

Trash, such as crabs, small stones, and paper cups, becomes lodged in the tube inlets. This reduces water flow and causes higher water outlet temperatures, which lower the exchanger duty. More significantly, the low tube-side velocity encourages precipitation of inorganic deposits and the lay-down of organic sludge.

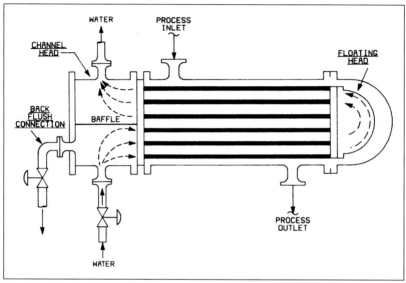

Fig. 16–1. Back-flushing a water cooler can unplug tubes.

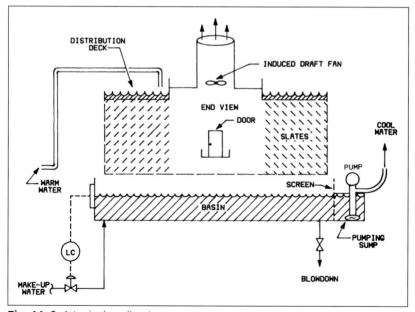

Fig. 16–2. A typical cooling tower.

Back-flushing is an effective method to open and clean plugged tubes. Figure 16–1 shows a properly installed back-flush connection. Note that the back-flush nozzle is the same size as the water-inlet nozzle (see chapter 12).

One should monitor the cooling tower basin water level when back-flushing an exchanger; do not empty this basin by back-flushing too long. Note that if the back-flushed water contains large pieces of debris, there is probably a hole in the cooling tower screen, located between the basin and the cooling water pump sump. It can easily be repaired without shutting down the pumps (see figure 16–2).

Air bumping

Injecting plant air into the cooling water inlet will agitate the tube interior and shake loose some deposits. Attach an air hose to the drain valve at the bottom of the channel head. If the water returning to the cooling tower becomes dirty, the air bumping is working. Air left in the cooling water return line will not hamper the cooling tower operation.

Back-flushing with air and water at the same time greatly increases the flushing action of the water. This is called air rumbling. For badly plugged exchangers, it will be necessary to close the cooling water outlet valve partially to force the air backward through the tubes. Some operators have injected sand-blasting grit in the back-flush water with favorable results.

Acid cleaning

After other methods of improving cooling have failed, acid cleaning should be considered. Certainly, this is an effective way of removing fouling deposits from tubes without taking the exchanger out of service. Unfortunately, refinery experience has shown the detrimental effects of acid cleaning on carbon steel piping used in cooling water service. There are a number of companies that specialize in acid-cleaning exchangers. To clean a water cooler, they will:

1. Determine the water flow through the exchanger. The operating engineer supplies the data, along with the tube material and suspected type of fouling deposit. This information is used to choose the type and amount of acid.

2. Pump acid from a truck to the exchanger inlet. The water circulation is maintained.

3. Inject a calculated quantity of a neutralizing chemical, such as caustic, at the exchanger outlet.

4. After acid washing (perhaps 30 minutes), return the cooling tower as a whole to a normal pH.

In one refinery, acid cleaning a single exchanger resulted in lowering the pH of the cooling tower pH to 3. Low-pH water is extremely corrosive to the welds made in carbon steel piping. In this instance, a dozen leaks in the underground cooling water piping resulted.

Regardless of the method employed, the operating engineer should check the heat-transfer coefficient before and after the cleaning. If repeated cleaning is necessary, avoid excessive use of acid cleaning, as eventually the cooling water piping will suffer.

Calculating Water Flow Rate

As an exchanger plugs, the tube-side pressure drop stays the same. The flow of water is diminished, however. To calculate the flow of cooling water:

1. Determine the shell-side duty in BTU/hr (Q).

2. Measure the tube-side temperature rise. Simply drain water from the inlet and outlet into a quart bottle with a glass thermometer (ΔT). The flow of water is then:

$$\frac{Q}{(\Delta T) \cdot 500} = GPM$$

Compare this number with the design water flow rate listed on the exchanger data sheet. If the water flow is low, and the pressure drop across the tube side is equal to or greater than design, the tubes are partially plugged. Low exchanger pressure drop indicates the cooling-water circulating pumps need to be overhauled.

Hydrocarbons Leaking into Cooling Towers

One day, an operating superintendent notes something peculiar going on in his unit's cooling tower. It appeared as if a geyser were erupting from the deck of one of the cooling tower's cells. Closer inspection showed that a jet of hydrocarbon vapor was blowing up through the water. The entire top deck of the seven-cell cooling tower was enveloped in butane vapor.

It was later determined that 4,000 B/D of isobutane were leaking through a hole in the floating head of a condenser into a cooling water system. The idea in troubleshooting hydrocarbon leaks is to find them while they are still small. Large leaks can ignite and burn down the cooling tower. There are four indications of hydrocarbon leaking into a cooling tower:

- **Biological growth is promoted.** Slime accumulates on the cell decks or the rate of chlorination needed to maintain a residual chlorine concentration is increased. (Amine or other organics leaking into the cooling water will have the same effect.)
- **A gas test meter** shows combustibles are emanating from the cooling water return-line discharge distributor.
- **A heat haze rises from the decks.** This is not a heat haze at all, but rather high concentrations of evaporating light hydrocarbons.
- **A shaking cooling water return line,** which moves because of expanding vapor in the lines.

Which exchanger is leaking?

Having determined that hydrocarbons are entering the cooling water return header, one will want to find which exchanger is the culprit. Open the ¾-inch vent on top of the channel head (refer to figure 16–2). Then, without immersing the probe in the flowing water, check for combustibles with a gas test meter.

If a leaking water cooler cannot be isolated without shutting down an entire process unit, it is a common refinery practice to let it leak. However, if you can see a hydrocarbon haze rising from the cooling cell decks or if the gas test meter shows a localized concentration in the explosive range, the leaking exchanger should be removed from service without delay.

Warm Cooling Water

An exchanger with a high water exit temperature is suffering from either plugged tubes, a low water pump discharge pressure, or increased exchanger duty. On the other hand, a high water-inlet temperature is indicative of cooling tower deficiencies. (Refer to figure 16–2, which depicts a typical induced-draft cooling tower.)

To determine if there is anything wrong with the cooling tower, check the wet bulb temperature. Tie a piece of wet cloth around the end of a glass thermometer and twirl it until a minimum temperature is observed. The water pumped out of the cooling tower basin should be 5°F–15°F warmer than the wet bulb temperature. Check the vendor's design specification for the expected approach temperature to the wet bulb temperature. A typical performance curve supplied by one vendor is shown in figure 16–3. If the water is warmer than predicted by the performance curves, there is something wrong. Do the following:

1. Walk through the inside of the tower (there is a door in the end of most cells). See if the slates (also called fill) are deteriorated.

2. Try to redistribute the water flow to each cell so that the height of water on each cell's deck is the same. There are block valves provided on each cell water-inlet line for this purpose.

3. Check each deck for large holes. Water pouring through large holes is not effectively cooled by the up-flowing dry air.

4. See whether the distribution holes on the deck are plugged. Is there a large amount of water spilling over the sides of all the cell decks? If the deck is covered with a thick slime, increase the rate of chlorine addition until a residual chlorine concentrate is observed. Do not worry about the white foam that will appear; this is normal. Sometimes it may be necessary to clean debris manually from the holes in the deck. If, after cleaning the decks and rebalancing flow to the cells, water is still spilling over the side, the cooling tower itself is probably overloaded.

High exchanger outlet temperature

If the water effluent from a cooler is unusually warm but the inlet temperature is about normal, check the pressure on the discharge of the cooling water circulation pump. Low cooling water pressure is sometimes due to deterioration of the cooling water pump's impeller or wear rings (see chapter 11). A badly fouled screen on the inlet to the pumping sump (see figure 16–1) will also reduce the efficiency of the pump. If the level in the pump side of the sump is more than a few feet below the water level in the basin, the screen should be cleaned.

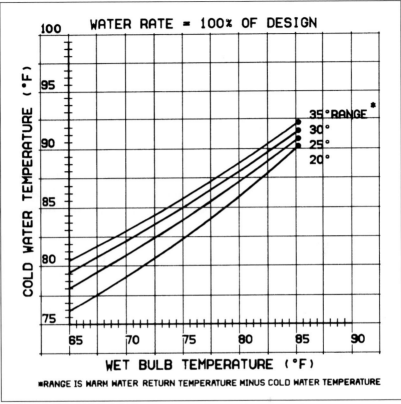

Fig. 16–3. A typical performance curve for a water cooling tower.

Water-Side Fouling

A cooling tower treatment program has three objectives: 1) Minimize hardness deposits on exchanger tubes; 2) minimize biological growth in cooling water; and 3) reduce exchanger corrosion. Formulating a detailed treatment program is best left to the professionals in the business—the companies that supply the treatment chemicals. Included in the price of these chemicals is a charge for technical service. There is a simple method to secure first-class technical assistance from a chemical vendor: ask one of his competitors to review the unit's treatment program and recommend an alternate. The chemical vendors do not market any wonder ingredients, so do not hesitate to switch to another supplier if technical assistance is not satisfactory.

Biological growth

The chemical vendor will supply a biocide additive. The refinery will be responsible for adding chlorine. Controlled addition of biocide and chlorine will limit biological growth and prevent the buildup of slime in the exchanger tubes. Be careful to avoid excessive use of biocide. In one refinery biocide killed the "bugs" in the activated sludge section and was therefore responsible for a major upset at the effluent treating plant.

Excessive use of chlorine leaches out the lignite content of the cooling tower's wooden components. Therefore, maintaining a continuous high residual concentration of chlorine reduces the useful life of a cooling tower. If one attempts to control biological growth solely with biocide, however, the "bugs" in the cooling tower will mutate and become resistant to the biocide.

pH control

The pH of cooling water naturally increases due to evaporation. Sulfuric acid is added as required to control the pH at 6.5–7.0. An excursion of high pH (9–10) will cause a rapid increase in inorganic fouling deposits in the exchanger tubes.

On one unit the automatic pH controller failed and excess sulfuric acid was charged to the cooling tower. The low-pH water caused numerous leaks in the cooling water piping. Only when water was seen to be spraying out of the pipes did operating personnel realize they were circulating 4-pH water. Substantial quantities of caustic were injected to the cooling tower basin to neutralize the jetting water.

Unfortunately, too much caustic was used and the water's pH jumped to 9. The iron, which had been dissolved through the corrosive action of the low-pH water, came out of solution. The cooling water turned brownish-red. Rapid exchanger fouling was the end product of this incident.

Poor pH control will undermine any vendor's water treatment program. Corrosion follows from too much acid addition, and fouling is a consequence of insufficient acid injection or caustic spills into the cooling water system.

To protect exchangers from the effects of low-pH water, some refiners have an alarm that sounds in the control room when the pH falls below a preset level. When the alarm goes off, acid is blocked away from the cooling tower. Then the basin blowdown rate (see figure 16–1) is increased to a maximum, consistent with the capacity of the water make-up control valve. Neutralizing caustic should be added as a last resort, and then only with extreme care.

Cycles of concentration

The operating engineer will be under constant pressure from the environmental engineer to maximize a cooling tower's cycles of concentration. The cycle of concentration is a measure of the amount of cooling water lost to the sewer, calculated as follows:

$$\frac{\text{calcium in circulating cooling water}}{\text{calcium in make-up water}} = \text{cycles of concentration}$$

As the circulating water evaporates, its calcium concentration increases. Usually, a cycle of concentration of four is fine, with six indicating outstanding control of cooling water spilled to the sewer. The higher concentration of calcium promotes hardness fouling of exchanger tubes. However, a proper water treatment program will control this fouling at an acceptable level.

Monitoring Exchanger Fouling

Maintaining technical records is a large part of an operating engineer's job. When it comes to troubleshooting a problem of reduced cooling capacity, records of historical heat-transfer coefficients are valuable. In addition to pinpointing when fouling suddenly increased, a plot of heat-transfer coefficient with time permits rational planning of an exchanger cleaning program. The heat-transfer coefficient is calculated as follows:

$$U = \frac{Q}{A \cdot \Delta T}$$

where:

Q = Heat duty, BTU/hr

A = Exchanger surface area, ft^2

ΔT = Temperature driving force

The temperature driving force is calculated as follows:

$$\Delta T = \frac{(To - ti) - (Ti - to)}{\ln \dfrac{(To - ti)}{(Ti - to)}}$$

where:

To = Hot-side outlet temperature

Ti = Hot-side inlet temperature

to = Cold-side outlet temperature

ti = Cold-side inlet temperature

ln = Natural log

The ΔT calculated above is for a true countercurrent flow. A correction factor for deviation from countercurrent flow is needed when checking a multitube pass exchanger. Procedures and charts for calculating this correction factor may be found in the TEMA standards book.[1]

Evolution of Air from Cooling Water

The condenser shown in Figure 16–4 is elevated 40 feet above the cooling water supply and return headers. Especially on older units, condensers are elevated for efficient drainage of the condensate into the reflux drum. The cooling water supply must then flow uphill by 40 feet. As shown in figure 16–4, the combined effect of the rise in elevation and the exchanger ΔP results in a cooling water outlet pressure of 15 inches of mercury. Also, the water has been heated from 70°F to 120°F. Heating the water and reducing its pressure reduces the solubility of air in the cooling water. This is logical as the water was just in contact with air in the cooling tower. The air flashes out of the water. This swells the volumetric flow leaving the water side of the condenser outlet.

The generation of bubbles of air in the condenser tubes and in the water outlet line chokes off the flow of water. Any dissolved light hydrocarbons in the circulating water makes the problem worse.

I attached a hose to connection B (figure 16–4) and ran the hose down to grade. Air and water flowed out together from my hose into a bucket of water.

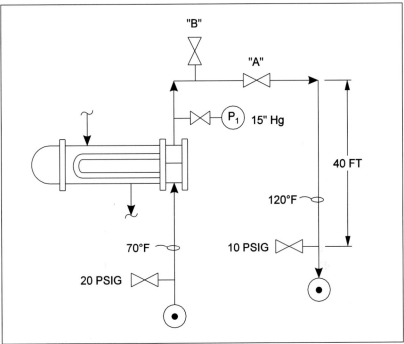

Fig. 16–4. Loss of cooling water flowing due to evolution of air at vacuum conditions.

The air-encumbered water flow is reduced by the air and raises the water outlet temperature. This also reduces the process heat transfer rates. To restore the heat transfer condenser efficiency, you should partly close valve A. It is not desirable to close this valve enough to eliminate the partial vacuum at P_1. In my experience, closing valve A so that the water pressure at P_1 reaches about 3 to 5 inches of mercury is optimum. By optimum I mean that the water outlet temperature drops to 3°F to 5°F. Closing valve A further (Figure 16–4) will increase the water outlet temperature. Of course, each system is different and the optimum pressure at P_1 can only be found by field experimentation.

A True Story from Lithuania

I was working on a condenser problem in the Mazaikiu Naphtha Refinery. The operators were restricting the tower top reflux rate to a splitter because of the following rational:

- During the summer, cooling water gets hotter.
- Then the cooling water outlet temperature from the splitter's overhead condenser rises above 50°C.
- Above 50°C (122°F), the lay-down of hardness deposits (i.e., calcium salts) inside the condenser tubes is accelerated.
- The hardness deposits reduce the cooling water flow through the tubes.
- The lower water flow raises the water outlet temperature.
- The higher water-side outlet temperature raises the rate of calcium salt deposits inside the tubes.
- A positive feedback loop has been created. That is, the problem feeds upon itself.
- To prevent the creation of this positive feedback loop, the splitter reflux rate must always be kept low in the summer.

The problem was lack of cooling water flow through the tubes. I checked the water inlet pressure. It was 3 bars (45 psig). This was close to the 3½-bar cooling water pump discharge pressure. I checked the cooling water outlet pressure. It was 2½ bars. The cooling returned to a water basin at zero bar as shown in Figure 16–5. The water supply and return lines were the same length and diameter. Why would the supply line ΔP be ½bar (i.e. 3½ bar minus 3 bar), but the return line ΔP be 2½ bar (2½ bar minus atmospheric pressure)? The only visible difference was that the return line dipped underneath a dirt road before it discharged into water basin.

I checked the pressure of the water just upstream of where the line passed under the road. The pressure was 2 bar. How could I lose 2 bar of pressure (29 psi) in just 20 meters of line, when I had only lost ½ bar (7 psi) in 1 kilometer through the same diameter line?

That evening the road was excavated. Hidden 2 meters below the surface of the road was a full line size isolation valve that was 90% closed. As the carbon steel valve had been buried for two decades, it could not be opened.

The maintenance division hot tapped two connections on either side of the stuck 400 mm valve. A 300-mm bypass pipe was flanged up into place. When the bypass was opened, the pressure drop across the stuck isolation valve dropped to about ½ bar.

You might think—as I had—that removing half the frictional loss through the system would increase the water flow by about 40% (i.e., flow varies with the square root of pressure drop). But this was not the case. Sadly, the observed water flow (based on a flow meter on the P-1 pump discharge, shown in Figure 16–5), only increased by 15%. But why?

The problem was that we were operating well out on the P-1 pump curve. Thus, a 15% increase in water flow caused the pump discharge pressure to drop by about 1 bar. Strangely, none of the operators in Lithuania, some of whom were there for the plant startup by the Russians 20 years ago, recalled ever seeing the buried valve.

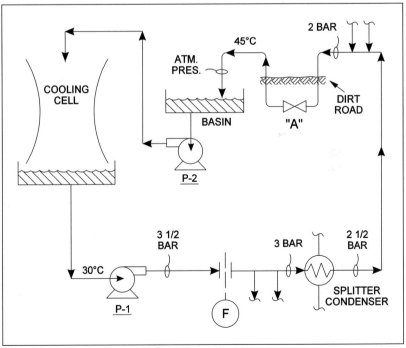

Fig. 16–5. Valve A was buried underneath a road and mostly closed.

Troubleshooting Checklist for Water Coolers

Plugged Tubes
Back-flushing
Air-bumping
Acid-cleaning versus piping corrosion
Upsetting cooling tower pH due
  to acid-cleaning
Sand-washing

Hydrocarbon Leaks
Watch for hydrocarbon haze
Rapid biological growth on cooling
  tower cell decks
Check for gas with test meter on
  cell decks
Vibrating cooling water lines
Check vent on exchanger channel
  head for leaks

Cooling Tower Deficiency
Check wet bulb temperature
Inspect interior for damaged fill
Eliminate large holes in
  distribution decks
Redistribute water to individual cells
Unplug distribution holes
Increase chlorination rate
Use biocide

High Exchanger Water Outlet Temperature
Cooling water pump deficiency
Plugged cooling tower screen
Plugged exchanger tubes
Plugged floating head

Water-Side Fouling
Hardness deposits
Poor pH control
Biological deposits
Insufficient chlorine addition
Routine tabulation of heat-
  transfer coefficients
Review vendor's
  treatment program
Cycles of concentration

Reference

1. *Standards of Tubular Exchanger Manufacturers' Association,*
   4th ed., 1959.

Alarms and Trips

An upset on a process unit can cause a shutdown if prompt, corrective action is not taken. Even worse is the possibility that the unit may self-destruct if an abnormal condition is allowed to continue.

An alarm alerts the operators that something has gone awry in the process. A trip shuts down an endangered piece of equipment when it is too late for human intervention. When a trip fails to function, severe damage to a refinery process unit can result.

The process operating engineer is often given the job of investigating why such an emergency condition has shut down a refinery unit. This assignment will include formulating a plan to prevent similar incidents. Understanding how alarms and trips function, what can go wrong with them, and how they are tested onstream is a prerequisite to troubleshooting and preventing unit accidents.

Steam Turbine Trips

During the Nazi invasion of Russia in World War II, Soviet engineers were instructed to destroy turbogenerators before retreating. First, the trip mechanism was wired so that it would stay latched. Next, the turbine was run up to maximum speed. Finally, the pump circulating lube oil to the turbine bearings was switched off.

Such a sequence of similar events can occur accidentally, and the refining industry has a significant history of major equipment failures because of them.

Trip mechanism

Steam turbines are designed to trip off due to overspeed, excessive vibration, low suction pressure, loss of lube-oil pressure, and so on. Turbines are tripped off by releasing the tension on a spring. Normally, this spring holds a trip-throttle valve open in the steam inlet line.

The lever that maintains tension on the spring is called the trip. This trip is latched when the turbine is rotating. When the turbine is in trouble, a mechanism unlatches the trip and the spring immediately slams the trip-throttle valve shut. A problem arises when the trip-throttle valve sticks. Typically, solids in poor-quality steam build up, fouling the valve. When the trip unlatches, the throttle valve may not close.

Exercising the valve parts is the way to prevent sticking. You can manually close the trip-throttle valve a half turn without activating the trip. If this is done once a week, sticking can often be avoided.

The best prevention of a malfunction of the trip-throttle valve is periodic testing. On one large alkylation unit's refrigeration compressor, the following procedures were implemented every month:

1. The lube-oil pressure to the mechanism activating the trip-throttle valve was intentionally bled down.
2. If the throttle valve itself didn't slam shut, it was manually closed with the handwheel until it broke loose.
3. The trip was immediately re-latched, and the throttle valve wound back open.

Running without refrigeration for a few minutes didn't affect the alkylation unit. For processes that cannot tolerate such an outage, the turbine throttle valve may be wired open or gagged. The gag prevents the throttle valve from closing all the way during testing. However, never gag open a trip during normal operations.

Emergency lube-oil pumps

Frequently, a major unit shuts down when a compressor trips off. The cause may be low lube- or seal-oil pressure. The pumps providing the lube- and seal-oil pressure are typically spared (backed up). A steam turbine-driven pump is normally backed up by a motor-driven spare. The motor-driven pump should come on when the lube-oil pressure drops below a preset point.

How does one check the motor-driven lube pump for operability? First, note that the local pump control switch has three positions: On or Hand, Auto, and Off. The control switch should normally be set on Auto. Briefly switch the pump to On to determine if the pump itself is working. Then turn the switch back to Auto.

Now slowly reduce the turbine driven lube oil pump discharge pressure by turning the governor speed controller counterclockwise. Maintain this pressure well above the compressor trip pressure but below the lube-oil pressure switch's set point. You can then see if the motor-driven lube- or seal-oil pump will come on when it should.

In practice, back-up emergency equipment that is not routinely tested will eventually become nonfunctional. An alarm or trip—or, for that matter, any item of refinery equipment—will become unreliable through disuse.

Liquid Levels

A boiler generating 600-psi steam was destroyed when a laborer, cleaning up around the boiler's feed water pump, accidentally pushed the pump's stop button. The water level in the boiler started dropping. The low-level alarm sounded in the control room. An operator was duly dispatched to investigate. He reported that the liquid level glass was either completely full or empty (i.e., he could not see any interface). The chief operator went out to check for himself. Shortly thereafter, a boiler tube ruptured and high-pressure steam escaped into the fire side.

The flow of fuel to the boiler should have been stopped automatically when the water level fell to a dangerous level. Unfortunately, the low-level trip failed to function. (Many newer boilers come equipped with backup low-level trips.) The operating engineer investigating this incident determined that this trip had not been tested for years. He set up the following program to prevent a recurrence. First, the fuel gas control valve was locked into its normal operating position with its hand jack. Then the low-liquid-level trip pot (see figure 17–1) was blocked in and drained down. Finally, an operator verified that the pneumatic signal to the fuel-gas control valve fell to zero. (This valve was AFC—air failure closes.)

Many trip valves are separate from control valves. They can often be jammed open with a bushing prior to testing the trip circuit for operability.

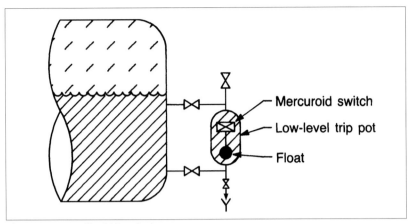

Fig. 17–1. Low-liquid-level trip.

A routine shutdown of a piece of equipment offers an excellent opportunity to test alarms and trips. For example, if a turbine is to be shut down for maintenance, its speed can be carefully brought up to see if the overspeed trip is operative. It is no exaggeration to state that if an overspeed trip is inoperative, the turbine should not be run.

High-liquid-level alarms

Some of the more common functions of a high-liquid-level alarm in a refinery are:

- Protecting a compressor from slugs of liquid carried over from a knockout drum.
- Alerting operators that a boiler's steam drum is overflowing.
- Protecting gas-fired heaters from slugs of liquid hydrocarbon because the fuel-gas knockout drum is overfilled.

Such alarms are easy to check. First, individually blow down the taps connecting the alarm pot to the vessel. Having determined that these are unplugged, block the taps in. Next, remove the plug from the top of the alarm pot so the pot is vented to the atmosphere. Then crack open the bottom tap and slowly fill the pot to overflowing. This should activate the alarm in the control room.

Unfortunately, periodic routine checking of high-level alarms in this manner is uncommon. What usually transpires is that a high level in a process vessel goes undetected until a unit upset occurs.

The operators then observe that the alarm has not functioned. Since it is better late than never, a work order is issued for the instrument mechanic to investigate the cause of the failure.

Distillation Tower Overpressured

A depropanizer was designed to operate at 270 psig on a corrosive stream with a high fouling tendency. It was protected from overpressuring by a relief valve set at 300 psig. On one occasion the operators found that the depropanizer feed pump could not maintain flow to the tower. An operator went outside to investigate. He found that the tower pressure was 450 psig.

The cause of this near disaster is shown in figure 17–2. Tower pressure was held by adjusting steam flow to the reboiler. A single pressure transmitter signaled both the high-pressure alarm and the pressure control valve (i.e., the reboiler steam inlet valve).

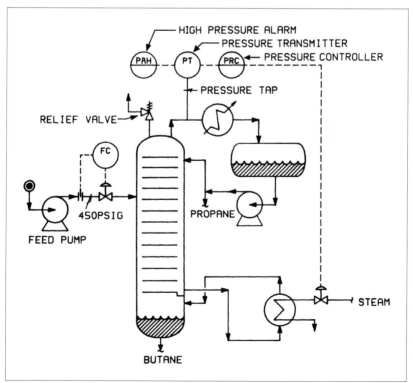

Fig. 17–2. Poorly located pressure alarm.

The increased tower boil-up overpressured the tower. Since the high-pressure alarm depended on the same tap as the pressure controller, no alarm was sounded. The relief valve, which was set to pop open at 300 psig, did not open even at 450 psig. Corrosion products had completely plugged the relief valve. (A rupture disk is one common method used to protect relief valves from plugging.)

As the tower pressure rose, the feed pump, shown in figure 17–2, was backed down. Only when the depropanizer feed flow was lost did the operators suspect a problem. A recurrence of this potentially serious incident was prevented. First, the tap for the high-pressure alarm was switched to the reboiler return line. This tap was blown out once a week to ensure that it was not plugged. Also once a week, the high-pressure alarm transmitter was pressured up to 300 psig with bottled nitrogen to see if the alarm would actually sound.

This incident illustrates a concept generally applicable in troubleshooting alarm and trip features: The sensing element for an alarm or trip must be separate from the sensing element used for control purposes.

Many refiners will actually alarm from the same transmitter generating the recorded signal. For example, in figure 17–2, the high-pressure alarm was activated from the pressure control valve transmitter. Surprisingly, this is not uncommon.

Automatic Temperature Shutdowns

A malfunctioning high-temperature furnace trip caused a major tray upset in a large ethylbenzene fractionator.[1] Figure 17–3 shows the tower control system. When the circulation pump briefly failed, the flow of hot vapor to the tower was interrupted. The tower temperature controller was called for more heat, and furnace fuel firing consequently increased. At this point, the automatic furnace shutdown system should have tripped off the furnace fuel due to the excessive furnace outlet temperature. This trip failed to function.

After a few minutes, circulation was again established. However, the heater tubes were not allowed time to cool, and an extremely high vaporization rate resulted. The high vapor rate produced a pressure surge that dislodged the tower trays.

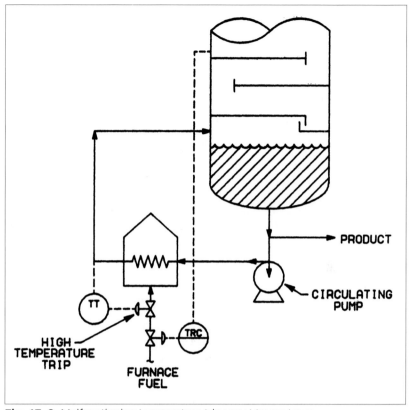

Fig. 17–3. Malfunctioning temperature trip upset tower trays.

This incident could have been prevented by a routine check of the high-temperature trip. A straightforward procedure to do this is:

1. Temporarily place an object in the fuel-gas trip valve that will jam it halfway open when it is activated.
2. Unscrew the thermocouple used to activate the trip from its thermowell.
3. Connect a portable TI to this thermocouple.
4. Expose the thermocouple to a source of heat sufficient to obtain the fuel-gas trip temperature. Check this temperature with a portable TI.
5. Observe that the fuel-gas trip valve is actually activated.

This procedure leaves no doubt that the temperature-trip system will work in an emergency.

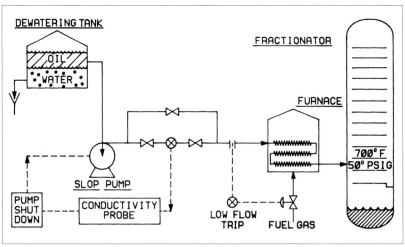

Fig. 17–4. Water slugs trip pump to protect fractionator.

Water Shot Damages Trays

All refineries generate a certain amount of heavy, wet hydrocarbons of variable composition called "slop." The oil recovered from refinery sewers is the most common source. The slop is dewatered and rerun. A typical processing scheme is shown in figure 17–4.

A major problem with rerunning wet hydrocarbons is unexpected slugs of water that appear in furnace feed. Such slugs of water are only partially vaporized in the preheat furnace. Upon entering the hot fractionator flash zone, the water suddenly flashes. Pressure surges, powerful enough to upset the fractionator trays, are a frequent consequence.

A common method to guard against this type of incident is a conductivity probe, installed as shown in figure 17–4. Water or a hydrocarbon-water emulsion conducts electricity far better than oil. Increased electrical current flow through the conductivity probe indicates the presence of water. A signal from the probe can then be used to shut down the slop pump.

The operating engineer troubleshooting tray damage in a fractionator exposed to slugs of water should check the conductivity probe trip for operability. If no such trip exists, installation of such a device should be considered. Testing a conductivity probe trip is simple.

1. Block in and bypass the trip.
2. Remove the conductivity probe from the charge line.

3. Immerse the probe in a bucket of water.

4. Restart the slop pump as soon as it trips off.

Losing flow for a few seconds won't significantly upset downstream equipment, provided one has taken the precaution to bypass any low-flow trips. The exception to this is the radiant tubes in a furnace, which might coke up or overheat quickly when flow is lost.

Low-Flow Trips

Centrifugal pumps are subject to mechanical damage when they lose suction pressure or when the discharge flow is blocked. Furnace tubes may overheat and fail when liquid flow through the tubes is greatly reduced. Low-flow trips protect process equipment against these failures. This trip may be used to shut down a pump or block off fuel to a furnace. To test an orifice-type low-flow trip, proceed as follows:

1. As previously described, temporarily prevent the low-flow trip from shutting down other parts of the process by use of hand jacks, gags, etc.

2. Blow clear both the high- and low-pressure legs running from the orifice taps to the flow meter (see figure 17–5).

3. Open the orifice taps' bypass valve. The indicated flow will then drop to zero, and the trip should be activated.

Paddle-type devices are sometimes used for low-flow trips. These have a somewhat lower reliability than orifice-flow trips.

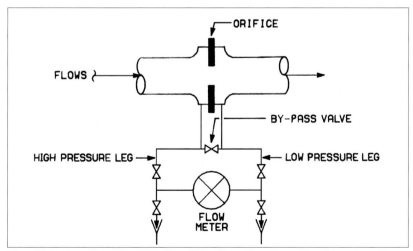

Fig. 17–5. Open bypass to test low-flow trip.

Eliminating Unnecessary Trips

The operating reliability of many process units is impaired because of defective trips. For example, the flame scanner on a furnace will shut off the fuel gas when the scanner no longer "sees" the flame. Too often, the flame is still on, but the scanner's window has become dirty.

There is a great temptation for operating personnel to bypass such trips permanently. Refinery management, however, takes a dim view of abandoning safety trips that are supposed to protect both lives and equipment.

On a sulfur recovery plant, excessive tripping of the reaction furnace feed was reducing unit reliability. The process engineer was asked to investigate. He found that all of the tripping incidents were due to the low-air flow trip. When the compressed air flow dropped below 1,000 SCFM, both the hydrogen sulfide (i.e., fuel) and air flow would automatically be stopped. The trip was designed to prevent the formation of a combustible mixture in downstream vessels when air flow to the reaction furnace stopped. With no air, there would be no combustion in the furnace.

In reality, however, there was never any loss of combustion until the trip shut off the air and H_2S. The indicated low air flows were just instrument and meter-related malfunctions. Therefore, minor instrument problems were causing major process problems by activating the low-air-flow trip.

But should this trip be permanently removed? The determining factor in answering this question is the response taken by the operators when both air and hydrogen sulfide flows were interrupted. They did not reestablish combustion by relighting the pilot light, as the designers envisioned. This would have taken many hours and so was impractical. The operators proceeded as follows:

1. All unit trips were temporarily bypassed.
2. Air and hydrogen sulfide flows to the reaction furnace were manually reestablished.
3. The hot bricks in the furnace were used to reinitiate combustion.

The process engineer concluded that the response of the operators to the false low-air flow trip was creating the very hazard that the trip was supposed to prevent, i.e., the accumulation of a combustible mixture in downstream vessels.

If, in actual refinery practice, a trip is creating a hazard that the designer did not anticipate, the process engineer may have to recommend discontinuing the use of the trip to management. The alternate course of improving the reliability of the trip to a satisfactory level may not always be feasible.

Lights Out on Alarm Panel

Operators occasionally deactivate alarms that go off too frequently and which they do not consider important. This is done by pulling a card inside the control-room alarm panel. To determine if any alarms have been illicitly deactivated, push the test button on the control-room board. Any alarms that do not light up have had their cards pulled or need new bulbs. An alarm that will not go off is a potential hazard.

On the other hand, alarm lights that are almost always on also create an indirect danger. The appearance of an alarm panel always lit up leaves the impression among the newer operators that some alarms can be ignored. If an alarm light remains on, have its card pulled until the reason for its being routinely illuminated is rectified. The philosophy for current panel alarms on a computer console is the same as for the older pneumatic control board.

A word of caution

Whenever a trip is to be tested by initiating a nonroutine operation for the first time, the engineer should visualize the result of a malfunction and use appropriate safeguards to avoid equipment damage or injury to personnel. Look before you leap; management may not want to give you a second chance.

The outstanding example of this advice is the disaster at the former Soviet Union's Chernobyl nuclear power plant. The plant engineer disconnected the emergency reactor cooling water backup system to run a test. The reactor overheated, which caused the control rods to jam. The result is a huge wildlife refuge in the middle of the Ukraine.

Troubleshooting Checklist
for Alarms and Trips

Steam Turbines

Exercise trip-throttle valve

Watch for solids in driver steam

Gag throttle valve and unlatch trip

Test lube-oil backup pump

Liquid Levels

Cross-check gauge glass level with
　control room level

Drain low-liquid-level trip pot to test

Blow out liquid-level pot taps

Fill high-level alarm pot to test

Pressure

Blow dirt from pressure taps

Locate pressure alarm on its own tap

Locate pressure recorder
　on separate tap

Test pressure transmitter
　with bottled N_2

Temperature

Impress trip millivolt valve
　at thermocouple head

Check thermocouple with
　portable potentiometer

Heat thermocouple
　to activate trip

Eliminate Unnecessary Trips

Check operator response
　to unnecessary trip-outs

Evaluate effect on process
　if trip is absent

Operate with alarm lights
　normally off

References

1. C. H. Kilgore, "Report on Ethylbenzene Tower Explosion—
   Tenneco Oil Company," presented at API Meeting, Operating
   Practices Committee, Salt Lake City, Oct. 10, 1979.

section **3**

Practical Problems

"The price we pay for success is the willingness to risk failure."

~ Michael Jordan

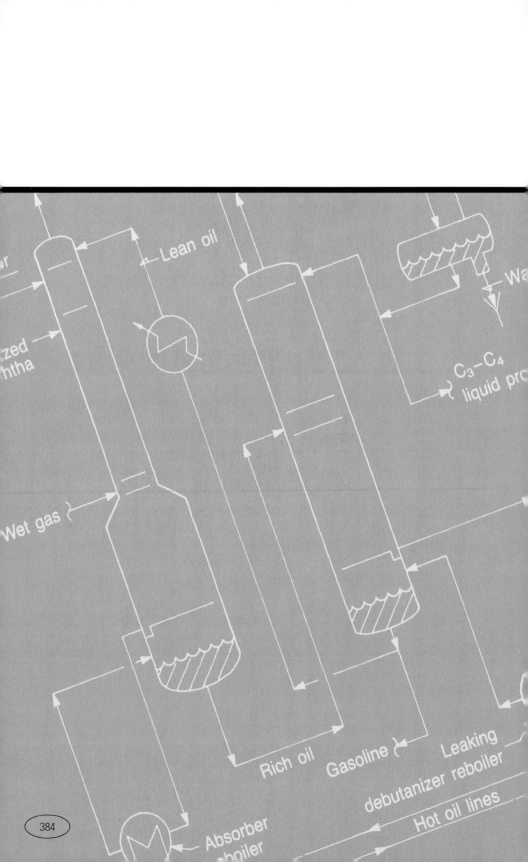

Additional
Distillation Problems

T he distillation of light hydrocarbons is an integral feature of almost every refinery gasoline-producing unit. A typical product slate from such a plant might be:

- Propane for LPG
- Isobutane for alkylation plant feed
- Normal butane for gasoline-vapor pressure control
- Light naphtha for blending into gasoline or isomerization unit feed
- Heavy naphtha feed to a reformer unit

Each of these products has a specification to meet. In addition to consistent quality, each stream should be run down at a steady rate. Products transferred between units at erratic rates upset the operations of the downstream unit.

Reboiler-Induced Foaming and Flooding

Flooding of distillation tower internals is the usual cause for reduced fractionation efficiency. In theory, distillation efficiency should start to decline at 85% to 90% of flood. However, many columns begin to exhibit flooding characteristics at 60% to 70% of flooding capacity.

Field observations reveal a common cause of premature flooding in hydrocarbon fractionation to be high foam levels generated in the bottom of towers served by circulating reboilers. These high levels cannot be observed in the ordinary manner and so tend to elude detection.

Identifying the Incipient Flood Point

The term *incipient flood* is that point in a trayed tower's operation when the spray height of liquid from the tray below begins to impinge on the tray above to the extent that entrainment reduces fractionation efficiency. Incipient flood in a packed column is that point in the column's operation at which liquid hold-up increases to an extent that reduces fractionation efficiency.

From an operator's point of view, incipient flood is observed as follows:

- An increment of reflux and reboiler duty does not improve fractionation and may even worsen the split.
- The temperature difference between the bottom and the top of the tower is reduced as the reflux and reboiler rates are increased, and the tower's temperature profile becomes erratic.
- A small, slow reduction in tower pressure reduced fractionation efficiency.
- A small increase in reboiler duty causes a disproportionately large increase in the tower ΔP, which also becomes erratic.

Experienced plant operators often refer to this situation as the column's "prime point," "jugging," or "optimum point." My term for these symptoms is *incipient flood*. For many refinery naphtha splitters, debutanizers, or depropanizers, the incipient flood point (i.e., the capacity at which the tower works best) should be reached at 80% to 90% of the calculated jet flooding capacity. Many towers, however, reportedly work best at a capacity of only 60% to 70% of flood.

High liquid levels cause flood

My field experience has shown that the most common cause of flooding in mechanically intact towers is fouling. A close second is high bottoms liquid levels. For towers served by forced circulation reboilers, a high liquid level will cause the tower to flood.

When the liquid level in the bottom of a column rises to the reboiler return nozzle, the liquid in the bottom of the column is forcibly lifted by the reboiler vapors. The entrained liquid is blown against the underside of the bottom tray. Since flooding will progress up a tower, a liquid level covering the reboiler return nozzle will cause the entire fractionator to flood.

For most once-through and many circulating thermosiphon reboilers (see figures 18–1 and 18–2), a high liquid level covering the reboiler vapor return nozzle will retard thermosiphon circulation and cause a precipitous loss in reboiler duty, instead of tower flooding.

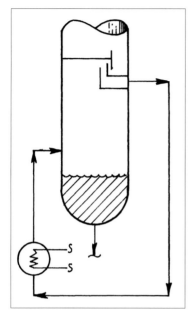

Fig. 18–1. A once-through thermosiphon reboiler.

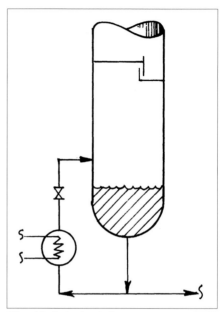

Fig. 18–2. A circulating thermosiphon reboiler.

Effect of foam

Systems subject to low surface tension (de-ethanizers, refrigerated absorbers) or particulates (amine regenerators, sour water strippers) are known to flood prematurely due to foaming. However, many other relatively clean, higher surface-tension systems also appear to cause premature flooding.

Foam-induced flooding has been considered a problem occurring on tray decks or inside packed beds. Certainly this is correct. For many columns, however, it is a high foam level formed in the bottom of the tower which causes premature flooding. When this foam level rises to cover the reboiler vapor return nozzle, flooding results.

This foam is generated in the reboiler. The amount of foam formed is a function of the system surface tension, particulates, the reboiler duty, and most importantly, the circulation rate through the reboiler.

You may wish to perform the following experiment to dramatize this effect. Bring a pint of water to a rapid boil. Add a cup of particulates to the boiling water (finely crushed iron sulfide; cocoa; coke fines; etc.). The resulting foamover is an example of what happens to the reboiler effluent when corrosion products circulate through the reboiler.

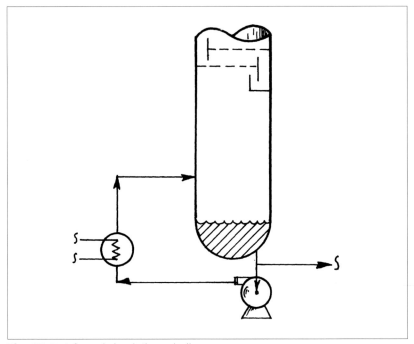

Fig. 18–3. A forced circulation reboiler.

My field experiments on a forced-circulation reboiled absorber (see figure 18–3) and a circulating thermosiphon reboiled depropanizer indicate that increases in the rate of liquid circulation through a reboiler also promote the formation of increased foam levels in the bottom of a tower. Apparently, the extra turbulence imparted to the boiling fluid by the increased liquid circulation accelerates foam formation in the reboiler.

External indications of foam

Figure 18–4 indicates split liquid levels—a sign of foam in the bottom of a fractionator.

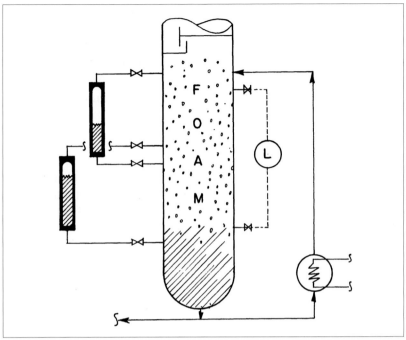

Fig. 18–4. Split liquid levels are a sign of foam.

Foam itself will not appear in a gauge glass assembly, or level-trol. When foam flows into a gauge glass, contact with the surface of the level assembly causes the foam to separate into liquid and vapor. Thus, only flat liquid can be observed in the gauge glass. The density of this liquid ranges from two to three times the density of the foam inside the tower. Alternately, the foam level inside the tower is two to three times higher than the liquid level in the gauge glass.

If foam entirely spans the level taps of a gauge glass, the settled liquid observed in the glass will not indicate the foam level in the tower; rather the level in the gauge glass will indicate the density of the foam as follows:

$$\frac{L_1}{L_2} = \frac{D(foam)}{D(liquid)}$$

where:

L_1 = Liquid level in gauge glass above lower tap

L_2 = Distance between taps

$D(foam)$ = Density of foam

$D(liquid)$ = Density of liquid

The multiple liquid levels shown in figure 18–4 and so often observed in operating units cannot, of course, indicate multiple or split liquid levels in the column. They simply indicate foam above the top tap of the upper gauge glass. The progressively greater levels in the lower gauge glasses show that the foam density progressively becomes greater in the lower portions of the column. Anyone who drinks a glass of beer with a head is familiar with this phenomenon.

Preflash towers

A common example of foam formation in the bottom of a fractionator which induces flooding occurs in a crude preflash tower. In this case, stable foam accumulates in the bottom of the column. Figure 18–5 shows a portion of a preflash tower. Once the foam level rises to the feed-inlet nozzle, the trays flood and black distillate is produced.

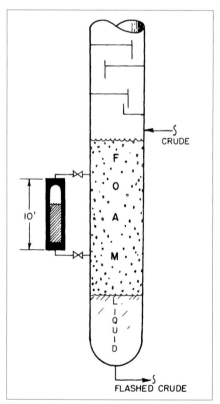

Fig. 18–5. Foam formation in the bottom of a crude preflash tower.

To observe the existence of foam above the top tap of the level controller, we proceeded as follows:

1. A pressure gauge was placed at the base of the tower.

2. The liquid level was lowered by 3 feet (i.e., 30%), based on the observed liquid level on the gauge glass.

3. The pressure indicated at the base of the tower dropped by 2 psi.

4. The specific gravity of the crude oil at the process temperature was 0.76. Therefore 2 psi of head loss equated to 7 feet of liquid actually pumped out of the tower.

This means that the foam level in the tower (assuming that the bottom 10 feet or so of the tower contained settled liquid) truly dropped by 7 feet. The density of the foam between the level taps dropped about 30%. The density of the foam inside the tower between the level taps before the experiment was started was:

$$D(foam) = \frac{3.0\,ft}{7.0\,ft} \times 0.76 = 0.33 \text{ sp gr}$$

While this procedure proves the existence of foam above the top tap of the gauge glass, it cannot find the true foam level inside the tower until the foam level is pulled down between the taps of the gauge-glass assembly.

Many refiners resort to employing gamma rays or a neutron source (K-Ray) to find the actual foam level in the bottom of such columns and then use a silicon defoamer chemical to control the foam level.

Flood Control

A tower served by a forced-circulation reboiler (as shown in figure 18–3) was fractionating poorly. A pressure-drop survey indicated flooding. The liquid level in the gauge glass was well below the reboiler vapor-return nozzle. The calculated percentage of flood was 70%.

The reboiler duty and reflux rate were not altered. The reboiler circulation rate was not changed. The indicated liquid level on the control panel was reduced from 58% to 32%. The tower ΔP dropped by 40% and the fractionation efficiency vastly improved. This experiment proved that foam initially was covering the reboiler vapor-return nozzle. It was the high foam level, and not any tray deficiency, that had caused the tower to flood. The indicated liquid level on the control panel was far below the true froth level in the tower. Lowering the liquid level an arbitrary amount restored fractionation efficiency.

The circulating thermosiphon reboiler shown in figure 18–2 served a propane-butane splitter that was flooding prematurely. Without a lowering of the liquid level or a change in other process parameters, the split between these hydrocarbons was greatly improved. The circulation rate through the reboiler was reduced by throttling on the reboiler effluent flow. This diminished agitation in the reboiler and thus minimized foam formation. As further proof, the outside shift operator reported that the usual tower bottoms "split" level had vanished, and only a single liquid level could now be seen in the gauge glasses.

Towers that lose capacity when the source of feedstock is altered are likely suffering from foam-induced flooding. My experience indicates that corrosion products (iron sulfide) are the usual culprits. A variety of extraneous chemicals (surfactants, products of amine degradation, oil field "soap sticks," pipeline flow promoters, anti-foulants), and any finely divided particulate matter all can cause foaming and flooding in distillation towers.

In summary, whenever a distillation tower's performance is improved by lowering the bottom's liquid level, foam formation should be suspected. If the apparent liquid level is well below the reboiler vapor-return nozzle before the level is reduced, the existence of a thick froth layer in the bottom of the column is most probable. Reduction in reboiler circulation is one proven method to diminish the foam height in the bottom of a tower. Injection of a silicon defoaming chemical is also effective in fighting foam.

Instrumentation Malfunctions

The purpose of a feed-forward control scheme is to improve the stability of a process. For instance, a sudden increase in feed rate will automatically increase the steam to the reboiler. A conventional controller would wait for the reboiler temperature to drop before bringing up the steam rate.

On one distillation tower (an isobutane-normal butane splitter), an advanced feed-forward control system was installed. The new controls were an instant success. However, after a few days the operators reported that gremlins had entered the new system and were playing malicious tricks. The butane splitter, without any prior notice, would suddenly go completely wild. Both the reboiler steam and reflux rates would change erratically.

After a few weeks, the mystery was solved. Whenever an electrical outlet in the control room was used, the feed-forward system was affected in the same way that running an electrical appliance interferes with TV reception.

Note: If a tower cannot be operated successfully in manual, regardless of operator time and attention, simply tuning the controllers will not help.

A more subtle but frequently encountered problem is attributed to plugged orifice-flow meter taps.

A debutanizer that was controlled by a feed-forward system operated perfectly. One day the debutanizer began to die; that is, the reflux and steam flows, the tower pressure, and the temperature all started to decline. An alert operator noticed the debutanizer's impending demise. He switched the reboiler steam flow control from automatic to manual and thus manually restored the steam flow back to its normal rate. This move resuscitated the tower.

The debutanizer's problem appeared to be chronic. The symptoms reappeared with greater intensity and frequency. After several weeks of unstable operations and 100,000 barrels of off-spec butane, management decided to shut the tower down for repair. The nature of the planned repairs was never specified.

Suddenly, a long-time shift operator hit on the solution: The orifice taps on the reboiler steam flow meter were plugged. The taps were drilled out. The tower's control immediately reverted to its former smooth operation. An investigation revealed what had happened.

The feed-forward control logic was designed to reduce the reboiler heat input when the debutanizer feed rate dropped. The measured steam flow was relayed back to the feed-forward controller. If the measured steam flow did not decline in step with the reduced feed rate, the controller signaled the steam flow control valve to continue closing. Then the plugged orifice taps caused a misleading steam flow rate to be transmitted back to the feed-forward controller. This controller continued to throttle the reboiler steam, trying vainly to reduce the measured steam flow rate. Of course, the actual heat input to the reboiler dropped precipitously. With reduced boil-up, the reflux flow dried up. As the tower cooled off, its pressure also fell.

Excess Feed Preheat

A distillation tower's reflux rate must correspond to the tower's heat input. The heat input is the sum of the feed enthalpy and the reboiler duty. For a tower run at constant reflux rate, the reboiler duty will decline as the tower feed enthalpy rises. This relationship is summarized as follows:

$$\Delta \text{ reflux duty} = \Delta \text{ reboiler duty} + \Delta \text{ feed enthalpy}$$

When the reboiler duty declines, the stripping vapor rate on the trays between the reboiler and feed point also drops. This reduces the ability of the tower to strip the light key component from the bottom product.

Too much feed preheat, therefore, can make it impossible to meet bottoms product specifications. That's why experienced operators hold a constant feed temperature. Unfortunately, for certain systems a very large increase in percent vaporized (and thus enthalpy) occurs for a small change in temperature. As an example, consider a tower with a feed composition of 2% propane, 72% butane, 24% pentane, and 2% hexane. A few degrees extra preheat in this tower's feed caused a large increase in the amount of butane vaporized. This led to a considerable drop in stripping vapor rate and a consequent increase in butane in the bottoms pentane product.

The tower's feed preheat source was exhaust steam of a variable pressure. Flow of steam to the preheater was neither metered nor controlled. A technical investigation revealed that as exhaust steam pressure (and condensing temperature) rose, the ability of the tower to fractionate declined. The problem was resolved by installing a flow control loop on the exhaust steam to the preheater. The tower instability—which had originally been thought to be an instrumentation failure—proved to be a process problem.

Slug Flow in Risers

Erratic distillation tower operation may be due to the uneven flow of a vapor-liquid mixture in a vertical run of pipe. An example of such an arrangement is the reflux drum-condenser combination pictured in figure 18–6.

The reflux drum is elevated above the condenser. The effluent from the condenser consists of a vapor-liquid mixture. The line between the condenser and drum is called a riser.

If the velocity in the riser is too low, the two phases will separate. A head of liquid will build up and create back pressure against the distillation tower. Periodically, a slug of liquid will be pushed through the riser and relieve the back pressure. The riser then gradually refills with liquid. This type of slug flow causes fluctuating tower pressure. Other instances of slug flow in distillation service are thermosiphon reboiler outlets and partially vaporized feed lines.

If one finds that increasing the rate or temperature in a riser improves the operating stability of a tower, slug flow should be suspected. To confirm this suspicion, calculate the linear velocity in the riser. A rough rule of thumb is above 25 ft/sec, slug flow is unlikely; below 15 ft/sec, slug flow is likely.

To overcome slug flow instability, some distillation columns are equipped with dual risers. A small-diameter riser is used at low throughputs. At higher rates, the larger diameter riser is put into service.

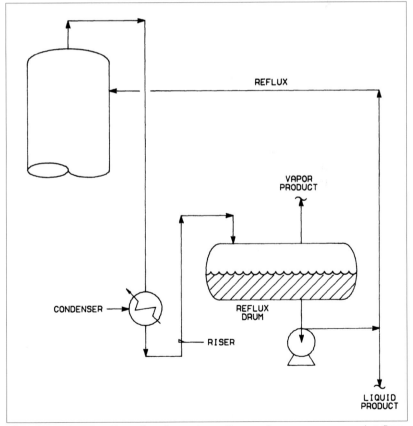

Fig. 18–6. Elevating the reflux drum above the condenser may cause slug flow in the riser.

Hydrocarbons in Steam

The troubleshooter should always be prepared to encounter the unexpected. In a refinery the integrity of any system is suspect. Even the composition in an ordinary low-pressure steam header should be questioned.

On one unit a depropanizer was reboiled with 30-lb steam. One morning a small feed preheater, also using 30-lb steam, was put in service. Soon after, the ability to control the depropanizer pressure was lost. Figure 18–7 illustrates the process flow.

Both the tower pressure and reflux rate declined simultaneously. Steam flow to the reboiler became extremely erratic. The overhead and bottoms products went off specification. At first the operators thought they had experienced an instrument failure. All the controllers were switched from automatic to a manual mode of control. This did not help.

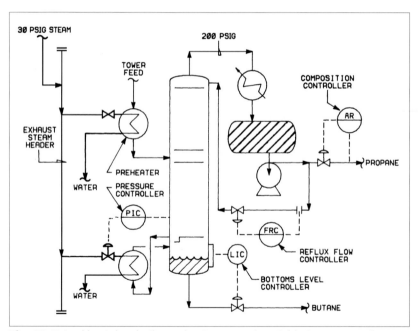

Fig. 18–7. Leaking tubes in the preheater caused unstable tower operation.

As a rule, if stable operations cannot be restored by running a tower on manual, it is safe to assume that there is a fundamental process problem and not simply an instrument failure.

At this point the process operations engineer was consulted. Observing that the reboiler steam flow was the most erratic process variable, he checked the reboiler.

He noted that the reboiler duty was partially restored when the vent on the top of the channel head (i.e., the steam side) was cracked open. He checked the vented steam with a gas test meter and found that it contained several percent hydrocarbons. Concluding that the reboiler had developed a tube leak, he advised management to shut down the tower to repair the reboiler.

The depropanizer was taken out of service and the reboiler disassembled. It was found to be in perfect condition. The reboiler was put back together and the tower returned to service. The tower immediately resumed unstable operating characteristics. What was to be done next?

The operating engineer obtained a sample of the vented gas from the channel head and submitted it to the lab for gas chromatograph analysis. The lab results were 15% propane, 30% isobutane, 50% normal butane, and 5% pentane.

The engineer recognized this component distribution as being identical to the depropanizer feed composition. He then recalled that the tower instability problems started on the same day that the feed preheater was put online. The operating engineer had the steam-inlet block valve to the preheater closed. Within 10 minutes, steam instead of gas began blowing out of the reboiler vent; the steam flow to the reboiler lined out and the depropanizer pressure began rising. An hour later the tower operation was smooth and stable.

The preheater had several leaking tubes. Liquid hydrocarbon had backed into the exhaust steam header through these damaged tubes and flashed to a vapor. The light hydrocarbons flowed into the depropanizer reboiler along with the low-pressure steam. The propane and butane were, of course, much too volatile to condense. They filled up the reboiler tubes and thus reduced the ability of the reboiler to transfer heat between the condensing exhaust steam and the depropanizer.

The point of this story is that the integrity of any piece of process equipment should never be taken for granted. The tubes in heat exchangers are prone to leak. This point is emphasized by the following incident.

Heat Integration Causes Upset

In most large refineries, the principal light-ends distillation units are associated with a crude unit, cracking plant, coking unit, or naphtha reformer. The reason for this is energy conservation.

For example, in an FCCU's main fractionator, large amounts of heat must be removed to keep the fractionator in heat balance. This is done by circulating a hot-oil stream (called a pumparound) drawn from an intermediate tray through a series of heat exchangers (see figure 18–8).

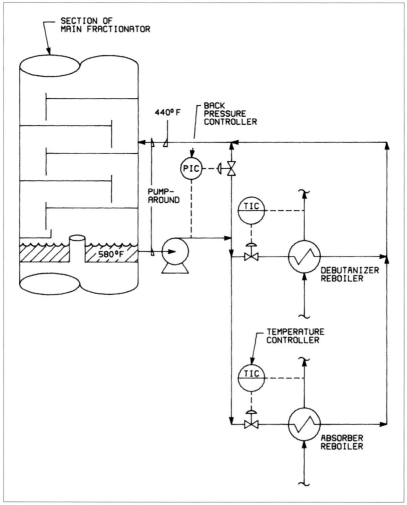

Fig. 18–8. Hot oil from the fractionator supplies heat to gas-plant reboilers.

By supplying heat to reboilers from a hot pumparound stream, waste heat from the main fractionator is gainfully utilized. On the other hand, heat-integrating two process units creates the opportunity for some interesting troubleshooting assignments. The following story is not a myth; it really happened.

One morning as I sat quietly at my desk in corporate headquarters, the boss dropped by to see me. He had some unpleasant news. One of the company's refinery managers was planning to visit our office to discuss the quality of some of the new plants that had been built in his refinery. As an example of how not to design a unit, he had chosen a new gas plant for which I had done the process design. The refinery manager had but one complaint: "The gas plant would not operate."

I was immediately dispatched to the refinery to determine which aspect of my design was at fault. If nothing else, I should learn what I did wrong so as not to repeat the error.

Upon arriving at the refinery, I met with the operating supervisors. They informed me that, while the process design was fine, the gas plant's operation was unstable because of faulty instrumentation. However, the refinery's lead instrument engineer would soon have the problem resolved.

Later, I met with unit operating personnel. They were more specific. They observed that the pumparound circulating pump (see figure 18–8) was defective. Whenever they raised hot oil flow to the debutanizer reboiler, the gas plant would become destabilized. Reboiler heat-duty and reflux rates would become erratic. Most noticeably, the hot-oil circulating pump's discharge pressure would fluctuate wildly, indicating cavitation. They felt that a new pump requiring less net positive suction head was needed (see chapter 11).

Both these contradictory reports left me cold. Anyway, the key to successful troubleshooting is personal observation. So I decided to make a field test.

When I arrived at the gas plant, both the absorber and debutanizer towers were running smoothly but not well. Figure 18–9 shows the configuration of the gas plant. The debutanizer reflux rate was so low it precluded significant fractionation. Also, the debutanizer pressure was 100 psig below design. Only a small amount of vapor, but no liquid, was being produced from the reflux drum. Since the purpose of the gas plant was to recover propane and butane as a liquid, the refinery manager's statement that the gas plant would not operate was accurate.

As a first step, I introduced myself to the chief operator and explained the purpose of my visit. Having received permission to run my test, I switched all instruments on the gas-plant control panel from automatic to local/manual. In sequence, I then increased the lean-oil flow to the absorber, the debutanizer reflux rate, and the hot-oil flow to the debutanizer reboiler.

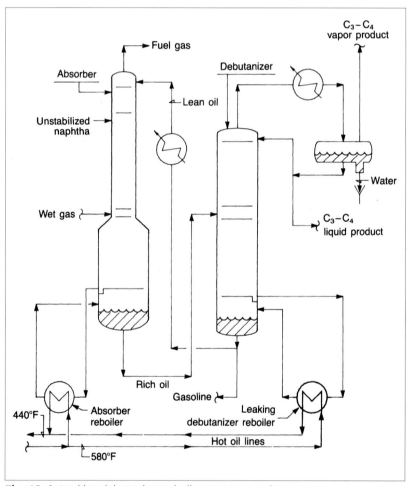

Fig. 18–9. Leaking debutanizer reboiler upsets gas plant.

The gas plant began to behave properly. The hot-oil circulating pump was putting out a steady flow and pressure. Still, the plant was only producing a vapor product from the debutanizer reflux drum. This was because the debutanizer operating pressure was too low to condense the C_3–C_4 product. By slowly closing the reflux drum vapor vent valve, I gradually increased the debutanizer pressure from 100 psig toward its design operating pressure of 200 psig.

Suddenly, at 130 psig the hot-oil flow to the debutanizer's reboiler began to waver. At 135 psig, the debutanizer pressure and the hot-oil flow plummeted. This made absolutely no sense. How could the debutanizer pressure influence hot-oil flow?

To regain control of the gas plant, I cut reflux to the debutanizer and lean-oil flow to the absorber. I was now back where I started. The thought of impending failure loomed.

I repeated this sequence twice more. On each occasion, all went well until the debutanizer pressure was increased. By this time it was 3 A.M. Was it also time to give up and go home?

Just then, I noticed a commotion at the main fractionator control panel. The operators there stated that the fractionator was flooding again—for the third time that night. The naphtha production from the fractionator had just doubled for no apparent reason.

In every troubleshooting assignment there always occurs that special moment, the moment of insight. All of the bits and pieces fall into place, and the truth is revealed in its stark simplicity.

I cut the debutanizer pressure back to 100 psig and immediately the flooding in the main fractionator subsided. The operators then closed the inlet block valve to the hot-oil side of the debutanizer reboiler and opened up a drain. Naphtha poured out instead of gas oil. This showed that the debutanizer reboiler had a tube leak.

Whenever the debutanizer pressure reached 130 psig, the reboiler pressure exceeded the hot-oil pressure. The relatively low-boiling naphtha then flowed into the hot oil and flashed. This generated a large volume of vapor that then backed hot oil out of the reboiler. The naphtha vapors passed on into the main fractionator and flooded this tower. Thus, the cause of the gas plant instability was neither a process design error, instrument malfunction, nor pumping deficiency. It was a quite ordinary reboiler tube failure.

Composition Causes Instability

Distillation columns are designed to fractionate between light and heavy key components. For example, the feed to one large butane splitter was:

none key component	5% propane
light key component	40% isobutane
heavy key component	45% normal butane
none key component	10% pentane

A well-designed tower should be able to handle wide swings in the ratio of light to heavy key components. However, large increases in composition of the none key components will upset the operating stability of many towers. The process engineer should be able to identify such upsets as being external to the tower.

The best troubleshooting tool to identify instability in light-ends distillation towers due to variability in feed composition is an online gas chromatograph. If reliable gas chromatographs can be obtained from both the overhead and bottoms products, they should be used to control the tower's operation directly. For a tower subject to swings in none key components composition, closed-loop analyzer control will go a long way to improving fractionation stability. The following example amplifies this point.

Temperature controller fooled

For the butane splitter cited previously, the critical specification was not more than 10% normal butane in the isobutane overhead product. This requirement was met by controlling the splitter top temperature to 140°F. The isobutane product would then have the following composition: 11% propane, 79% isobutane, and 10% normal butane.

Now suppose the refinery crude unit that contributes feed to the butane splitter suddenly increases the propane content of its butane product. Assume this change raises the propane content in the splitter's feed by 20%. If the tower top temperature is maintained at 140°F, the isobutane product composition would be 13% propane, 66% isobutane, and 21% normal butane.

In effect, the light none key component propane is pulling the heavy key, normal butane, up the tower. After all, the distillate product must be at its dew point when it leaves the tower as a vapor.

The normal butane content of the isobutane product will now considerably exceed the 10% specification.

To operating personnel, it appears as though the splitter controls have suddenly failed. The process engineer should, however, recognize this as a communications problem between the crude unit and the butane splitter control centers.

Condensing capacity exceeded

The tower was limited by condensing capacity; that is, the heat exchangers used to condense the reflux and distillate were marginally inadequate. Whenever the propane content of the overhead would increase, the bubble-point temperature of the liquid in the reflux drum would drop. This made it more difficult to condense the tower overhead vapors. As the splitter was only equipped to make liquid products, the inability to condense even a small fraction of the overhead vapors upset the tower. Here is the way this happened.

First, the reflux drum liquid level dropped as the uncondensed vapors accumulated in the drum. Next, the operators reduced the reflux rate to prevent the reflux pump from running dry. Then, the splitter pressure rose rapidly, and the relief valve popped. This effectively vented the lighter hydrocarbons from the tower and allowed the operators to regain control.

This incident teaches that a distillation tower producing only a liquid overhead product must have excess condensing capacity available. Even if for a moment only a small fraction of the overhead cannot be condensed, control of the tower will be entirely lost.

Steam Reboiler Condensate Seal

The majority of refinery light-ends distillation towers are reboiled with steam. Likely, the use of steam reboilers predates the petroleum industry. Thus, one would think that engineers would have developed reboiler design to a fine science.

Nothing could be further from the truth. Steam reboilers are a major cause of tower instability, and the problems seem related to one function: draining the condensed steam out of the reboilers' channel head (see figure 18–10). Either the condensate does not drain freely from the reboiler and backs up over the tubes, or the condensate seal is blown and heat-transfer efficiency is impaired.

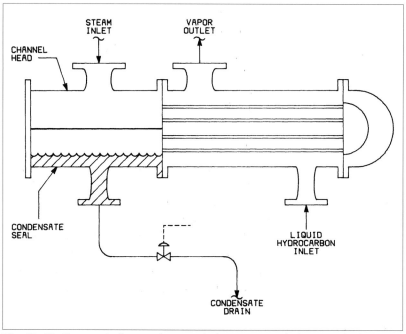

Fig. 18–10. Steam reboiler condensate drainage system is improperly designed.

The idea of the condensate seal is quite familiar to operators. For engineers lacking field experience, it is a surprising concept. Figure 18–10 shows a water level in the channel head. This is the condensate seal. Once this level drops out of the channel head and steam starts blowing through the control valve, the reboiler heat-transfer duty is drastically reduced.

If an operator opens the control valve too far, he can entirely drain the channel head. As the condensate level in the channel head falls, more tubes are exposed to the condensing steam. This increases reboiler duty. However, when the condensate seal is blown, reboiler duty falls so sharply that it upsets the tower operation.

Marginal Condensate Drainage

On one reboiler using low-pressure steam, the flow of steam to the reboiler was cycling. The repeated swings in reboiler duty destabilized the tower's operation. Repeated checks of the controls and instruments did not reveal any deficiency.

This problem turned out to be a simple hydraulic phenomenon. The controls on the tower are shown in figure 18–11. Note that the steam supply pressure is 30 psig, and the condensate collection header pressure is 20 psig. Steam flow is controlled by throttling on the reboiler steam inlet line.

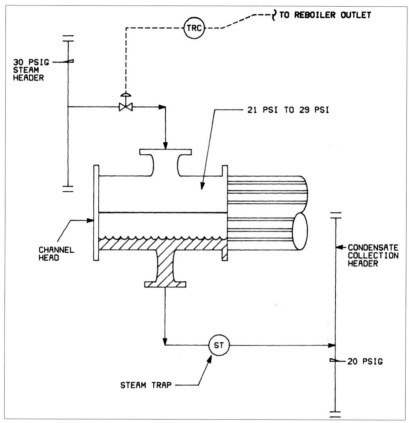

Fig. 18–11. Variable-channel head pressure promotes uneven condensate drainage.

The reason for the cycling of steam flow was revealed by putting a pressure gauge on the channel head. The following chain of events was then observed:

1. The temperature recorder controller (TRC) holding the reboiler outlet temperature signaled the steam-inlet control valve to pinch back.
2. As the control valve closed, the pressure in the channel head dropped to 21 psig.

3. When this pressure dropped, it became difficult for the steam condensate to drain from the channel head. The condensate level rose and submerged several rows of tubes.

4. The submerged tubes were not available for heat transfer. As a consequence, reboiler duty, and thus the reboiler outlet temperature, fell.

5. The TRC then opened the steam-inlet control valve, and the channel head pressure rose rapidly to 30 psig.

6. The higher channel head pressure forced the condensate level down and quickly increased reboiler duty to an undesirably high level. The reboiler outlet temperature then jumped, and the cycle was repeated.

The dynamics of this system was such that it could never be lined out by tuning instrumentation. In the end, the control valve on the inlet steam was removed. The reboiler outlet temperature was then controlled by backing condensate up into the channel head and submerging heat-transfer tubes. This was accomplished by installing a control valve on the condensate line downstream of the existing steam trap. The revised arrangement ensured that the pressure in the reboiler channel head was a constant 30 psig. At this pressure, there was never any difficulty in draining condensate into the collection header.

Tower Flooding

When a distillation column's trays become completely overloaded, the tower becomes inoperative, not just unstable. Liquid may be carried over the top of the tower due to tray downcomer flooding or a sharp drop in tray efficiency (jet flood) can result.

A marginal flooding condition will, however, result in an unstable tower operation. Unexplained swings in bottoms product rate or bottoms level likely mean the trays are alternatively priming and dumping. A sudden drop in the temperature difference between the top and bottom of a tower, followed by a return to a normal temperature profile, indicates a marginal flooding condition. The best way to troubleshoot tower flooding is with a differential pressure survey across the trays (see chapter 19).

Thermosiphon Reboiler

Fluctuating reboiler duty will have the same effect on a distillation tower as pumping the gas pedal on a car (that is, unpleasant changes in acceleration). A typical thermosiphon reboiler configuration is shown in figure 18–12. The density difference between the liquid-filled reboiler inlet line and the vapor-liquid mixture in the outlet line drives the process-side fluid through the reboiler.

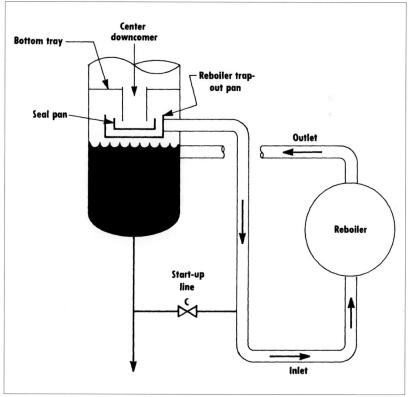

Fig. 18–12. Submerging the reboiler return nozzle causes erratic tower operation (courtesy *Oil & Gas Journal*).

When the tower bottom's liquid level rises to the reboiler return nozzle, thermosiphon circulation is inhibited. Covering and uncovering this nozzle is one cause of fluctuating reboiler duty, as reduced circulation through the reboiler will cut heat transfer.

Sometimes, thermosiphon reboilers seem to die at low loads. Instead of a gradual reduction in heat output, the reboiler suddenly seems to stop working. This is because the low-percent vapor in the reboiler return line does not result in very much density difference between the reboiler inlet and outlet lines. Without the density difference driving force, thermosiphon circulation stops and heat transfer is reduced.

Varying heat output from a thermosiphon reboiler may also be related to insufficient liquid flow. A leaking trapout tray will permit only a part of the liquid flowing down the tower to be collected and drawn off to the reboiler. This problem is easily recognized by observing the reboiler outlet temperature. When this temperature falls as reboiler duty rises, liquid flow to the reboiler is certainly increasing. Conversely, a high outlet temperature combined with a low duty indicates the reboiler is running dry.

Turndown Problems

Many towers equipped with grid, valve, or sieve trays do not operate efficiently at low feed rates. This is due to tray-deck leakage. As the pressure drop of the vapor flowing through the grid or sieve holes falls below the weight of the liquid on the tray deck (as determined by the height of the weir), the tray will start to leak.

As the liquid is now running down through the sieve holes, the flow of liquid over the weir stops. The liquid level on the tray deck (as shown in figure 18–13) falls below the top of the weir. This unseals the bottom edge of the downcomer from the tray above and permits vapor to bypass the tray deck and flow up the downcomer. With vapor flowing up the downcomer, liquid backs up on the tray above. The greater weight of liquid on the tray deck encourages more leakage through the sieve holes.

Figure 18–13 illustrates how sagging tray decks can aggravate the problem of low vapor flows. Also, the preceding discussion is almost as applicable to valve trays as to sieve trays. Valve trays most certainly leak at low vapor rates regardless of any vendor claims to the contrary.

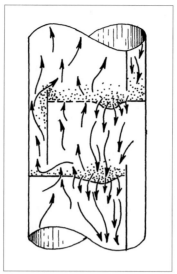

Fig. 18–13. Low vapor rates unseal downcomers.

In summary, operators too often blame tower instability problems on instrument malfunctions. Quite often, such instability is an early warning sign of a fundamental process problem or a mechanical failure. Remember, if one cannot run a tower in the manual control mode, instrument tuning will not help.

Troubleshooting Checklist
for Distillation Problems

Instrument Malfunctions

Wrong feedback-feed
  forward control
Electrical interference
Tune instruments
Switch from auto to manual control

Reboilers and Preheaters

Excessive feed preheat
Reboiler tube leak
Hydrocarbons in reboiler
  steam supply
Loss of thermosiphon circulation
Steam condensate
  drainage restricted
Condensate seal blows out

Other Problems

Trays overloaded-tower floods
Slug flow in feed riser
Slug flow in reflux-drum inlet flow
Variable amounts of
  nonkey components
Flooded condenser short
  of condensing capacity
Foam-induced flooding
Unsealing downcomers
Surfactants in feed

Fouled Trays

The accumulation of deposits on tray decks is a subject important enough to merit its own chapter. Dirty trays cause problems by promoting premature flooding and reducing tray efficiency. A tower that has begun flooding at lower reflux rates probably needs to be washed. The most common form of flooding associated with plugged trays is indicated by increased reflux rates having no effect on a tower's bottom temperature, flow rate, or reboiler duty. If it is possible to raise the reflux rate by 20%–30% without a noticeable change in the tower's heat and material balance, the top of the tower is completely full of liquid. Each incremental barrel of reflux flows overhead in the vapor line and directly back to the reflux drum. Such a situation is due to plugged trays.

Propylene-Propane Splitter

In one instance, a tower fractionated an olefin-rich stream from a refinery FCCU into a propylene concentrate and LPG propane. The unit had recently been revamped to reduce the propylene's product moisture content. Afterward, the trays below the feed point were found subject to premature flooding, for example, the trays would flood at 60% of their design capacity. The operators believed this was due to faulty design, while the designers contended that better operators were needed.

As the reflux rate was increased, the differential pressure gauge reading across the bottom trays would slowly rise. At 48,000 B/SD reflux, the pressure drop would start increasing exponentially. This meant that liquid was accumulating on the tray decks and backing up the downcomers. The tower was starting to flood.

Figure 19–1 shows this tower along with the tray pressure-drop data. The tower internals consisted of Linde high-capacity sieve trays. The holes in the sieve decks were 3/16-inch diameter. Sieve trays with small holes are intended for clean services because they are prone to plugging.

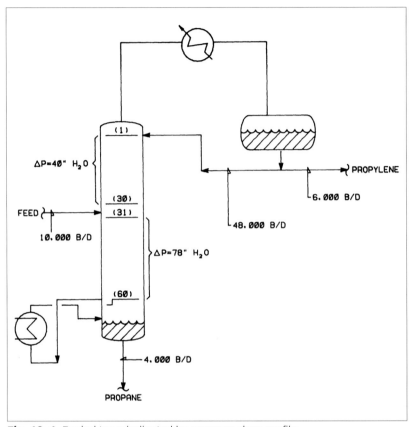

Fig. 19–1. Fouled trays indicated by pressure-drop profile.

An analysis of the pressure drop proved that the tower was not poorly designed or operated but that the trays were fouled. The calculated pressure drops at current operating rates were compared to the measured tray pressure drops (table 19–1).

Table 19–1. Comparing calculated versus observed ΔPs shows bottom half of tower has plugged trays.

	Calculated	Observed
Reflux rate	48,000 B/SD	48,000 B/SD
ΔP for trays above feed point	36" H₂O	40" H₂O
ΔP for trays below feed point	36" H₂O	78" H₂O

For trays above the feed point (tray 31), the calculated pressure drop was close to the observed value. For the trays below the feed point, however, the measured pressure drop was double the calculated value. (One should consult with the tray vendor for the actual correlations needed to calculate tray pressure drop. Often, the design tray ΔP is listed on the vender's original data sheet. However, for emergency use, a very approximate method is given in the Appendix.)

It was concluded from the table that the bottom half of the tower had to be washed. The unit was shut down, and the propylene-propane splitter opened for inspection. The feed tray was covered with a thick deposit. Lower trays were also badly plugged. A component analysis of the deposit indicated it to be largely potassium (K).

Dowell (a company that specializes in chemical cleaning refinery process equipment) was then called upon to wash the tower. They circulated a 5% inhibited HCl solution through the splitter. The unit was brought back on-line. Design reflux rates (72,000 B/SD) were reestablished at a normal tray pressure drop. Table 19–2 summarizes the steps to follow when acid washing a tower.

Table 19–2. How to clean tower trays by acid washing.

1. Obtain a sample of fouling deposit and see if it will dissolve in HCl.
2. Fill the tower with a 5%–10% HCl solution with 1% corrosion inhibitor and a surfactant agent.
3. Circulate the acid from top to bottom, monitoring the acid strength.
4. Circulate for one shift.
5. Rinse the tower with water and drain.
6. Refill the tower with diluted KOH solution; circulate and drain.
7. Water wash again before inspecting to make sure the tower is clean.

Note: Acid cleaning of refinery equipment is typically done by outside contractors.

What had happened? The propylene-propane splitter had formerly run for many years without difficulty. After the unit was revamped, the sieve decks had plugged.

As part of the revamp project, a KOH dryer was installed upstream of the splitter. A small amount of dissolved KOH was carried out of the drier vessel by the splitter feed. Once in the tower, all of the KOH in solution dropped out as a dry powder and plugged the sieve decks.

Amine Regenerator

This tower was used to steam-strip hydrogen sulfide from an amine solution. This is a somewhat corrosive service. Thus, iron sulfide particles (products of corrosion) accumulate in the circulating amine solution. The regenerator was equipped with stainless steel valve trays. It had been onstream for three years before the trouble started.

The first sign of a problem was an increasing concentration of amine in the reflux drum water (see figure 19–2). Over a period of months, this concentration increased from 1% to 20%. In itself, this change did not impair operations. The next indication of trouble was increasingly erratic flows of both the overhead product (hydrogen sulfide gas) and the bottoms product (lean amine). As time went on, this became a big problem, especially at high feed rates. Finally, the tower began to massively carry over. Without warning, the liquid level in the reflux drum would jump up. Stripping steam and amine feed to the regenerator was reduced to restore the reflux drum level.

After a few months of fighting the recurring tower flooding, operating management decided to shut down for inspection. When the manways were opened, everyone expected to see the feed tray full of iron sulfide. However, both the feed tray deck and downcomer were clean. Luckily, it was decided to investigate further. Crawling down through the tower, we found that the lower trays were encrusted with a reddish solid. The maximum amount of fouling was on the fourth tray below the feed point. At the bottom of the tower, the trays were again clean.

The deposits were so deep on the deck of the fourth tray that the valves were almost covered. Thus, vapor passage was restricted and a high pressure drop resulted. This high pressure drop on the fourth tray caused the regenerator to flood.

Note that a restriction to vapor or liquid flow on one tray will cause the tower to flood above that tray. If, as is often done, only the trays conveniently located at manways are inspected, a plugged tray can be missed. The process engineer should not be averse to crawling through a tower to make a thorough personal inspection.

In the case of the amine regenerator, no explanation could be found for the fouling distribution. The solids (consisting mostly of iron and sulfur) were soluble in hydrochloric acid. The trays were hydraulically cleaned. Typically, an amine regenerator tower ought to be routinely acid washed (without opening) every few years.

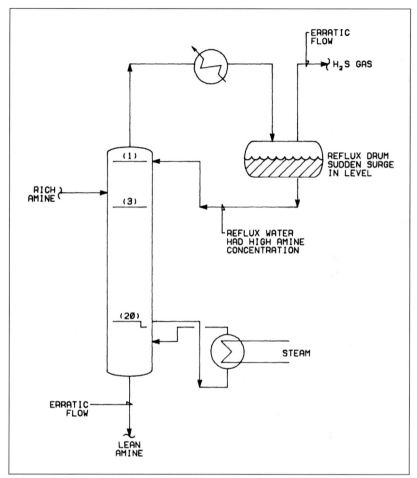

Fig. 19–2. Signs of flooding in an amine regenerator.

Gas Absorber

In many refineries, operating supervision on individual process units changes every few years. The shift operators themselves may be new to the unit or may have short memories. Thus, long-established operating procedures on older units are sometimes lost.

In this case a new superintendent on a gas plant noted that the butane content of the absorber off-gas (see figure 19–3) was 8%. Since his unit's principal function was to recover liquid butane from wet gas feed, he initiated a process design study to reduce the butane in absorber off-gas from 8% to 3%. While the process engineer was simulating the tower operation on a computer program, someone recalled that the absorber used to do a better job of recovering butane. An investigation was begun. It was determined that the current lean-oil circulation was far below historic rates. Since the amount of butane recovered in an absorber is a function of the lean-oil circulation rate, it looked as if the cause of the poor butane recovery had been determined.

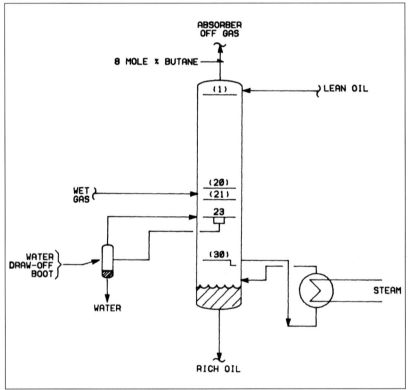

Fig. 19–3 Plugged absorber trays prevented adequate lean-oil circulation, and butane recovery suffered.

Lean-oil rates were being restricted to prevent carryover of liquid hydrocarbons into the absorber off-gas. That is, the tower was flooding at an abnormally low liquid load. Finally, an old hand recalled that formerly it had been standard practice to water wash the absorber every year or two. Coker gas, which was a major component of the wet gas feed, was rich in ammonia; ammonium salts would deposit on the absorber trays. After a good wash with clean steam condensate, which removed the water-soluble deposits, the absorber was returned to service. Lean-oil circulation was tripled; the butane content of absorber off-gas dropped from 8% to 2%.

In general, this type of trouble may be readily identified on absorbers by: 1) reviewing historic unit data on parameters affecting absorber efficiency (pressure, temperature, lean oil, and gas flows); 2) calculating the percent of flood and tray pressure drop at current operating conditions; and 3) measuring the absorber tray pressure drop in the field and comparing it against the calculated value.

Alkylation Depropanizers

Many alkylation units use auto refrigeration to cool their reactors. Effluent refrigeration commonly used on Stratco units also falls within this classification. To maximize isobutane purity in the refrigerant recycle stream, circulating refrigerant is depropanized.

Occasionally, towers in this service must be washed to remove sulfate or acid coke deposits on the tray decks. The wash liquid is accumulated in the reflux drum and then pumped to the top tray with the reflux pump. The wash liquid can then be pumped from the bottom of the tower and recirculated back to the reflux drum.

A tower in need of washing will not always exhibit overt signs of flooding (except for an increase in tray pressure drop.) The alkylation depropanizer would gradually lose fractionating ability. That is, with 5% propane in the bottoms product and feed and reflux rates held the same, the butane content in the tower overhead would increase. When the butane content of the propane product exceeded LPG specs, it was time to wash the tower. This was a simple procedure. The trick was to dry the tower out prior to start-up.

Alkylation depropanizers are exposed to acid containing hydrocarbons during normal operation. As long as the depropanizer is dry, these hydrocarbons remain noncorrosive. Upon contact with moisture, free acid is formed and the tower is speedily attacked by corrosive weak acid.

Isomerization and polymerization processes using $AlCl_3$ as a catalyst also have this problem. Water washing is occasionally required to clean tower trays, but a cup of free water left in the tower after washing causes a year's worth of corrosion on start-up.

Dehydrating light-ends towers during start-up is discussed in chapter 20. It is a skill that the operating engineer will find well worth acquiring.

HF Alkylation Iso-Strippers

The feed to this tower is usually located close to the top tray. Both the feed tray and the three or four trays located below this point are subject to plugging, fouling, and corrosion. Evidence of a plugged feed tray is several percent alkylate in the isobutane recycle stream. Replacement of the usual ½-inch-sieve-hole trays with a "Nutter V-Grid" type tray (LVG) or a ¾-inch-hole sieve deck, has diminished plugging in this service. Monel tray decks and especially Monel downcomers are recommended.

In some services, corrosion to the overhead reflux condensers contaminates the reflux flow with particulates. Alternately, caustic may back up into the reflux drum through the distillate product line. This can happen during a unit upset. Either way, the tower top tray can become plugged while all the trays below remain open.

In this case only the top tray will flood, and a tray ΔP survey will not reveal the problem. To recognize this difficulty, one need only observe what happens when the reflux rate is increased.

The plugged top tray will prevent the reflux from cascading down to the lower trays. The liquid reflux will just overflow into the condensers and circulate back to the reflux drum. The tip-off to this problem is that neither the reboiler duty nor the bottoms temperature is affected in the normal way by raising reflux. The tower's heat balance appears as if the reflux rate had never been increased. This is not much different from the signs of normal tower flooding, except that the ΔP on all but the top tray is not excessive.

Often, a tower that has been flooding prematurely is opened for inspection. Fouling deposits on the tray decks are suspected of having caused the flooding. However, no such deposits are observed on the tray decks. Why?

In the course of making a tower ready for entry, we ordinarily steam out, or water wash, the column. In so doing, we may remove the offending deposits. Then, when the tower is opened for inspection, we find the trays are clean. We then erroneously conclude that our theory of flooding due to tray-deck fouling was incorrect.

Most tray deposits in refineries are insoluble in water and cannot be vaporized with steam. Such deposits are typically iron sulfides which will auto-ignite when dry at ambient temperatures. Thus, they must be chemically removed prior to making the tower safe for entry. In the course of this necessary safety procedure, the reason for opening the tower in the first place is eliminated. Failure to remove the iron sulfide can cause evolution of sulfur dioxide, which is quite fatal at concentrations of 1,000 ppm or less.

Cleaning Fouled Tower Internals

In a refinery, trays are typically fouled with iron sulfide deposits if there is hydrogen sulfide present in the process flow. This iron sulfide is dangerous in the sense that it will auto-ignite when dry and exposed to air at ambient temperatures. I used to remove such deposits by circulating a dilute hydrochloric acid solution.

In the 1980s, it became popular to circulate a "Zyme solution" to clean towers; this solution is less corrosive than dilute hydrochloric acid, but of course far more expensive.

One relatively newer method to clean trays as well as beds of packing is offered by the Refining Technology Corporation. They are located in Corpus Christi, Texas. One of my clients, Suncor, has reported excellent results using their services to clean a coker fractionator with trays and a bed of structured packing using their program. Two separate chemical systems are employed. Their "Permanna System" is used to dissolve pyrophoric iron (i.e., $Fe(HS)_2$, iron sulfides). Their "Quick Turn System, removes heavier hydrocarbons.

Structured packing has presented a real safety problem when not cleared properly of iron sulfides. Not only will the iron sulfides and hydrocarbons catch fire when the contaminated packing is exposed to air during a turnaround but the packing itself will melt and will certainly be ruined. Please recall that the pyrophoric iron corrosion deposits are insoluble in either hydrocarbons or water.

Troubleshooting Checklist for Dirty Trays

Tower Floods
Exponential increase in tray
  pressure drop
Calculated versus observed ΔP
Fractionation drops as reflux
  rate is increased
Reflux drum level jumps
Lean-oil circulation rate reduced

Fouling Deposits
KOH or NaOH entrained in feed
Iron sulfide particles
Ammonium salts
Acid coke in alkylation service
Check intermediate trays
Use V-grid trays
Water-soluble deposits

Washing Procedure
Circulate inhibited HCl solution
Water wash from reflux drum

Dehydrating Light-Ends Towers

The novice operating engineer may believe that his career objectives can best be accomplished by communicating with management or demonstrating an ability to organize complex projects. This is false. Skill in dehydrating light-ends towers on start-up is the primary criterion for a promotion to the executive suite. To some, this may seem an exaggeration. But to those who have witnessed the corrosion damage wrought by water in distillation columns, the importance of dehydrating light-ends towers cannot be overemphasized.

Moisture left in towers will also accumulate on the tray decks and reduce tower capacity. Then again, certain fractionators must make a bone-dry product. Water left in these towers on start-up can cause off-spec production for weeks.

Alkylation Depropanizer

The material distilled in some towers becomes super-corrosive when exposed to moisture. Some depropanizers in sulfuric-acid alkylation service are quite susceptible to moisture-initiated corrosion. One such depropanizer had a long history of corrosion failures. Most frequently, the overhead condenser tubes would start leaking. When this occurred, propane vapors rose from the cooling tower. Often, the effects of corrosion were less subtle; clouds of hydrocarbons would burst forth from leaking lines.

After a time it became obvious that we were forming weak sulfuric acid in the tower on start-up. On one occasion several days after such a start-up, we opened the drain on the case of the spare depropanizer reflux pump. Weak acid (it turned pH paper red) poured out.

Weak acid is death to carbon steel. We investigated further. Small amounts of weak sulfuric acid were found in nearly all of the cold, low, or dead-ended locations. With traces of sulfuric acid in the tower, numerous leaks soon developed in the carbon steel tubes of the overhead condenser. This forced a shutdown.

Several start-ups later, we finally developed a procedure to dehydrate the tower effectively:

1. We assumed the depropanizer had become wet while it was out of service.

2. A sketch was made of the bleeders (i.e., small drain valves) that were to be opened to check for water (figure 20–1). Note the location of the bleeder on the floating head of the overhead condenser. This particular drain valve had been overlooked in all previous attempts to dry out the tower.

3. After the tower had been freed of air by steaming (a standard refinery procedure), it was pressured to 60 psig with nitrogen. All bleeders were then repeatedly blown down to remove the bulk of the water from the tower.

4. An external source of dry, acid-free butane was routed to the depropanizer via a rubber hose. Steam was lined up to the tower's reboilers, and the reflux pump started to maintain level in the reflux drum (this is called putting a tower on total reflux).

5. The depropanizer was run on total reflux for two or three days. The bleeders located on the reboiler and tower bottoms dried up quickly. The drains in the overhead system dehydrated very slowly.

6. The last bleeder to dry up completely was, of course, the one that had always been overlooked on the condenser floating head.

7. The normal acidic depropanizer feed was charged to the tower. The depropanizer was run for the next day at a minimum operating pressure and maximum temperature to boil off residual moisture quickly.

Refinery management had difficulty understanding why it took four days to start up a tower that used to be put online in just four hours. They were, however, pleased with the greatly reduced rate of corrosion failures on this depropanizer.

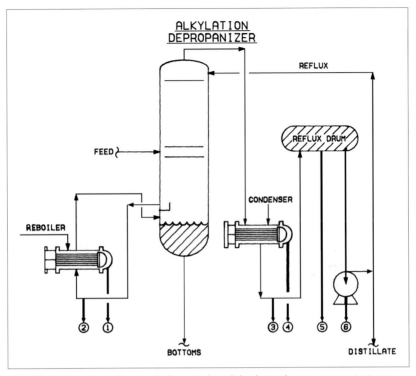

Fig. 20–1. Numbers denote drains used to dehydrate the tower on start-up.

Water-Removal Trays

In services where the hydrocarbon feed normally contains dissolved moisture, water must continuously be drawn off the tower. Often, all of the water in the feed stream will boil overhead; then it can be collected in the reflux drum and easily be drawn off. Less frequently, water will leave with the bottoms product. Problems arise when the water, which behaves as a nonideal component in hydrocarbon systems, condenses out inside the tower. This happens because the water gets trapped between the top and bottom tray.

The tower top temperature and gross volume of overhead vapors are not high enough to force the moisture in the feed to be evaporated overhead. The tower bottom temperature and boil-up rate are too high to permit all of the water in the feed to escape with the bottoms product. Thus, this water is trapped inside the tower.

Of course, at steady state, the water must eventually leave the tower, but not before a separate water phase is formed. The water phase accumulates on the tray decks, possibly as high as the tray weir. Indications of free water in a lights-ends fractionator are:

- Reduced tray efficiency
- Increased tray pressure drop
- Unusual reboiler temperature
- Extensive failures of reboiler tubes due to corrosion
- Corrosion to the tower shell and trays

The presence of free water in a tower can be predicted by calculation. First, assume that there is no appreciable entrained water in the hydrocarbon feed. If there is a significant amount of entrained water, try to estimate the amount by catching a large sample of representative hydrocarbon feed—a difficult task, especially with butane and lighter components.

Water dissolved in hydrocarbon can be measured by the Karl-Fischer water analysis technique. If this type of analytical method is not available, the moisture content of a hydrocarbon stream that has been in contact with a free-water phase is calculated as follows:

1. Convert the hydrocarbon feed rate to moles per hour.
2. Using figure 20–2, find the mole fraction of water dissolved in the hydrocarbon.
3. Multiply values obtained in the steps above to find moles of water in the feed.

Next, determine if this quantity of water can be drawn overhead in the tower. This is calculated as follows:

4. Find the tower top temperature, the reflux drum temperature, and the tower pressure.
5. Calculate the mol/hr of reflux and the mol/hr of net distillate product.
6. Calculate the maximum mol/hr of water that can be taken overhead.

$$W_o = \frac{(VP_w)}{(PT)}(R+D)$$

where:

W_o = Maximum mol/hr water that may evaporate overhead

VP_w = Vapor pressure of water at the tower top temperature, psia (look this up in a steam table)

PT = Tower pressure, psia

R = Moles of reflux

D = Moles of net overhead product

7. Calculate the mol/hr water put back into the tower with the reflux. This is done by multiplying R by the corresponding mole fraction of water obtained in figure 20–2.

8. Subtract the mol/hr water obtained in step 7 from W_O found in step 6. This is the maximum amount of water that can be driven overhead.

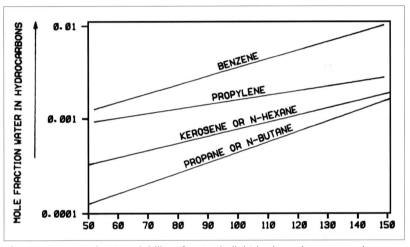

Fig. 20–2. Approximate solubility of water in light hydrocarbons assuming hydrocarbon is in equilibrium with a free-water phase.

Water Trapout Tray

If both calculation and observation confirm that water is settling out inside a tower, the operability of the existing trapout tray should be checked. If no trapout tray exists, one should be installed at the earliest opportunity.

Figure 20–3 illustrates a typical water trapout tray scheme. The trapout pan itself is too small to make a decent separation between water and hydrocarbon. A mixed phase is drawn off and flows to a small drum (boot) located at grade. In the boot, a good separation is made between the two phases. Water is drawn off; the hydrocarbon flows back to the tower through the recirculation line.

The flow of liquid circulating through the boot shown in figure 20–3 needs to be adjusted to maximize water production. The block valve located on top of the boot should be closed one turn per hour until the water drawoff rate is maximized. Too low a circulation rate will force water to overflow the internal sump in the tower. Too great a circulation rate will retard water settling in the boot.

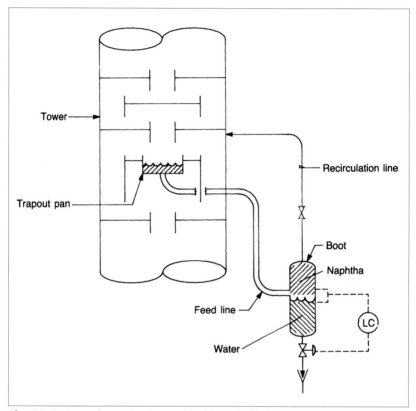

Fig. 20–3. Removing water trapped inside a distillation tower.

In most services, the water boot LC will have to be run on manual. If there is more than one water draw-off nozzle, only leave one open to the boot at a time. I once flew halfway around the world to Durban, South Africa, to tell a client that when two valves on the same tower are lined up to the boot, liquid from the upper drawoff will back up into the lower drawoff nozzle and bypass the boot.

The driving force producing the flow of liquid through the boot is the density difference between the liquid in the feed line and the recirculation line. If either the circulation line or boot bottom nozzle plugs, water removal efficiency will be impaired. Blowing nitrogen up the recirculation line will usually clear any blockage.

A large hole in the trapout pan will also reduce the capacity to remove water. That is why such pans should be checked when a tower is opened for inspection.

A gas plant reboiled absorber is the most common example of a tower that requires a water trapout tray (figure 20–4). Water accumulates in such absorbers because they are not refluxed. In general, towers that run at low reflux ratios are most subject to internal water accumulation.

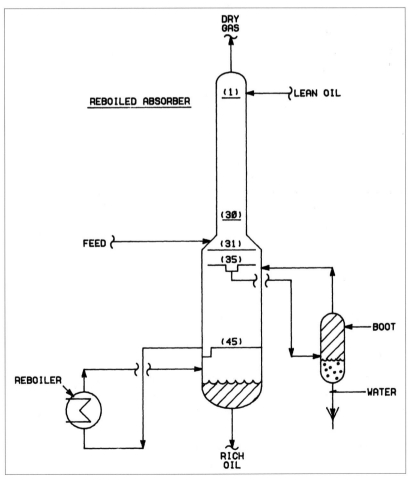

Fig. 20–4. Tray 35 is used to remove trapped water.

Some absorbers have a water draw-off in the region of tray 20 (figure 20–4). This permits rapid dehydration of an absorber during start-up. These draw-off trays seem to dry up during normal operations. It's good but nonessential to have these start-up water draw-off trays.

Sometimes, a distillation column will begin losing efficiency due to water accumulation on the trays or perhaps the bottoms product has turned cloudy because of moisture. If there is no trapout tray, try cooling off the tower's feed. This will reduce the moisture content of the hydrocarbon.

Procedure Checklist for Tower Drying

Reasons for Dehydrating Towers on Start-up
Tower in alkylation service
Unit used $AlCl_3$ or HCl
Critical water spec on products

Drying Procedure
Locate all low-point drains
Blow out drains with nitrogen
Put tower on total reflux
Maximize temperature
  and minimize pressure
  for several shifts

Indications of Free Water on Trays
Low tray efficiency
High tray ΔP
Reboiler tube leaks
Corrosion to tower shell
Dew-point calculation for overhead

Water Trapout Tray Failure
Plugged circulation line
Hole in trapout pan
Water drains plugs
Circulation rate too high
Using two draw-offs at the same time

Vapor-Liquid Separation

The simplest process operation carried out in a refinery is the separation made between vapor and liquid. The driving force to effect this separation is gravity, with the rate proportional to:

$$\text{Settling rate} = K \left(\frac{p_L - p_v}{p_v} \right)^{1/2}$$

where:

K = K factor (see table 21–1 on page 432)

p_L = Density of liquid

p_V = Density of vapor

This relationship, along with the volume of vapor to be processed, is used by the designer to calculate the diameter of most process vessels (including distillation towers) in all refineries. The separation will usually take place in an empty vessel called a knock-out (KO) drum. Although it is hard to imagine a simpler operation than vapor-liquid separation, failure of this function accounts for many of the accidents and unit upsets that plague petroleum refineries.

The common term applied to such failures is liquid carryover. The most frequent cause of carryover from KO drums is that the drum fills with liquid and runs over. Operators euphemistically refer to these incidents as foamovers, excessive entrainment, boilovers, and so on—the implication being that human intervention could not have prevented the incident. Although such things do happen, the process operating engineer ought not to be confused by these statements. Most likely, no one was watching the level in the KO drum and it simply ran over.

It can never be repeated too often that it is difficult to locate a liquid level visually inside a process vessel. Methods to find and control levels have been discussed in other sections of this book, but one should recognize that operating personnel do not always have the time or initiative to go outside and check levels.

For instance, on one unit, an absorber was recovering propane from fuel gas. The bottom level control on the absorber failed and the tower filled with liquid naphtha. The naphtha carried over into a KO drum and filled that too. The gas from this unit flowed into the refinery fuel gas collection system that also contained a large KO drum. These facilities were also filled with naphtha.

Finally, the fuel gas was distributed throughout the refinery to a number of furnaces. At each furnace, there was a fuel-gas KO drum. These drums also filled up. Only when naphtha was observed raining down from the furnace burners and a number of fires had started did operating personnel notice that the fuel-gas system was full of naphtha.

Inlet Nozzle Submerged

The liquid level in a vertical KO drum need not completely fill the vessel before massive liquid carryover ensues. Once the level rises over the feed-inlet nozzles, entrainment of liquid in the vapor outlet increases rapidly. That is why most vertical KO drums are designed to maintain the maximum liquid level several feet below the feed inlet nozzle.

There are, of course, some rather fundamental reasons why KO drums carry over other than just high levels. For instance, the ordinary vortex is a source of trouble.

Vortexing

A liquid draining through a nozzle can initiate a vortex if the liquid level is low and the drainage rate is high. This relationship is quantified in figure 21–1.[1] If calculations indicate a KO drum is carrying over due to vortexing, a simple modification to the nozzle inside the tower will break the vortex.

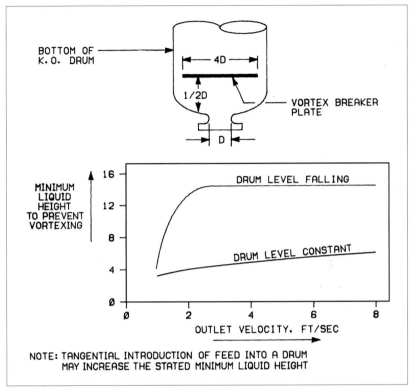

Fig. 21–1. Vortexing can cause liquid carryover.

Entrainment

A persistent carryover problem, as indicated by a constant accumulation of liquid in downstream equipment, is likely due to entrainment of small droplets of liquid in the gas stream. For example, the small amount of liquid that is withdrawn daily from the case drain of certain compressors is due to entrainment.

Stokes's Law is a rigorous method to calculate entrainment rates:

$$V = \frac{2gr^2 (p_L - p_V)}{9(vis)}$$

where:

   V = Rate of a droplet settling under the influence of gravity

   g = Gravitational constant

   vis = Viscosity of vapor

   r = Radius of the droplet

Using Stokes's Law, one could calculate the quantity of liquid entrained in a vapor stream. However, the particle-size distribution (r), which is one of the parameters used in applying Stokes's Law, is rarely known in refinery applications. Therefore, a number of rules of thumb have been developed by designers for sizing KO drums, summarized in table 21–1.

If the operating engineer comes up against a situation where carryover increases as a function of gas rate, he should use table 21–1 as a guide to decide if the problem is entrainment. On the other hand, the calculated velocities predicted by this table should not be regarded as representing an ultimate process limit.

Table 21–1. Rules of thumb for qualitatively predicting entrainment for vertical knock-out drums.

	K Factor	
	Without Demister	With Demister
Little entrainment, low pressure service	0.15	0.30
Little entrainment, high-pressure (1,000+ psi) H_2 service	0.10	0.20
Moderate entrainment	0.25	0.45
Severe entrainment	0.50	—
Live-steam deentrainment	—	0.20

Demisters

If a carryover problem has been traced to entrainment, the use of a demister will probably help. The qualitative effect of a demister on entrainment is also shown in table 21–1. A demister is a pad placed in front of the KO drum outlet nozzle, as indicated in figure 21–2. The droplets of liquid impinge and are coalesced on the Brillo-like fibers of the demister pad.

Pressure drop across a clean demister is negligible for moderate pressure services. For example, air blown through a 4-inch demister at 10 ft/sec would exhibit a pressure drop of only 1 inch of water. For vacuum services, even this small drop could be a problem. A detailed quantitative treatment on the use of demisters is presented by O. H. York in the June 1963 issue of *Chemical Engineering Progress*.[2]

Some operating superintendents dislike demisters. Parts of the pad may break loose and be drawn into a compressor suction or become lodged in downstream piping. In fouling service, the demister can coke with solids and restrict vapor flow from the KO drum and actually promote entrainment of liquids.

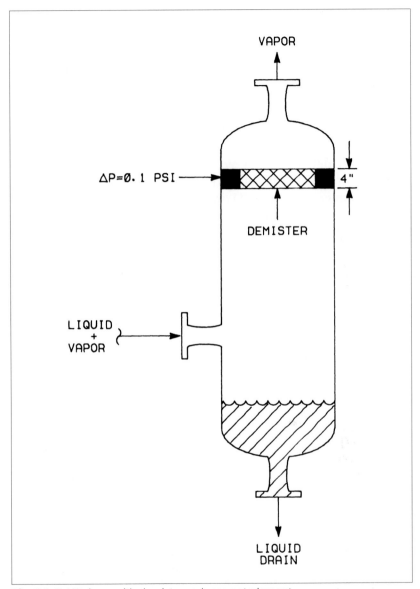

Fig. 21–2. KO drum with demister reduces entrainment.

It may seem contradictory, but demisters in many services do not do a good job of removing a fine mist at low velocities. This is because the droplets of liquid must impinge on the demister fibers with a certain momentum to coalesce properly. If a unit suffers from increased entrainment at reduced throughputs, it is probably due to this phenomenon.

Centrifugal separators

A handy device used to retrofit existing KO drums for reduced entrainment is a centrifugal separator. This is a relatively small pot that is mounted above the KO drum on the vapor line. A centrifugal motion is imparted to the vapor as it passes through this pot. The liquid droplets are swirled against the sides of the centrifugal separator where they coalesce and are drained out the bottom by means of gravity flow through a loop seal back to the knock-out drum. These devices only require a small (few tenths of a psi) pressure drop. They are not meant to handle slugs of liquid and are useful only for smaller vapor flaws.

Foaming

The most powerful tool that the troubleshooter brings to his job is not his technical education but the simple experiences of everyday life. For instance, the tendency of dirty liquids to foam is well known to all of us.

Particulate matter reduces the surface tension of liquids and thus promotes foaming. A common example of this in a refinery is the flooding of amine scrubbers due to iron sulfide particles in the circulating solution. A system which utilizes a circulating liquid that suddenly exhibits a carryover problem is probably subjected to foaming. Check the circulating liquid for particulate content. Examples of this are sulfuric acid recycle in alkylation units and closed hot-oil reboiler systems on gas plants.

Distillate-hydrogen systems

Much of the distillate (furnace oil, diesel oil, kerosene, etc.) produced in a refinery must be hydrodesulfurized. Part of this process involves a low-temperature (about 100°F–140°F), high-pressure (several hundred psig) separation of hydrogen-rich gas and the desulfurized liquid.

Field experience has shown that the drum in which this separation is made has a tendency to carry over distillate into the hydrogen-rich gas. The gas is scrubbed with amine to remove hydrogen sulfide. The oil in the gas is picked up by the amine and flows to the sulfur recovery plant. On one unit such an incident resulted in a fire at the tail end of the sulfur plant.

This carryover problem has been shown to be due to foaming. Whenever hydrogen is being flashed from a cool, heavy stream (distillate, gas oil, decanted oil), foaming should be suspected as the cause of carryover.

For one plant, the problem was corrected by replacing a vertical KO drum with a larger horizontal vessel. This provided more liquid surface area upon which the foam could break. This was done because the rate of foam dissipation is roughly proportional to the liquid surface area. Any mechanical device that provides additional disengaging surface will also speed that rate of foam dissipation. Foam can also be broken by direct contact with hot (250°F) steam-heated pipes. At least one commercial installation has been built that utilizes this concept.

Water causes foaming

A KO drum in a crude preheat train was operating at 300°F and 30 psig. The light vapors flashed off were condensed as a separate naphtha product. Periodically, this normally water-white product would turn brown due to contamination with crude oil. The operating temperature of the KO drum was above the boiling point of water at 30 psig. Occasionally, a shot of water would reach the KO drum. Mixing water and oil at temperatures above the water's boiling point caused the oil to foam. This problem was identified by correlating an increase in the water content of the condensed naphtha, with the brown color observed in the naphtha.

Rapid depressurization

This phenomenon needs little amplification-just open a warm bottle of beer. The same type of foamover has been observed in process units.

During a unit start-up, it often becomes necessary to reduce the pressure in a KO drum. Even after a unit is lined out, a reduction in operating pressure in a vessel may be required. As the liquid in such a vessel is almost always at its bubble point, the reduction in pressure will initiate boiling. If the pressure is reduced quickly, the boiling action can be very violent and lead to a large carryover of foam from the vessel.

Anyone who has ever operated a delayed coking unit can testify that large amounts of a very heavy liquid can be foamed over by depressuring a coke drum too rapidly (see chapter 2). It follows that a constant operating pressure can stop a carry-over from a KO drum if the drum is foaming over due to fluctuations in pressure.

Detecting foam levels

Note that foaming in a vessel cannot usually be detected with a single gauge glass. The liquid in the glass is usually flat and will therefore have a greater density than the frothy liquid in the vessel. If the liquid in the vessel consists entirely of foam, it will actually be about twice as high as the liquid level indicated in the gauge glass. (As a rule of thumb, a foamy hydrocarbon has about one-half the density of the same hydrocarbon after it has settled down.) However, sometimes foam density is only 10% of that of the clear liquid.

One commercially proven method to follow foam levels in difficult services is with a gamma-ray level detector. This device, in effect, looks through the vessel wall and finds the foam level by detecting changes in density inside the vessel. Gamma-ray level detection has proven to be quite valuable in black-oil services. Neutron back-scatter level detection is even more common.

Another indication of foam is split liquid levels. Assuming a tower is equipped with multiple-gauge glasses and that each set of gauge glasses taps is connected directly to the vessel, foaming will be indicated by multiple liquid levels in the gauge glasses. The gauge glasses are simply measuring the density of the foam in terms of the flat liquid specific gravity in the glass. (See chapter 18.) Unfortunately, if the gauge glasses are all connected to a stilling well or external liquid chamber, they cannot be used to detect foam.

Defoamer

Carryover due to foaming can be controlled by injecting a small amount of silicone antifoam agent. A variety of such chemicals are on the market and a specific one, effective for many services, can usually be found. However, for most refinery applications, continuous use of antifoam chemicals would be prohibitively expensive. Also the additive may have negative effects on certain down-stream processes. If the wrong defoaming chemical is used, foam formulation can be enhanced. This happened on a distillate fuel desulfurizer when a silicone defoamer, normally used to control foams in a coking unit, was tried. When the correct defoamer is used, concentrations of as little as 1 ppm are effective.

One successful application of a silicone antifoam chemical reported was in a visbreaker fractionator. At high conversions the tower bottoms tended to carry over into the normally clear distillate product. This situation was brought under control by injecting 10 ppm of an ARCO silicone defoamer into the vapor space above the fractionator bottoms.

Unfortunately, silicone defoamers often wind up contributing to the silicone content of reformer naphtha feed. The silicone accumulates on the reformer's hydrodesulfurizer pretreater catalyst. This shortens the pretreater's run length and therefore precludes the use of silicone defoamers in most refinery applications, unless the silicone is used in small concentrations. Frequent use of defoamers will promote foaming in amine systems.

Troubleshooting Checklist
for Vapor-Liquid Separation Problems

High Liquid Level
Inlet nozzle submerged
Operator fails to check
  level visually
Vortexing

Entrainment
Calculate allowable velocity
Demisters reduce entrainment
Check for damaged demisters
Centrifugal separators

Foaming
Particulates accumulate in closed
  circulating systems
Distillate-hydrogen separation
Insufficient foam breaking surface
Break foam with hot pipes
Water shots in hot oil
Rapid depressurization
Gamma-ray foam detectors
Defoaming chemicals
Split liquid levels

References

1. F. M. Patterson, "Vortexing Can Be Prevented in Process Vessels and Tanks," *Oil & Gas Journal*, Aug. 4, 1969.

2. O. H. York and E. W. Poppele, "Wire Mesh Mist Eliminators," *Chemical Engineering Progress*, June 1963.

Refinery Metallurgy for Novices

To a refinery craftsman, the difference between a chemical, mechanical, or civil engineer is unimportant. If a person is an engineer, he ought to be qualified to make technical decisions relating to all refinery problems. It is not surprising then that welders and pipefitters will often ask the process operating engineer which materials should be used when an emergency repair must be made.

Without guidance, refinery maintenance forces will usually replace corroded pipes, leaking exchanger tubes, or worn pump parts with material of the same metallurgy. This practice leads to repeated failures. The operating engineer should have a sufficient knowledge of metallurgy so that he or she can select better, if not the optimum, replacement materials when an expert is unavailable for consultation.

This chapter introduces refinery material selection nomenclature. The pitfalls inherent in the use of certain expensive alloys are discussed. A few examples are cited, showing how you can get by with ordinary carbon steel by making minor adjustments to process conditions.

Whenever a corrosion failure occurs, the operating engineer ought to determine the nature of the environment that initiated the failure. Armed with this information, the metallurgical engineer can select the right replacement alloy.

Metals Commonly Used in Refinery Service

A refinery is constructed largely of plain carbon steel, which is easily cut and welded. It is the lowest-cost metal available. Thus, most refineries will stock a wide variety of carbon steel piping and accessories. One should not expect that the refinery maintenance division will be able to fabricate new piping sections quickly or retube an exchanger bundle with any material other than carbon steel.

- **Cast iron** is used for pumps, in underground sewers, and alloyed with silicone, for handling sulfuric acid.

- **Admiralty** (a copper alloy in the brass family) is widely used in water-cooled heat exchangers.
- **Chrome steels** contain between 1% to 9% chromium. They provide better protection against high-temperature sulfidic corrosion and oxidation and superior strength (at temperatures above 650°F) when compared to carbon steel. They are used widely to fabricate tubes for process heaters.
- **Stainless steels** come in a wide variety of alloys. Although in general their properties are far superior to carbon steel, selecting the wrong alloy for a given service may lead to a serious failure. Most modern distillation tower trays are fabricated from stainless steel containing 12% chromium. (410 S.S.)
- **Titanium, Monel, Hastalloy, Carpentor 20, and HK-40** are some of the other common but very expensive metals used in refinery service.

Stainless Steel Nomenclature

A total of 18 different stainless steel alloys are used in the petroleum industry. In general, stainless steels provide good corrosion resistance in environments containing sulfur and high strength in elevated temperature service. On the debit side, stainless alloys are subject to failures that carbon steel would resist. This is especially true when chlorides are present. For simplicity, the various alloys are divided into two groups: magnetic and nonmagnetic.

- **Magnetic.** These are the 400 series stainless steels. They contain between 11% to 18% chromium, but no nickel. Types 410 (12% chrome) and Type 430 (17% chrome) are most commonly used.
- **Nonmagnetic.** These are the 300 series stainless steels. They contain between 17% to 30% chromium and substantial amounts of nickel. Type 304 (18% chrome, 8% nickel) is the general-purpose alloy of this group. The 304 stainless is also commonly referred to as 18–8. Type 316 (18% chromium, 10% nickel, and 2.5% molybdenum) provides better pitting resistance than Type 304 and resistance to naphthenic acids.

The 300 series offers excellent corrosion protection. Type 430 has good resistance to corrosion, but Type 410 provides corrosion protection only in mild environments. All stainless steels suffer from pitting corrosion in the presence of chlorides. The 300 series also is subject to stress-corrosion cracking by chlorides and polythionic acids.

Field identification of pipe

Carbon steel pipe will have a brownish or rusty color. Three hundred series stainless piping will resist atmospheric oxidation and maintain a dull silver color. Four hundred series steel (9%) chrome looks very similar to carbon steel in service. Check the stainless piping with a magnet. Chrome-nickel alloys (300 series) generally will be nonmagnetic, while chrome stainless steel (400 series) will be magnetic.

Table 22–1 summarizes symbols used to identify material types. These symbols are usually permanently marked on piping, flanges, and valve bodies at the time of fabrication.

Table 22–1. Field Identification of Steels.

Material Mark	Symbol	Material Mark	Symbol
302	△	347	⌷
304	☐	405	⸤ ⸥
304 - ELC	⌐ ⌐	410	☐
309	⊂⊃	430	⌑
310	⌑	MONEL	◇
316	8	INCONEL	<
316 - E:LC	◡◡	NICKEL	N
317	⌂	NI-O-NEL	U
319 - ELC	3	HASTELLOY "B"	B
321	⋈	HASTELLOY "C"	C

Corrosion Resistance

An operating superintendent was once asked what his primary responsibility was. He replied, "To keep the hydrocarbons inside the pipes and vessels." Corrosion of process lines and equipment is the major source of leaks in a refinery. If it were not for continuous and varied corrosion, most of the technical and maintenance employees in a refinery could be eliminated.

The choice of materials selected to replace corroded piping or vessel internals depends on the type of corrosion and the availability of suitable alloys.

How are corrosion rates measured and what rate is too high? Corrosion rates are reported in mil/year. A mil is 1/1000 of an inch. For example, a section of carbon steel piping handling 500°F high-sulfur diesel oil may corrode at a uniform rate of 10 mil/yr. If the pipe's wall thickness is ¼ inch (250 mil), the average wall thickness would be reduced to ⅛ inch after 12 years of service.

A corrosion rate that reduces a pipe wall to half its original thickness after 10 to 15 years is a reasonable target for common processes such as crude running, coking, fluid catalytic cracking, gas plants, alkylation units, and so on. The following section details the causes of corrosion in a petroleum refinery.

Sulfur

High-temperature sulfur corrosion is the most common cause of uniform thinning of process equipment. Figure 22–1 shows the effect of temperature on corrosion rates.[1] This figure may be used for oils derived from medium or high-sulfur crudes, in the absence of a substantial hydrogen partial pressure. The data are best thought of as the minimum corrosion rate because mechanisms such as pitting and crevice corrosion can greatly accelerate failure frequency. High vapor velocities (above 200 ft/sec) in piping and furnace tubes can produce accelerated corrosion attack due to droplet impingement.

Pitting

Often, a hydrocarbon leak starts out as a pinhole. Inspection of the inside of a pipe that has experienced such a failure will frequently show most of the pipe wall to be quite thick. However, small pits in the metal have ruined the pipe. Pitting occurs when the protective surface oxide or sulfide layer is removed in small areas. Once started—if the environment is right—the attack can accelerate because of differences in electrical potential between the large area of passive surface versus the small area of the active pits.[2]

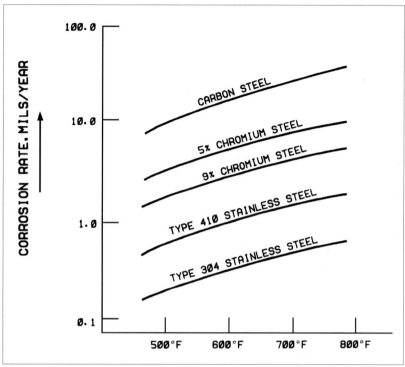

Fig. 22-1. Corrosion rates due to sulfur in hydrocarbons (adapted from McConomy).

Using the common Type 304 stainless steel (often called 18–8) in low pH chloride service (such as crude tower overhead condensers) will likely lead to severe pitting. Stainless steels that contain several percent molybdenum (Types 316 and 317) reduce pitting tendency in many environments. Stagnant cooling water left in stainless-steel heat exchanger tubes during shutdowns will lead to rapid, severe pitting.

Carbon steel piping exposed to low-pH water with SO_2 or SO_3 present is also subject to pitting. This has been noted on several sulfuric acid alkylation units. Type 316 stainless is recommended to resist pitting in this service.

Weld leaks

Corrosion problems on process units often first show up as weld leaks. More precisely, the failures occur in the heat-affected zone adjacent to the weld. The welding procedure lowers corrosion resistance at the metal grain boundaries in the heat-affected zone.

Stress relieving and improved welding techniques will not necessarily mitigate the problem. Use of low-carbon steels or Type 321 stainless (contains titanium) and Type 347 stainless (contains columbium) for replacement piping subject to frequent weld leaks will minimize the rate of failure.

Naphthenic acid

Organic acids present in crude oils are collectively called naphthenic acids. The concentration of these acids in different crudes is highly variable, and their corrosive effect on unprotected crude units can be very severe.

The neutralization number (mg of KOH to neutralize 1 g of crude) is a measure of its naphthenic acid content. A neutralization number above one indicates that naphthenic acid attack in the crude unit is going to be a problem. In addition, naphthenic acids will cause product problems. For example, virgin kerosene produced from certain crude oils contains high concentrations of naphthenic acids and, as a result, will not meet haze tests.

Severe naphthenic acid corrosion occurs primarily in vacuum distillation towers at temperatures in the range of 550°F to 650°F.[3,4] In piping, there will be an increase in general corrosion in direct proportion to an increase in neutralization number.

Absence of scale and severe pitting of the steel, as if it were caused by erosion, often is evidence of naphthenic acid corrosion. It should be noted, however, that this type of damage also is produced by droplet impingement at high velocities in the absence of naphthenic acids.

A crude unit which has metallurgical protection for running high-sulfur crudes is not protected against naphthenic acid attack. For example, naphthenic acids attack 12-Cr stainless steel and carbon steel with equal vigor. Type 316 or, preferably, Type 317 is required to resist naphthenic acid corrosion, but even these alloys will pit in severe cases. Other materials effective against naphthenic acid corrosion are aluminum or alonized steel.

Probably the most cost-effective metallurgy to use in vacuum towers exposed to naphthenic acid attack is type 316(L) S.S. The (L) denotes low carbon content. Recent operating experience has shown that 316(L) is quite a bit more resistant than type 316. While type 316(L) is not quite as rugged as type 317, it is a good deal cheaper and usually more readily available. Type 304 has, on the other hand, shown itself to be marginal in a vacuum tower exposed to naphthenic acid attack. High sulfur content actually helps suppress corrosion due to naphthenic acids.

H₂S

Hydrocarbon desulfurizers produce an environment with significant partial pressures of both hydrogen and hydrogen sulfide. This renders the metal more prone to high-temperature sulfur corrosion than in the absence of hydrogen.[5] Figure 22–2 is drawn for an H₂S partial pressure of more than 6 psi plus a significant hydrogen pressure such as found in desulfurizers. Note that the addition of low percentages of chrome does little to reduce corrosion rates. Type 304 stainless steel (18–8) is the preferred alloy for this service.

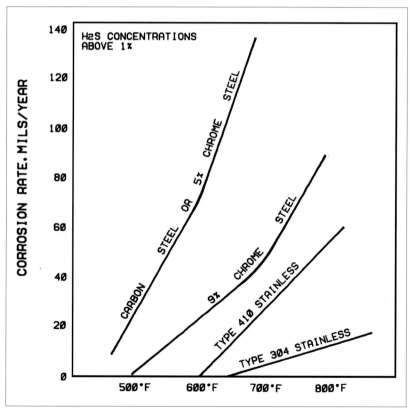

Fig. 22–2. Typical corrosion rates for H₂S/H₂ environments (after AISI).

Hydrogen blisters

Atomic hydrogen, formed from the dissociation of molecular hydrogen, may diffuse into steel surfaces. The hydrogen then recombines into its molecular form and becomes trapped inside the steel lattice. The subsequent pressure buildup results in blisters that eventually crack the steel, especially at the heat-affected zone of welded vessels.

This phenomenon is commonly encountered in high-pressure separator drums, absorbers, and strippers of vapor recovery sections in FCCUs. The presence of cyanides (due to incomplete water washing of cracked wet gas) activates steel surfaces and increases the rate at which atomic hydrogen enters the steel surfaces.

Gas scrubbers that remove H_2S by absorption in MEA solution are also subject to hydrogen blistering. When the MEA solution becomes overloaded with H_2S, hydrogen blistering is promoted at the bottom of the scrubber tower.

H_2 attack

High-temperature hydrogen attack of carbon steel occurs at temperatures above 500°F and hydrogen pressures above 100 psig. API Publication 941 provides guidelines and lists appropriate materials for this service.[6] Atomic hydrogen diffuses into and decarbonizes steel by combining with carbon to form methane, which then causes blistering.

Sulfuric acid

For acid used at ambient temperatures at strengths above 70% H_2SO_4, carbon steel is acceptable. Cast iron alloyed with silicon is best used for hot (150+°F), strong sulfuric acid. Weak sulfuric acid (less than 60%) destroys carbon steel piping at an amazing rate. In one instance, brand-new piping was rendered unfit for service after a weak sulfuric acid was accidentally charged through the piping for a single day.

Carpentor 20, an alloy containing large percentages of nickel and chromium, and Hastalloy B-2 provide excellent corrosion protection from sulfuric acid at any strength.

Ammonium hydrosulfide

Thermal processes such as coking, cat cracking, hydrodesulfurization, and hydrocracking all produce both NH_3 and H_2S. The resulting aqueous-phase ammonium hydrosulfide is antagonistic to carbon steel heat-exchanger tubes. Most of the ammonium hydrosulfide winds up in sour-water streams. Before disposal, the sour water must be steam stripped. The overhead condenser used for this stripper has an extremely corrosive environment. Titanium is preferred for this service.

Carbon dioxide

Concentrated CO_2 streams are encountered in ammonia- and hydrogen-producing plants. The carbonic acid formed aggressively attacks carbon steel. A more insidious problem is the formation of carbonic acid on the steam side of heat exchangers. As the steam condenses, traces of CO_2 in the refinery steam supply accumulate in the heat exchanger. The trapped CO_2 forms corrosive carbonic acid. If a leaking exchanger has corroded through from the steam side, the replacement tube bundle may have to be constructed from a chrome steel instead of carbon steel. Type 304 stainless steel is the preferred material against carbonic acid attack.

When a vent is left cracked open, CO_2 accumulations on the steam side of an exchanger will be eliminated. This practice effectively has reduced corrosion in many refinery steam reboilers, provided the vent is below the bottom channel head pass partition baffle.

Stress cracking

Exposed to the wrong environment, many alloys used in a refinery are subject to stress-corrosion cracking. Such cracks may propagate as fast as 0.05–10 mm/hr.[7]

Residual tensile stress resulting from fabrication is a major cause of stress-corrosion cracking. For example, welded carbon steel pipe can crack when exposed to a 10% caustic solution at 170°F or above. To avoid this failure, welded carbon steel piping sections that, even briefly, are exposed to caustic should be stress relieved.

Stress relieving is accomplished by heating the fabricated section in a special furnace. Alternatively, process equipment may be stress relieved in the field by electrical heating. The operating engineer should insist that replacement welded carbon steel piping be stress relieved if caustic can enter the piping system.

Admiralty cracking

An ammonia-rich gas stream was once accidentally charged to a gas plant. Shortly thereafter, leaks appeared in many of the tubes of the distillation towers' reflux condensers. The admiralty tubes (a copper alloy in the condensers) were cracked by the ammonia environment. Admiralty may also experience stress-corrosion cracking when exposed to concentrated amines.

Cracking stainless steel

Type 304 stainless steel, one of the most commonly used alloys in a petroleum refinery, is very vulnerable to chloride-induced stress-corrosion cracking. Since there is no way to stop stress-corrosion cracking once it starts, it is best not to expose Type 304 stainless or other similar alloys to a chloride environment.

Chlorides are routinely introduced into refinery streams primarily from two sources: salt in crude oil (most crude unit desalters are less than 95% efficient), and chlorides intentionally added to naphtha reformer reactors to promote catalyst activity. Although the amount of chlorides in a particular stream may be small, it is possible for the chlorides to concentrate at certain points. A reclaimer for an amine system is one such example. A rough guide to the time-temperature-concentration relationship to initiate cracking in Type 304 stainless is shown in Figure 22–3.

U-tube heat-exchanger bundles will have residual bending stresses. The bend may be vulnerable to chloride-induced stress-corrosion cracking unless the tubes are annealed to relieve the stresses. Type 304 stainless steel is also subject to stress-corrosion cracking when exposed to caustic solutions.

Polythionic acid cracking

Polythionic acid solutions can readily form at ambient temperatures in process equipment during shutdowns. Only the interaction of moist air and iron sulfide corrosion products is required.

The 300 series stainless steels—especially Type 304 (18–8)—are susceptible to rapid cracking by polythionic acid. Polythionic acid

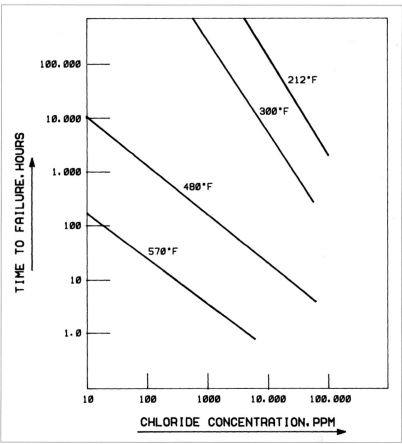

Fig. 22–3. Stress-corrosion cracking of type 304 stainless steel (18–8) in chloride solutions (after Lancaster).

cracking is a special form of stress-corrosion cracking in that it proceeds rapidly, even at ambient temperatures and moderate tensile stress. The 300-series alloys are made susceptible to this form of cracking by either operating at over 800°F, welding, or heat treating.

As iron sulfide is almost universally present in refinery process equipment, the danger of cracking such items as welded 304 stainless-steel piping during a unit outage due to polythionic acid attack is very real. Proper shutdown procedures are an effective way to reduce this hazard. In particular, circulating a neutralizing solution or purging with nitrogen during shutdown is recommended.

Type 321 or 347 stainless steel is one alloy in the 300 series that is far less prone to polythionic acid cracking than is 304 stainless steel.

The Dangers of Embrittlement

On a frigid winter day in the Midwest, a coking unit experienced an emergency shutdown. The coking heater was quickly cooled by natural convection. The heater's high (410 S.S.) chrome tubes had been operating at a 900°F metal temperature. After the shutdown, cold air rapidly chilled them to 0°F.

An engineering inspector entered the heater to hammer test the tubes. Hammer testing is a method to locate thin metal walls. A thin pipe tapped with a ball-peen hammer will resound with a dull, leaden sound. The inspector gave the first tube a firm blow. The tube cracked. Thinking he had located a defective tube, he continued hammer testing the tubes. Each tube fractured after being hit. After damaging a dozen tubes, the inspector realized that something was amiss and went off to seek expert consultation.

Rapid cooling had embrittled the chrome tubes. When struck with a hammer, the tubes fractured. Reheating the tubes to a moderate temperature (1,100°F to 1,200°F) and then slowly air cooling them would have removed the embrittlement and restored their ductility.

Loss of ductility is not a serious problem with carbon or low-chrome steels or with stainless steels containing nickel (300 series). However, when dealing with high-chrome steel pipe, the operating engineer should watch for embrittlement. In particular, when a unit has been shut down, any high-chrome piping that has seen service at temperatures in the 700°F to 950°F range must be treated with care.

Metallurgy in Extremely Rigorous Service

If money is no object, repeated equipment failures can be stopped by switching to expensive metals. In the long run, the expensive material may be far cheaper than living with repeated failures.

- **Titanium** is used in crude tower overhead condensers to replace carbon steel tubes. The titanium tubes withstand the effects of inadequate crude desalting and poor crude unit overhead-system pH control. Titanium tubes will be made of a thinner-gauge material than steel tubes. Thus, they are more subject to mechanical damage than are steel tubes.

- **Monel** is used in units exposed to wet HCl. Many polymerization and isomerization processes use aluminum chloride as a catalyst. When exposed to moisture, highly corrosive weak hydrochloric acid is liberated. In one coking unit, a combination of water in the fractionator reflux and high chlorides in coker feed combined to destroy 12% chrome trays. Replacement with Monel 400 (70% nickel, 30% cobalt) trays proved effective in resisting further corrosion damage.

- **Carpentor 20** is a good general-purpose alloy to use in sulfuric acid alkylation units. Field experience has shown it to be especially effective in high-velocity areas where acidic and wet hydrocarbon streams combine.
- **HK-40** is successfully used in very high-temperature service.

Alonizing

Alonizing is a process by which a steel surface is transformed into an iron-aluminum alloy. This is usually a cheaper alternate than using a high-grade alloy steel. Alonized coking heater tubes are reported to be less susceptible to coking. Alonizing reduces scale formation and thus reduces fouling of reactor catalyst beds. Occasionally, the alonizing job is incomplete and the base metal is left unprotected. This can cause the alonized process equipment to be vulnerable to rapid corrosion failure.

Plastic liners

Repeated failures in a piping system can be dangerous and discouraging. Sometimes one begins to wonder if the right metallurgy for the environment will ever be found. For this reason, many operating engineers try lined piping. A variety of plastic or Teflon internally lined pipe is available. Although it is quite expensive, the liners do seem impervious to the corrosive environments to which they are exposed.

However, in a number of services, internally lined piping has proved to be a failure. The lining has become separated from the inside of the pipe and restricted process flow. On one unit using extruded Teflon-lined pipe, the process flow was nearly stopped by the Teflon liner. Lined pipe does help, but it should not be considered a permanent substitute for finding the proper metallurgy for a system.

Heat-Exchanger Tube Fouling

Besides generally improved corrosion resistance, stainless steel tubes are less prone to foul in certain heat-exchange services. The stainless tubes maintain a smooth surface as compared to carbon steel tubes.

In high-temperature (700°F plus) hydrocarbon service, the smooth surface tends to reduce the buildup of coke. Other fouling deposits also adhere more readily to a roughened carbon-steel surface. A fluid catalytic unit slurry oil-to-fresh-feed pumparound exchanger is one service where replacing carbon-steel tubes with stainless steel has improved the overall heat-transfer coefficient.

Troubleshooting Checklist
for Corrosion Failures

Corrosive Environments

Sulfur causes uniform thinning

Pitting accelerated by
electrical potentials

Weld leaks at heat-affected zones

Naphthenic acid attack in crude units

Combined hydrogen sulfide
and hydrogen attack

Cyanides promote
hydrogen blistering

Weak sulfuric acid

Ammonium hydrosulfide
from cracking processes

Carbonic acid formed from
condensing steam

Stress Cracking

Rapid propagation of cracks

Carbon steel in caustic solutions

Stress relieving

Ammonia causes admiralty tubes
to crack

Chlorides initiate stainless
steel failures

Stainless U-tube bends vulnerable
to cracking

Polythionic acid cracking
during turnarounds

Embrittlement

High-chrome tubes lose ductility

Take care in hammer testing
when cold

Reheat to 1,100°F to 1,200°F

Protection from Corrosion

Titanium for crude-unit condensers

Monel in wet HCl service

Carpentor 20 for alkylation units

Alonizing coking heater tubes

Teflon-lined pipe

316(L) in vacuum towers

References

1. H. F. McConomy, "High-Temperature Sulfide Corrosion in Hydrogen-Free Environment," Proceedings of the American Petroleum Institute, Refining Division, Section 3, 1963.

2. Committee of Stainless Steel Producers, "The Role of Stainless Steels in Petroleum Refining," American Iron and Steel Institute, Bulletin SS-607-477-20M-HP, April 1977.

3. W. A. Derungs, "Naphthenic Acid Corrosion-An Old Enemy of the Petroleum Industry," *Corrosion*, December 1956.

4. J. Guitzeit, "Naphthenic Acid Corrosion in Oil Refineries," *Materials Performance*, October 1977.

5. A. S. Couper and J. W. Gorman, "Computer Correlations to Estimate High-Temperature H_2S Corrosion in Refinery Streams," *Materials Protection and Performance*, January 1971.

6. "Steels for Hydrogen Service at Elevated Temperatures and Pressures in Petroleum Refineries and Petro-Chemical Plants," API Publication 941, 2nd ed., June 1977.

7. Committee of Stainless Steel Producers, 1977.

8. J. F. Lancaster. "What Causes Equipment to Fail?" *Hydrocarbon Processing*, January 1976.

9. National Association of Corrosion Engineers, "Recommended Practice for Protection of Austenitic Stainless Steels in Refineries Against Stress-Corrosion Cracking," Standard PR-O1-70.

Unusual Noises and Vibrations

Many process equipment failures are preceded by unusual noises and vibrations. An alert operator can prevent damage to a vital refinery component by identifying such sounds. The troubleshooter may be asked to explain the cause of the noise and how to stop it. The origins of some of these problems are:

- Eroded valve internals
- Cavitating pumps
- Large pumps running at reduced capacity
- Surging centrifugal compressors
- Exchanger pass partition failing
- Flashing hydrocarbons in water lines
- Harmonic vibration in large exchangers
- Going fuel rich in a furnace
- Control valve screeching
- Water hammer in steam systems

After a few weeks on a unit, the operating engineer should be able to pick out unusual sounds from the general din. The cause of these sounds should be investigated as they invariably prophesy trouble for the refinery's process equipment.

Rotating Equipment

A large centrifugal compressor that has begun to surge will emit a deep, roaring sound at regular intervals of several seconds. A compressor in surge is operating in an unstable mode. Its rotor is sliding across the shaft radial support bearings and is pushing against and deforming the thrust bearing. This horizontal motion of the rotor will eventually damage the compressor's seals and bearings. Each surging sound indicates one more trip of the rotor sliding across the radial bearings. The louder the surge sounds, the more forceful and damaging is the horizontal motion of the rotor.

Compressors will eventually self-destruct because of surging. In general, a machine running at 3,000 rpm can withstand more surges than one rotating at 8,000 rpm. Regardless of its speed, one should immediately take corrective action upon hearing a compressor go into surge. Some of the causes and cures of surging are:

- **High discharge pressure.** Vent the receiver downstream of the compressor aftercooler to relieve the back pressure. Open up the water to the compressor aftercooler.

- **Low suction flow.** Open the antisurge valve. This allows gas from the discharge of the compressor to circulate back to the compressor suction.

- **High suction temperature.** Many installations will have the ability to inject a light hydrocarbon liquid upstream of the centrifugal compressor's suction. This liquid vaporizes and, thus desuperheats and cools the hot vapor flowing to the compressor suction. Typically, the vapor has overheated because the antisurge valve has been opened too far.

Cavitating pump

A centrifugal pump making a rattling noise is suffering from cavitation. This means that the suction pressure is insufficient to suppress vaporization inside the pump. The collapsing bubbles of vapor cause the rattling. The pump's seals, bearing, and impeller will be damaged if cavitation is allowed to continue.

The fastest way to stop cavitation is to reduce the flow by throttling on the pump's discharge. Then, raise the liquid level in the vessel from which the pump takes suction. If a steam turbine is driving the pump, slow down the turbine 100 or 200 rpm; this will usually stop the rattling.

Oversized pumps

Certain large centrifugal pumps will emit a periodic low-pitched sound when running at reduced capacity. Internal recirculation across the impeller is the cause of this noise. Damage to mechanical seal faces and the rotating assembly is the usual consequence of long operation at these conditions. The only operational change that will stop this sound is to increase the discharge flow.

Excessive vibration

Any piece of rotating equipment experiencing excessive vibration should be taken out of service before additional damage is done to its seals, bearings, or labyrinths. One quick way to determine the magnitude of the problem is to stand a nickel on its edge on the pump or compressor case. If the coin cannot maintain its balance, there is a good chance that vibration is too excessive to permit safe continued operation.

Heat-Transfer Equipment

One may expect rotating equipment to make odd sounds, but passive equipment such as heat exchangers should behave with greater decorum. That is possibly why noisy vibrations emanating from heat-transfer equipment are often such a mystery. Take, for example, the case of the vibrating waste-heat boiler.

Harmonic vibration

Many years ago, a furnace at the Amoco Oil Refinery in Whiting, Indiana, was retrofitted with a steam boiler to recover heat from hot flue gas. The gas was drawn through the boiler with a fan. Shortly after start-up, the operators noticed that when the fan ran at design speed, both the boiler and fan would loudly vibrate.

The natural reaction of the engineering staff was to overhaul the fan. The bearings were replaced, the rotating assembly rebalanced, and the fan itself repeatedly modified. Years passed and the trouble persisted. Field inspection of the boiler during turnarounds indicated that the intolerably loud noise was accompanied by tube failures at the baffles. Also, the observed shell-side pressure drop was higher than expected.

Finally, a novice engineer hypothesized that the problem was due to resonant vibrations induced by vortex shedding on the tubes. The baffle spacing was such that the frequency generated by the flue gas passing over the tubes became resonate.

Figure 23–1 shows the boiler arrangement. The young engineer further hypothesized that if dummy baffles were inserted between the existing baffles, the natural resonance of the boiler would be destroyed. This rather farfetched idea was tried with a great deal of skepticism. Much to everyone's surprise, the noise completely vanished.

The young engineer was Ron Cutshal. The skeptical manager was me. Ron took my position when I left Amoco in 1980. This is an outstanding example of how higher mathematics can be used to solve process problems. Energy savings were over 10 mm BTU/hr., for a tiny investment for Ron's dummy baffles.

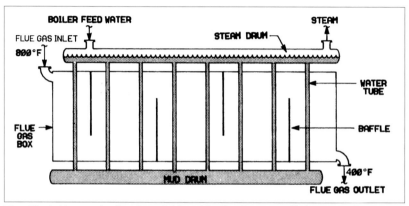

Fig. 23–1. Harmonic vibrations initiated by flowing gas will damage tubes.

Thumping furnace

To conserve fuel, the secondary air registers on a natural-draft process furnace should be partially closed. This will reduce the excess air and cut the loss of heat to flue gas. If the air registers are closed too much, the furnace will go air deficient. A firebox producing a drum-like thumping sound is not drawing in enough air to combust all of the available fuel.

To stop this noise, either throttle the fuel gas flow or open the secondary air registers. If you have just trimmed a furnace to minimize excess air, stay around for awhile and monitor the firebox to ensure that it does not start thumping. Air leakage into the convective section magnifies this problem.

Pass partition failure

The piping associated with one crude preheat exchanger started vibrating. After several days of operating with the vibration, flow of crude through the tube side of the exchanger suddenly stopped. Upon opening the exchanger's channel head, the pass partition (refer to figure 23–2) was found to be collapsed and lodged against the crude outlet nozzle.

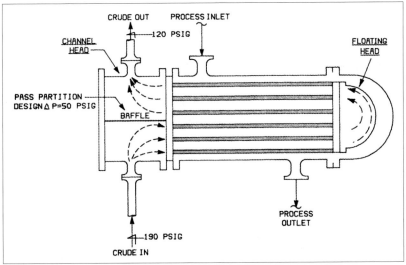

Fig. 23–2. Excessive tube-side pressure drop can bend pass partition.

The mechanical design had allowed for a pressure drop through the tube side of 50 psi; the operating pressure drop was 70 psi. Partially plugged tubes had greatly increased the normal crude ΔP. The pass partition had to withstand the high pressure differential. When it finally failed, it was pushed against the channel-head outlet nozzle.

Whenever an exchanger's piping is seen vibrating, the pressure differential across both the shell and tube side should be checked and compared to the design strength of the pass partition baffle.

Hydrocarbon leaks

Another cause of exchanger piping vibration is attributed to leaks. A light hydrocarbon liquid leaking into a lower pressure fluid will flash to a vapor. The sudden expansion generates pressure surges, which initiate piping vibration. Sampling the lower-pressure fluid for contamination with light hydrocarbons will reveal this problem.

Water hammer

To those who have lived in steam-heated apartments, water hammer is a well understood phenomenon. The loud clanging is caused by the rapid condensation of steam on cold metal or the mixing of steam with subcooled water.

A vertical steam reboiler, when operated at 20% of its design load, emitted bangs loud enough to awaken the dead. At 50% of design heat duty, it ran smoothly and quietly. A kettle reboiler produced a terrific steam hammer when it was first brought online. Once the contents had started boiling, it immediately quieted down.

A severe water-hammer problem is sometimes observed in a condensate collection system. When subcooled steam condensate and live steam mix in the same pipe, a strong water hammer will result. A malfunctioning steam trap, which allows steam to blow through into the condensate system, is often the culprit. Although the clamor generated by water hammer in steam systems may be quite deafening, refinery steam piping seems able to withstand a good deal of this type of punishment.

The cause of water and steam hammer is the same: sudden deceleration of water in an enclosed system. In the examples cited above, the rapid and localized condensation of steam accelerates a slug of steam condensate upstream by producing a localized area of low downstream pressure. The high-velocity water slams into a piping elbow or an exchanger channel head and produces the clanging sound associated with water hammer.

In a liquid-filled system, a sudden reduction in flow may also produce a banging sound. Again, the flowing velocity of water is dissipated in the form of a pressure surge which hammers against the piping system.

Screeching Control Valves

A control valve in good condition ought to give a positive shutoff. If a control valve is in a fully closed position but emits a whining or dull roaring sound, fluid is leaking past the control valve seat. If there is a large pressure drop across the valve (50 psi to 100+ psi), the noise can be unsettlingly loud.

A small stone or an old bolt can become lodged in the valve and prevent it from seating. The noise indicates that there is an extremely high-velocity fluid passing through the valve. Such a situation will erode the valve's internals if allowed to persist. Short of pulling the control valve out of service for repair, the best way to reduce the noise is to reduce the upstream line pressure.

If the normal operating position of a control valve is mostly closed and it makes a great deal of noise, pressure drop through the valve is excessive and its operational life will be shortened. A valve subject to severe vibration may be throttling a vapor-liquid mixture.

If vibration is diminished when the fluid temperature is reduced, this is almost certainly the problem.

Vibrating control valves

When a process-flow control valve is opened past a point, it may start to vibrate. The flow through the control valve becomes erratically low, and the pressure drop through the valve becomes erratically high. The vibrating sound coming from the valve indicates that the seat inside the control valve is loose. To prove this, leave the valve on automatic FRC and open up the control valve bypass a few turns. If the vibration stops, you can be sure the control-valve seat is loose.

Roaring block valves

The ability to locate a leaking block valve audibly is a skill that the troubleshooter will want to acquire. Hold your ear near the bottom of the valve and the sound of the rushing liquid will be quite distinct. Chances are, if you cannot hear any noise, the valve is not leaking through more than a few gpm.

Ordinary gate block valves are not intended for control purposes. If they are not left open more than two or three turns, the valve internals will erode due to the action of the high-velocity fluid. Then, when the valve is shut, it will leak through. In one service, 120 gpm of amine was found leaking through a closed 3-inch valve. The 300-psi pressure differential was sufficient to force this large volume through an eroded notch with a cross-sectional area much less than a dime.

Gurgling stripper

A side-stream steam stripper may sometimes be heard to emit a noise similar to water dripping into hot grease. This is, in actuality, a reasonable description of the gurgling's origin. Most often, a small amount of steam is condensing in the level gauge glass. The steam is drawn into the top tap from the stripper. As the steam condensate flows back into the stripper from the bottom-level tap, it boils. This can be stopped by the installation of an inert gas purge on the top tap of the level glass, to stop steam from entering the stripper.

Also, the stripping steam may be wet and the flashing water may be causing the noise. Wet steam can be identified by blowing the steam to the atmosphere through a ¾-inch bleeder. If the effluent from the bleeder stays white, right at the bleeder exit for 10 minutes, the steam is wet. Dry steam is invisible.

Troubleshooting Checklist for Noises and Vibrations

Centrifugal Compressor Surges
Rotor sliding across bearings
High discharge pressure
Low suction flow
High suction temperature

Pump Noise
Cavitation due to insufficient
 suction head
Throttle pump discharge
Raise liquid level
Internal recirculation

Heat-Transfer Equipment
Harmonic vibrations
Add deresonating baffles
Thumping process furnace
Exchanger pass partition failure
Light liquid hydrocarbon leaks
Water hammer in steam reboilers
Steam hammer in piping systems

Noisy Valves
Foreign object lodged in control
 valve seat
High fluid velocity
 damages valves
Block valves used for
 control purposes
Identifying leaking gate valves
Vapor-liquid flow into valve
Loose control-valve seat

Suggested Readings

Lowery, R. L., and P. M. Moretti. "Natural Frequencies and Dumping of Tubes and Multiple Supports." 15th National Heat Transfer Conf. AIChE Paper No.1, San Francisco, 1975.

Putnam, Abbott A. "Flow-Induced Noise and Vibration in Heat Exchangers." ASME Paper No. 64—WA/HT-21 (1964).

Walker, W. M., and G. F. S. Reising. "Flow-Induced Vibrations in Cross-Flow Heat Exchangers." *Chem. and Process Eng.*, November 1968.

Process Equipment

Better to light one small candle then to curse the darkness.
                    ~ Chinese proverb

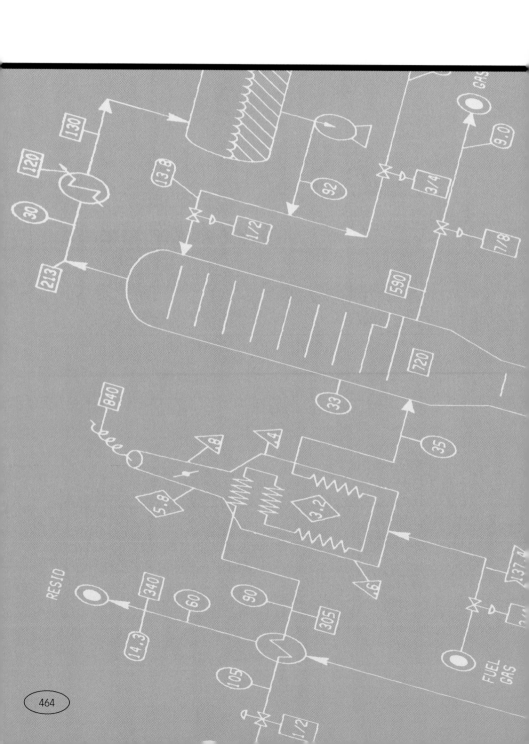

Natural Gas Drying

atural gas transported through common carrier pipelines must meet a moisture specification of 7 lb/MMscf of water. Gas is usually dried to meet this requirement by scrubbing with a concentrated glycol solution. Figure 24–1 shows a standard glycol contactor tower, generator, and pump.

Gas flows into the bottom of this tower where entrained water and naphtha drop out and are withdrawn under level control. The upflowing gas is contacted with the circulating glycol and dried. The glycol is pressured from the contractor to the regenerator, where it is heated to its boiling point to drive off water. Typically, 100 pounds of circulating glycol absorbs 3–4 pounds of water. After cooling, the reboiled glycol is pumped back to the contractor tower.

On the surface it would not seem possible that much could go awry with such a simple system. But, of course, the experienced process operator knows that it is only a matter of time before anything that can go wrong will go wrong. As a case in point, consider the operation of the glycol circulating pump.

This ingenious positive displacement pump is driven by expanding gas withdrawn, along with the wet glycol, from the contactor tower (see figure 24–1). The speed of this pump is set by a small valve that controls the amount of expanding gas emitted into the pump. An operator judges the amount of glycol circulation based on the audible strokes made by the pump's internals. The quicker the strokes, the greater the glycol circulation.

But suppose the pump has developed mechanical problems that reduce the volume of glycol normally pumped per stroke. Or perhaps the pump internals have deteriorated to the point that glycol circulation has stopped. Since glycol drying units are not normally equipped with flow meters on the circulating glycol, how can the process operator or the troubleshooting engineer recognize the problem?

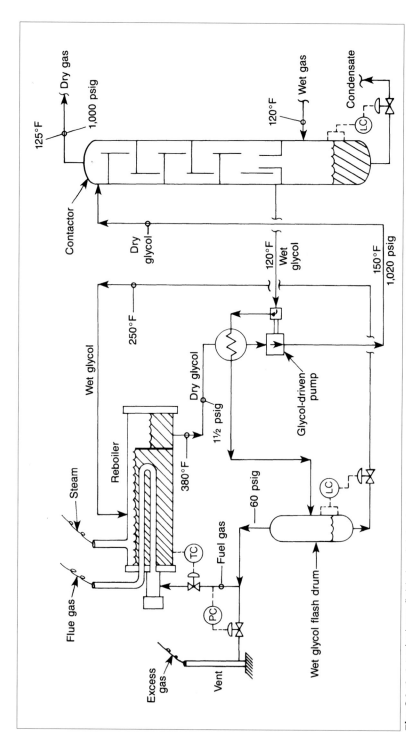

Fig. 24–1. A typical gas-field glycol dehydrator.

Glycol Pump Deficiencies

Our company's natural gas dehydration station was located in a picturesque section of the desert just south of El Gringo, Texas. When I arrived there on a Saturday evening to consult on an excessive moisture problem in our gas shipments, I was surprised to find the station deserted. Unlike petroleum refineries, natural gas processing equipment is designed to operate unattended.

The dehydration station consisted of six drying towers, each served by a dedicated glycol reboiler and pump. For the past two weeks the combined effluent gas from these six towers had become progressively wetter. Finally, the owner of the pipeline who received our gas drew the line: Either we dried our gas to the 7 lb/MMscf spec within two days, or we would have our connection to the pipeline blocked in. I had no idea which of the six parallel contactor towers was not drying the gas. Quite alone, and recalling tales of venomous snakes leaving their desert burrows at sunset, I began to investigate.

Indications of reduced glycol circulation

The first oddity I noticed was the noise from the vents associated with the individual reboilers. As figure 24–1 shows, the expanding gas used to drive the glycol pumps is also used as fuel to reboil the glycol. The excess gas not burned in the reboiler is vented under pressure control to the atmosphere. When the efficiency of the glycol pump is reduced due to mechanical problems, two factors act to increase excess gas venting:

- The reboiler firing rate drops because less glycol must be reheated.
- The amount of gas flowing from the tower to the glycol pump increases because there is less glycol liquid to restrict the flow of gas.

Hence, the net result of a reduction in glycol circulation rate due to reduced pumping efficiency is increased venting of excess natural gas. Of the six vents (one for each reboiler), only one was blowing hard. I also observed that the main burner on this particular reboiler was rarely on. *Note:* Temperature control on glycol reboilers works like your home heater—either full on or full off. Lack of firing on a glycol reboiler—that is, low reboiler heat duty—is another indication of a low glycol circulation rate.

The usual cause of glycol pump failure is deterioration of the O ring seals. Next morning, I requested that the suspect pump be overhauled. While this work proceeded, I continued my investigation.

Glycol regeneration temperature

The gas exiting the top of the contactor in figure 24–1 can be assumed to be in equilibrium with the reboiled—i.e., dry—glycol. The higher the glycol reboiler temperature, the dryer the glycol. The dryer the glycol, the dryer the treated natural gas. For most of the year in El Gringo, critical control of the glycol reboiler temperature gas was not vital. Relatively cool ambient temperatures maintained the top temperature of the contactor towers below 110°F. But now, in mid-July, this temperature was peaking at 122°F every afternoon. I checked my gas purification data book, [Kohl and Riesenfield's *Gas Purification* is an excellent data source for most types of glycol (Houston: Gulf Publishing, 1974)] and calculated that, for the 1,020 psig operating pressure of the contactors, it should be possible to meet the required moisture specification. My calculations were based on a reboiler temperature of 375°F. For triethylene glycol, the workhorse of the gas drying industry, the maximum recommended reboiler temperature to prevent thermal degradation of the glycol is 400°F. The six El Gringo dehydrator reboilers were all set to hold 375°F. But, by checking the actual reboiler temperatures with a calibrated thermometer, I determined that one of the reboilers was actually operating at 350°F as opposed to 375°F. This reduced temperature was sufficient to greatly increase the water concentration of the "dry" glycol, so that the moisture content of gas treated with this glycol stream doubled.

A simple recalibration of the reboiler temperature controller rectified this problem. Incidentally, operating a triethylene glycol reboiler at 375°F–400°F does not necessarily result in a noticeable increase in glycol degradation. The trick is to keep the glycol filters in good repair. Dirty glycol fouls the reboiler heat-transfer tube. This in turn causes hot spots on the heat-transfer surface, which accelerates thermal decomposition.

Leaking feed-effluent exchanger

The hot glycol from the reboiler is cooled by heat exchange with the wet glycol from the contactor. This heat transfer typically takes place in a double-pipe or plate-type exchanger. On one of the double-pipe heat exchangers, I noticed that the reboiled glycol was being cooled to a rather low temperature. I suspected that this could be an indication of a leaking feed-effluent exchanger. That is, cooler (120°F) wet glycol might be leaking into warmer (165°F) dry glycol. To verify my suspicions, I blocked in the dry glycol at the reboiler and at the suction to the pump. The appearance of a steady stream of liquid at an intervening bleeder confirmed that the feed-effluent exchanger was leaking. In effect, wet glycol was bypassing the reboiler and flowing straight back to the contactor tower.

After fixing the leak, this reboiler and the units that had suffered from an inefficient pump and a faulty temperature controller were put back online. The treated natural gas was checked and found to meet pipeline moisture specifications.

Further improvements

As a follow-up to the preceding problem, several other modifications were made to the El Gringo operation. To extend the life of the glycol pumps' O-ring seals, an aerial cooler, constructed from a section of finned-tube piping, reduced the dry glycol temperature by 20°F. Pumping the cooler glycol halved the amount of maintenance required on the glycol pump.

The composition of the glycol was also altered. A 50%–50% mixture of tetraethylene-triethylene glycol was substituted for the 100% triethylene glycol. This mixed glycol, while equally as effective for drying as its predecessor, is quite a bit cheaper than 100% triethylene glycol.[1] More importantly, it can be reboiled at a higher temperature to improve gas drying without encountering thermal decomposition.[2] Note that using pure tetraethylene glycol, while effective in a process sense, is much more costly than triethylene glycol.

Flooding Dehydrator Towers

The field supervisor's first indication of a flooded contactor tower is usually a report of excessive glycol loss. A check of a low-point bleeder on the gas pipeline downstream of the tower will show glycol. After refilling the glycol reboiler, the level in the reboiler gauge glass noticeably decreases after a few hours. This is a further indication of flooding. Of course, a dehydration system losing glycol this fast cannot dry natural gas on a continuous basis.

One simple explanation of such glycol losses is a leaking dry-gas-to-dry-glycol heat exchanger (figure 24–2). Note that the glycol pressure in this heat exchanger will be slightly higher than the gas pressure. To check for leakage, shut off and block in the glycol pump, block in the dry glycol at the contactor tower, and open an intervening bleeder between the pump and the tower. If gas does not blow out of the bleeder, the exchanger is not leaking.

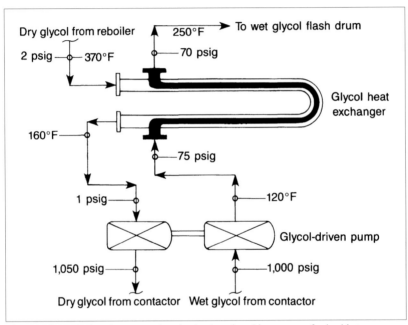

Fig. 24–2. Wet glycol may contaminate dry glycol because of a leaking heat exchanger.

Fouling versus flooding

A distillation column can flood due to tray damage, undersized liquid downcomers, high liquid level in the bottom of the tower, fouling, or excessive vapor velocity. Only the latter two difficulties are commonly encountered in natural gas conditioning. The troubleshooter should first check for flooding due to excessive vapor velocities. The following correlation may be used for trayed columns 2 feet or more in diameter with a standard 2-foot tray spacing:

$$V = 2.2 \left(\frac{[P \cdot 29]}{[T \cdot MW]} \right)^{1/2} \cdot D^2$$

where:

P = Tower pressure, PSIA

T = Tower temperature ($°F + 460°$)$°R$

D = Tower ID, feet

MW = Gas molecular weight

V = Volume of gas, MMscfd, that can be treated without excessive glycol losses

This equation is not intended for design purposes; rather it is based on field observations for towers exhibiting noticeable but tolerable glycol losses. These towers had been in service for some time and had been exposed to a moderate amount of fouling. If the actual volume (i.e., V) of gas exceeds the allowable volume as calculated above, you may be confident that an intolerable glycol loss is due to an excessive vapor velocity. Note that for sizing a new contactor tower, a coefficient of 1.8 in the above equation would be suitable.

Plugged trays

Drying towers in natural gas service can become rapidly fouled with drilling mud or formation fractionating sand (*frac sand*). The sand appears in the wellhead gas when the rate of gas production becomes excessive, and the sand is thus sucked out of the formation and into the well's tubing.

Drilling mud is found in natural gas for two reasons:

- A new well is not properly circulated and flowed-back to clear the drilling mud out of the production tubing prior to commissioning.

- During the drilling operation, excessive mud pressures are accidentally applied to the well, and the drilling mud is thus inadvertently forced into the producing formation. Some of this mud must eventually reappear in the downhole production tubing.

Not infrequently, a dehydrator loses its ability to dry gas from a field in which a new well has been put online. When this occurs, the culprit is invariably drilling mud plugging the contactor internals. For remote locations, one procedure that has proved to work is as follows:

1. A large water truck equipped with a pump to deliver about 50 psig, is sent to the site.

2. The dehydrator tower is blocked in and depressured. Both the tower inlet and outlet are disconnected from the gas piping. A special flange attachment, designed to mate up with a hose connection, is installed on the gas outlet line.

3. A 2-inch hose from the discharge of the truck's pump is connected to the dehydrator tower gas outlet line.

4. The pump is started and adjusted so that the pressure at the top of the tower—i.e., the water inlet—is about 5 psig. It is important not to apply too great a pressure because the trays could collapse.

5. Once the water draining from the bottom of the contactor tower appears clear, switch the water inlet to the bottom gas inlet. Overflow the tower until the water is again clear. The water overflow rate must be substantially higher than the normal glycol circulation rate to obtain enough liquid traffic to effectively wash the trays.

Why, you might ask, is it necessary to initially wash a badly fouled tower from the top down? A tray plugged with mud will severely restrict the flow of water. The resulting pressure drop may be sustained by the tray when it is pressed down onto the tray support ring; but the same pressure drop can rip the tray from its support ring when applied from the bottom of the tray.

In more accessible locations, it is a good practice to acidize a contactor tower after water washing. Acidizing consists of circulating an inhibited hydrochloric acid solution (typically 5% HCl) to the bottom of the tower with an acid truck. This is an effective method to clean contactors without promoting channeling of the gas flow through the trays. Acidizing is especially effective when iron scale deposits make up a portion of the fouling deposits. Including the acid disposal expense, acidizing a drying tower can cost $20,000–$50,000 (in 1985). When hydrocarbon deposits consisting of polymers formed in the glycol reboiler are the fouling component, a caustic wash, as opposed to acidizing, is in order. In the caustic washing procedure, a degreaser is also employed. A more thorough procedure is summarized in table 24–1.

Incidentally, once a tower has flooded to the extent that the liquid seal in the downcomers has been lost, the tower will continue to flood until the liquid seal has been reestablished. This must be done by interrupting the flow of gas through the tower, while glycol circulation is continued.

Table 24–1. Chemical cleaning a dehydrator contactor.

1. Circulate a solution of 9% sulfamic acid plus 1% citric acid plus 6% degreaser in hot water.
2. Circulate a 5% soda ash solution dissolved in water.
3. Circulate a 1% solution of whichever glycol is being employed.
4. Drain down and refill with concentrated glycol to normal liquid level.

After G. R. Daviet et al., "Switch to MDEA Raises Capacity,"
Hydrocarbon Processing, May 1984.

Dehydration Capacity versus Temperature

Three process requirements must be met for gas to be dried in a standard glycol dehydration unit:

1. The gas velocity through the contactor tower must not be great enough to entrain glycol into the dried gas. Theoretically, the entrainment of glycol does not interfere with drying. In practice, the continuous loss of glycol will knock a drying plant off-line as the unit's inventory of glycol disappears. Incidentally, it is not possible to measure the water content of gas containing a glycol mist.

2. The glycol pump must have the capacity to circulate enough glycol to absorb the water vapor contained in the natural gas. Of course, hotter gas can contain more water vapor. Increasing the gas temperature from 80°F to 100°F may double its water content.

3. The glycol reboiler must have a sufficient heat-duty capacity to regenerate the glycol at a high enough temperature to adequately dehydrate the gas.

As the temperature of the gas flowing through a dehydration contactor tower rises, its capacity will decrease as follows:

$$C_2 = C_1 \left(\frac{T_1 + 460}{T_2 + 460} \right)^{1/2}$$

where:

C_2 = Contactor capacity at temperature T_2, °F

C_1 = Contactor capacity at temperature T_1, °F

If a tower temperature increases from 80°F to 120°F, a tower's capacity will decrease by barely 3.5%.

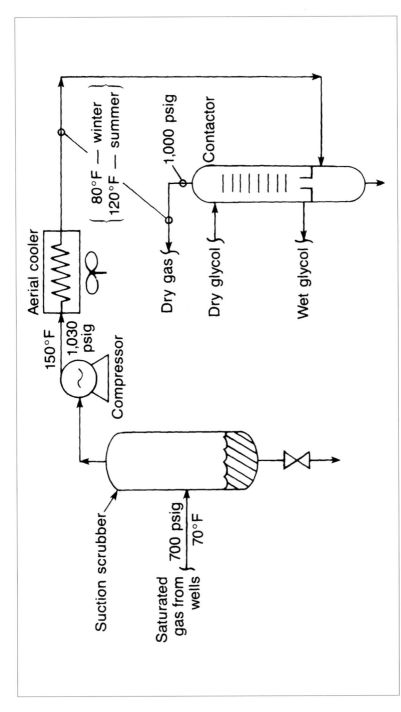

Fig. 24–3. Superheating gas through a compressor may not increase the required glycol circulation rate.

On the other hand, the amount of glycol circulation may or may not greatly increase as the gas inlet temperature rises. Figure 24–3 clarifies this point. A large booster compressor is serving a concentrated gas field. The gas produced from the wells enters the compressor's suction scrubber at a temperature independent of seasonal fluctuations. However, the aerial cooler on the compressor's discharge cools the gas to 80°F in the winter versus 120°F in the summer. Question: How much more glycol circulation is required to dry the gas? The requisite data to perform the calculation are given in figure 24–4.

At first glance, it would appear that three or four times as much glycol circulation is required. But remember that the 120°F compressed gas is not saturated with water vapor; it is really superheated. The compressed gas will have the same water content until it is cooled by the aerial cooler to below its dew point, in this case 79°F. If a contactor tower with 10–15 trays were employed, there would likely be no effect at all on glycol circulation requirement for the installation shown in Figure 24–3. For the typical six-tray contactor, industry correlations indicate that an additional 10%–30% of glycol circulation is needed; that is, far less than the 300%–400% required if the gas were saturated with water at the compressor discharge temperature.[3]

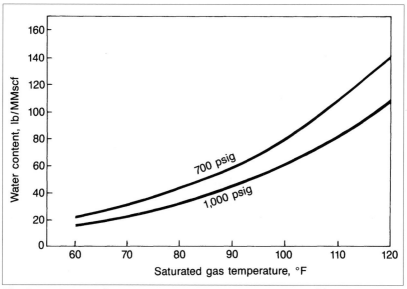

Fig. 24–4. Saturated gas temperature versus water content. (This curve applies only to saturated gas.)

Suppose, however, that the gas coming out of the ground is hot, perhaps 110°F. This gas, after compression and cooling to 1,000 psig and 120°F, would be saturated with moisture. Then, during winter operation, when the gas is cooled to 80°F, only one-third as much glycol circulation would be required as in the summertime. The condensed water corresponding to the difference in water content of 110°F, 700 psig gas versus 80°F, 1,000 psig gas would drop out in the bottom section of the contactor tower.

Glycol circulation vs tray capacity

If a 300% increase of glycol circulation is truly needed due to hotter saturated gas, will this increased liquid rate not affect the contactor's tray capacity? After all, in distillation column design, liquid flow rate over the tray weir is an important correlating parameter.

A glycol drying operation uses very small liquid rates in comparison to a distillation column. A typical 6-foot diameter drying tower may use 15 gpm of glycol flow, whereas a distillation tower of the same size may have 300 gpm of liquid flowing across its trays. Doubling or halving the glycol circulation rate does not appreciably affect the depth of liquid on the trays, and so it does not alter the trays' capacity. In simplest terms, a drying tower cannot be made to flood by speeding up the glycol circulation pump unless the glycol downcomers are partially plugged.

Tray design

To accommodate low liquid rates, trays of the design pictured in figure 24–5 are widely used in drying towers. The depth of liquid on the tray deck is such that the slots in the bubble caps are submerged. This forces the upflowing vapors to bubble up through the glycol. The depth of liquid on the tray is maintained by the height that the dual downcomer pipes protrude above the tray floor. The edges of these pipes are the equivalent of the straight outlet weir used on conventional distillation trays; the main difference is that there is very little height of liquid over the weir in glycol service due to the extremely low liquid rates.

Fig. 24–5. Typical bubble-cap tray (courtesy Smith Industries Inc., Houston, Texas).

Valve trays are used in lieu of bubble-cap trays in some drying columns. Valve trays are less expensive than bubble-cap trays and generally exhibit about 10% more capacity. In a practical sense, these advantages are outweighed by the superior turndown ratio of bubble-cap trays. I once had an occasion to run a field test on two drying towers operating in parallel. Both towers had the same number of trays. The bubble-cap trayed column dried its natural gas feed to 7.5 lb H_2O/MMscf at a flow rate varying from 70%–95% of design. The valve tray column produced 9.8 lb H_2O/MMscf at 65% of design capacity and 8.4 lb H_2O/MMscf at 90% of its design gas flow rate. I attribute this improved performance of the valve tray to reduced glycol leakage through the valves as the gas flow is increased. By contrast, a bubble-cap tray deck is leakproof.

The critical variable—reboiler temperature

Even if gas is superheated so that it does not contain any more moisture than a colder stream, it is much harder to dry. On one tower that was drying gas to 10 lb H_2O/MMscf, I tried to improve drying by doubling the glycol rate and halving the gas rate. The effect was nil. Only when I raised the glycol reboiler regeneration temperature by

10°F did the moisture content of the tower effluent gas diminish. It is all a matter of top tray equilibrium. That is, the moisture content of the dried gas cannot be any lower than the partial pressure of water in the glycol leaving the reboiler.

The temperature at which this equilibrium limit applies is the mixed temperature of the dried glycol and the wet gas plus the heat of condensation of the moisture removed with the glycol. Note that in this calculation the temperature of the gas is typically 25 times more significant than the temperature of the glycol.

As a rule of thumb, the glycol reboiler temperatures should be increased by 10°F for every 5°F increase in the equilibrium top tray temperature of the contact tower. Therefore, if you are drying gas from the discharge of an aerial cooler, you must raise your reboiler temperature by 10°F when the ambient temperature rises by 5°F to maintain a constant moisture spec in your dried gas. And remember, this is true regardless of the water content or flow rate of the wet gas.

So, we can see that the capacity of a contactor tower drying natural gas is not significantly reduced during hot weather. If, as the weather gets warmer, an operator neglects to increase the glycol circulation rate, it may appear to him that he must cut the tower's gas rate to maintain on-spec gas. Or he may run out of the reboiler heat duty required to heat the glycol sufficiently in hotter weather and also attribute this deficiency to excessive tower throughput. However, given sufficient glycol circulation of the proper moisture content, a contactor will properly dry natural gas until the trays in the tower flood.

Overheating glycol

The maximum recommended temperatures for continuous glycol reboiling are:

- Diethylene   340°F
- Triethylene   400°F
- Tetraethylene   430°F

Exceeding these temperatures is a self-defeating process. The glycol will begin to degrade. In this state it tends to foam in the contactor and cause premature flooding. A black, viscous glycol solution indicates that heavy hydrocarbons are carrying over in the gas to the contactor. A sweet, burnt-sugar smell, accompanied by a low pH and a dark but still transparent solution, signals that thermal degradation is occurring in the reboiler.[4] Salt laydown on the reboiler heater surface also produces the sweet smell indicative of glycol degradation.

The best way to improve drying when limited by the contactor top tray equilibrium is to use stripping gas in the reboiler. This is a patented process (U.S. Patent No. 4,179,328) and involves the injection of gas into the boiling liquid phase of the glycol reboiler. A horizontal sparger pipe is used to distribute the stripping gas, which reduces the partial pressure of steam in the reboiler and hence results in a drier glycol. The moisture content of the dried gas can be reduced by several lb H_2O/MMscf using stripper gas. Several scf per gallon of glycol circulated is a typical stripping gas rate. Unfortunately, all of the stripper gas is vented off the reboiler to the atmosphere and lost.

In summary, the essence of troubleshooting glycol dehydrators depends on differentiating between capacity and equilibrium problems. The glycol reboiler temperature and the pressure and gas inlet temperature to the contactor largely control drying equilibrium. The glycol pump, gas rate (on an actual volumetric basis), and the physical condition of the tower's trays determine the drying system's capacity limits.

I have written a separate book *Troubleshooting Natural Gas Operations* that discusses many gas field and processing problems. This chapter was partly based on this text. To order that book, email *norm@lieberman-eng.com*. The technology described is based on my experiences in Laredo, Texas in the mid-1980s. We produced gas from the H_2S-free Wilcox formation. I was in charge of 300 mostly depleted wells with production rates ranging from 30 to 300 thousand SCF per day.[5]

Troubleshooting Checklist for Gas Drying

Glycol Pump Deficiencies

Venting gas indicates reduced
  glycol circulation
Reduced reboiler firing
Deteriorated O rings
Minimize pumping temperature

Glycol Regeneration

Temperature
Must increase with hotter gas
Temperature controller calibration
Glycol filters avoid hot spots
Leaking feed-effluent exchanger
Tetraethylene versus
  triethylene glycol
Overheating leads to degradation

Flooding Dehydrator Towers

Glycol carryover
Plugged downcomers
High liquid level
Excessive vapor velocity
Fouled trays
Water washing drilling mud
  from a plugged tower
Acidize/degrease trays
Reestablished liquid seal
Foaming of degraded glycol

Dehydration Capacity

Circulation rate of glycol
Reboiler BTU rating
Effect of superheated gas
Contactor tray capacity
Glycol regeneration temperature
  versus contactor temperature
Stripping gas

References

1. R. J. Verritt, Manager, Glycol Product, KMCO Inc., Crosby, Texas, private communication to N. Lieberman, Jan. 25, 1984.

2. Silvano Grosso, "Glycol Choice for Gas Dehydration Merits Close Study," *Oil & Gas Journal,* Feb. 13, 1978, pp. 107–111.

3. Smith Industries Inc., Equipment Manual, "Section E: Dehydrators," Houston, Texas.

4. P. D. Hall et al., "Analytical Techniques Can Pinpoint Glycol Problems," *Oil & Gas Journal,* Sept. 24, 1979, pp. 176–188.

5. N.P. Lieberman, *Troubleshooting Natural Gas Operations, 2nd Ed.* Lieberman Publications, 2007.

Gas Compression— Centrifugal and Reciprocating

There are two general ways to compress gas in a refinery. Either a positive-displacement or a dynamic-type machine can be used. A reciprocating compressor is an example of a positive-displacement machine, whereas a centrifugal compressor is a dynamic machine. Centrifugal compressors are relatively:

- Expensive to purchase and install
- Energy-inefficient
- High capacity
- Low maintenance

By comparison, reciprocating compressors are:

- Cheaper to purchase and install
- Theoretically energy-efficient
- Low-to-medium capacity
- High maintenance

In practice, the maintenance effort required to keep reciprocating machines operating at their rated capacity and efficiency is so great that centrifugal compressors are preferred for most refinery services.

Centrifugal Compressors

These machines function as follows:

1. Gas flows into the suction of the compressor through the inlet guide vanes.
2. The first stage wheel (a rotating wheel attached to the drive shaft) accelerates the gas.
3. The velocity imparted to the gas by the first stage wheel is converted to pressure. That is, when the gas flows out of the wheel and into the stator or diffuser, the gas slows down. (The stator is a ring-shaped element fixed to the inside of the compressor case. It is stationary, hence the name "stator.") When a flowing fluid slows, its pressure goes

up. The higher-velocity gas exiting the compressor wheel decelerates in the stator. The kinetic energy imparted to the gas by the spinning wheel is converted to pressure in the stator.

4. Each set of a wheel and stator is called a stage. There may be three or four stages in a compressor case. Gas coming out of a compressor case is cooled before flowing to a second compressor case, which might also have three or four stages. All of the wheels in both cases are spun by the same shaft. The assembly of wheels and shaft is called the rotor.

The pressure developed by a centrifugal compressor is a function of:

- The number of wheels
- The speed of rotation
- The diameter of the wheels
- The volume of gas handled
- The density of the gas

As operating or technical people, we only need to worry about the last two factors. The effect of gas flow on pressure is shown in figure 25–1. Note that the vertical axis is labeled H_p. H_p is feet of polytropic head. Don't be confused by this term: It means feet of head for a compressible fluid. To relate feet of head to pressure increase, we multiply H_p by a number proportional to the density of the gas flowing into the compressor's suction:

$$\Delta P = H_p \cdot DV$$

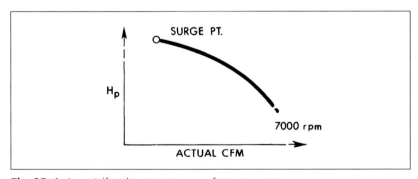

Fig. 25–1. A centrifugal compressor performance curve.

The DV (density of vapor at the compressor suction) is a function of:

- The molecular weight of the gas
- The suction temperature
- The suction pressure

Surge

Let's look at figure 25–2. The discharge pressure of the compressor is fixed by the absorber fuel gas pressure control (PC). If we start to raise the absorber pressure by closing this PC, the compressor will have to develop more differential pressure. That means it will have to produce more feet of polytropic head (H_p). As shown in figure 25–1 the compressor will be backed up on its curve. The volume of gas compressed will be reduced. The pressure in the wet gas drum will then rise. The other PC on the spill-back line will then start to close in order to reduce the volume of gas flowing back to the compressor's suction.

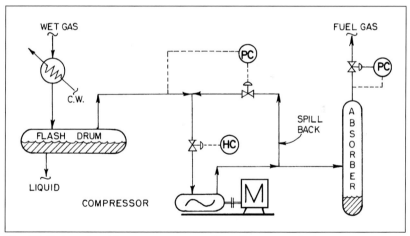

Fig. 25–2. A motor-driven compressor with anti-surge controls.

The increase in absorber pressure results in forcing the compressor to operate closer to its surge point shown in figure 25–1. If we raise the absorber pressure too much, the centrifugal compressor will stall or surge in the following manner:

1. The flow of gas across the rotor stops and reverses.
2. The gas flowing backwards pushes the rotor backwards.

3. The rotor slides back across its radial support bearings and slams against its thrust bearing. The purpose of the thrust bearing is to contain the axial movement of the rotor.

4. The compressor emits that horrifying sound characteristic of surge. Every surge corresponds to an impact of the rotor hitting the thrust bearing. Eventually, the thrust bearing is deformed and the axial movement of the rotor increases.

5. The rotor's wheels move closer and closer to the stators (which are fixed to the compressor case). When a wheel first touches a stator, both the wheel and the stator tear up and the compressor is wrecked.

Reducing the molecular weight of the gas will also cause a compressor to surge for the same reasons. I discussed this problem in detail in the section on FCCU flow reversals (chapter 7).

Stopping surge

A surging compressor is really attempting to overproduce polytropic head. As the operator, you are asking it to produce more polytropic head than it is really capable of developing. Each surge is a distress call for help: "Reduce your demand for polytropic head or I will die!"

To save the compressor we can reduce our demand for polytropic head by:

1. Raising the set point on the spill-back PC. This will raise the compressor's suction pressure by spilling more gas into the suction and allow the compressor to run out on its curve (see figure 25–1).

2. Lowering the absorber pressure.

3. Increasing the molecular weight of the gas. The denser the gas, the lower the Hp required to generate a given differential pressure (see chapter 7).

Overloading the driver

Let's assume we have a constant-speed, motor-driven compressor. The compressor is starting to surge, or stall, or slip, or become unstable. The operators pinch on the cooling water to the wet-gas condenser. The flash drum is slightly heated and the molecular weight of the gas to the compressor suction goes up 25%. The compressor now begins to develop more differential pressure for the same H_p. The flash drum pressure drops and the spill-back PC opens. As the spill-back valve opens, the volume of gas compressed increases.

Compression work (or amp load on the motor driver) is proportional to the volume of gas compressed. Therefore, any increase in the density of gas to the compressor will force the spill-back open, increase the number of moles flowing to the compressor and thus increase the driver horsepower requirement. We might even cause the motor to trip off due to over-amping.

Suction throttling

There are two extremes in operating a centrifugal compressor. At one end, we have surge, caused by low gas density and high discharge pressure. At the other end, we have driver overload, caused by high gas density and high discharge pressure. If we have a situation where the gas density is too high and the driver is over-amping, we can throttle on the hand control (HC) suction valve shown in figure 25–2. As we pinch on this valve, the flash-drum pressure will go up and the spill-back PC will start to close. With fewer moles of gas to compress, the amp load on the motor will fall. Of course, we are moving toward surge when we do this (see figure 25–1) so we can only pinch the HC valve so far. Naturally, we may use this technique to save electricity on a routine basis.

The quickest, simplest way to stop a compressor from surging, or stalling, is to lower the discharge pressure. If your car is starting to stall as you drive up a steep hill, try driving up a less steep hill. If your compressor is starting to stall at a 200-psig discharge pressure, try lowering the discharge pressure to 120 psig.

Centrifugal compressors are a complex subject. I have dealt with rotor fouling in *Troubleshooting Natural Gas Processing* (Lieberman Publications), and with design considerations in *Process Design for Reliable Operations*, Second Edition (Lieberman Publications). In this book, for use of centrifugal compressors in refrigeration service, see chapter 10. Chapter 7 discusses FCCU wet gas compressors.

Surge in variable-speed machines

When a variable-speed centrifugal compressor starts to surge, should its speed be increased or decreased to stop the surging?

Figure 25–3 shows a set of performance curves for a centrifugal (not an axial) blower. This type of blower is simply a single-stage, low-head, centrifugal compressor. It is a common piece of equipment in refinery sulfur plants.

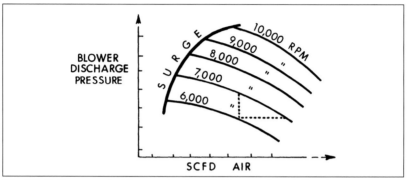

Fig. 25–3. Surge control is complex in variable-speed centrifugal compressors.

I noticed one evening in Texas City, during the 1980 strike, that I could precipitate surging (or unstable discharge pressure and air flow) by either speeding up or slowing down a blower. Referring to figure 25–3, assume I increased the blower speed from 6,000 rpm to 7,000 rpm. Would I find my new operating point by moving vertically up the dotted line (and thus closer to surge); or would I find my new operating point by moving horizontally across on the dashed line (and thus away from surge)?

If I held the blower discharge flow constant as I increased the speed, only the discharge pressure would increase, and I would move closer to surge. On the other hand, if I held the blower discharge pressure constant as I increased the speed, only the discharge flow would increase and I would move away from surge.

Depending on the relationship between pressure drop and flow caused by the equipment and controls downstream of the blower, an increase in blower speed could either cause a blower to surge, or stop it from surging. The shape of the surge curve, which is unique to each compressor, also influences the tendency of a machine to surge as its speed is altered.

There are a variety of computer software programs available to optimize the speed, suction throttle valve position, and spill-back valve position to minimize driver horsepower and keep a compressor safely away from its surge point while still controlling suction pressure.

Reciprocating Compressors

For many applications, reciprocating compressors (recip) are preferred over centrifugal compressors. For example, a recip will often be selected to compress hydrogen-rich gas. A centrifugal compressor would need many stages to boost the low molecular gas; while a reciprocating machine compresses light reformer off-gas as readily as 30 mol wt coker wet gas.

A reciprocating unit is (at least in theory) more energy-efficient than a centrifugal compressor. However, a centrifugal will preserve its efficiency better in a fouling service than a recip. In general, reciprocating compressors are cheaper to purchase and install than centrifugals. Also, damage to a centrifugal compressor will likely cost a great deal more to repair than damage to a recip. Finally, on-stream factors for centrifugal compressors in ordinary refinery service approach 95%; while reciprocating units might average 75%.

A simple example of a reciprocating compressor is shown in figure 25–4.

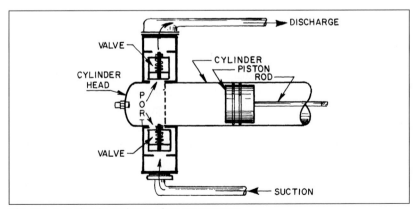

Fig. 25–4. A simple reciprocating compressor.

Evaluating lost compression horsepower

The first step in troubleshooting reciprocating compressors is to quantify the extent of the problem. How much compression work has actually been lost? An approximate rule of thumb is:

$$HP = n\left[\left(\frac{P_2}{P_1}\right)^{0.25} - 1.0\right] \cdot \frac{T_1}{520}$$

where:

  n = gas, MMscf

  P_2 = Discharge pressure, psia

  P_1 = Suction pressure, psia

  T_1 = Suction temperature $(460 + °F)$, °R

  HP = A number proportional to compression work

The next step is to decide if the lost compression work is due to an engine deficiency or a compressor problem. To ascertain that a gas-engine driver is not limiting compression work, the following questions should all be answered in the affirmative.

1. Are all engine exhaust-gas temperatures running below maximum?
2. Is the compressor running at its rated speed?
3. Is the fuel gas manifold pressure below maximum? (At a constant speed, the engines torque is linearly proportional to the fuel gas manifold pressure.)
4. Are all unloader pockets closed?

Next, check the unloader pockets. An adjustable head-end unloader is a device used to reduce the capacity of a cylinder, without reducing the compressor's efficiency. Figure 25–5 illustrates the function of an unloading pocket. Increasing the clearance between the piston and the cylinder head will reduce the volume of the gas compressed per stroke.

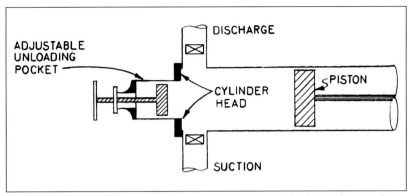

Fig. 25–5. An unloading pocket reduces engine load and volumetric capacity.

Unloader failure

Most large reciprocating compressors are equipped with pneumatically operated, automated unloaders. A malfunctioning unloader remains in an open position and thus reduces the capacity of the compressor. To identify this problem, proceed as follows:

1. Set the compressor to run at a constant speed.
2. Close the suspect unloader pocket and note the effect on the engine's fuel gas manifold pressure.
3. If the fuel gas manifold pressure does not increase, the unloader pocket did not really close, and it is probably broken. (***Note:*** For motor-driven machines, the motor amp loading gives the same indication as the fuel gas manifold pressure.)

Using this technique, I once discovered that one end of a compressor's two double-acting cylinders had a defective unloader. This failure reduced the capacity of the affected cylinder end by 40% and hence reduced the compressor's capacity by 10% (i.e., 40% × 25%).

Compression Work versus Temperature Rise

A rule of thumb for compression troubleshooting is that the theoretical increase of gas temperature due to compression is linearly proportional to a compression horsepower. A useful application of this rule of thumb is the approximation:

$$T_2 - T_1 = \left[\left(\frac{P_2}{P_1} \right)^{0.25} \right] - 1$$

where:

T_1, T_2 = Suction and discharge temperatures, °F

P_1, P_2 = Suction and discharge pressures, psia

Note that the temperature rise is independent of compressor speed, unloader configuration, or gas volume; it is only a function of the compression ratio and of course compression inefficiency. Figure 25–6 can be used to calculate the theoretical temperature increase for compressing natural gas. The 0.25 exponent assumes that the ratio of specific heat is 1.33, a typical value for natural gas.

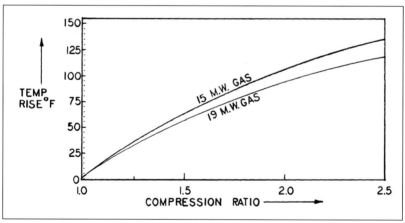

Fig. 25–6. Theoretical temperature rise due to compression.

Table 25–1 shows that for the compressors discussed, the temperature rise for the individual cylinders varied from 28°F for the No. 1 cylinder crank end to 42°F for the No. 2 cylinder crank end. The key point of this table is that compression efficiency varies inversely with temperature rise. As both the suction and discharge pressures were the same for all cylinder ends, the reason for the variable temperature rise was different efficiencies of compression. Because the work performed by the piston at each cylinder end was about the same (except for No. 2 cylinder head end which had a bad unloader), the observed temperature increases were inversely proportional to the gas flows. This means that if the No. 1 cylinder crank end was moving 30 MMscfd of gas, then the No. 2 cylinder crank end was moving only 20 MMscfd and the No. 1 cylinder head end was moving 23 MMscfd.

Table 25–1. Discharge temperatures of a two-cylinder, double-acting reciprocating compressor.

Compression End	Suction Temp., °F	Discharge Temp., °F	Temp. Rise, °F	Relative Efficiency
No. 1 cylinder crank end	60	88	28	100%
No. 1 cylinder head end	60	95	35	75%
No. 2 cylinder crank end	60	102	42	67%
No. 2 cylinder head end	60	90	30	93%

Cylinder temperature measurement technique

If individual thermowells are not available, you can still use the above technique to determine the relative compression efficiency of individual cylinder ends. A contact thermocouple may be used to measure the surface temperature of the compressor discharge valve. It is the relative temperature rise of the compressed gas that is of interest.

The compressor valve inefficiency corresponding to the excessive discharge temperature from the No. 2 cylinder crank end, could have been due to a variety of problems:

- Leaking piston rings
- Late compressor discharge valve closure
- Suction compressor valve leaking
- Late suction valve closure
- Discharge valve leaking
- High valve losses due to excessive flow

One way to discriminate between the possible problems is to perform a Beta Scan survey of the compression cylinders. A Beta Scan or Indicator Card is a pressure-volume diagram describing the actual compressor cylinder end performance. The pressure inside the cylinder is plotted against the piston position. A piston position of 0% corresponds to the piston position closest to the cylinder head. A perfect pressure-volume diagram is shown in Figure 25–7: This is the Carnot Cycle. Figure 25–8 shows Beta Scans for several maladies affecting compressor valves.

The Beta Scan plot obtained from the No. 2 cylinder crank end is shown in figure 25–9. This plot shows the valve installed in this cylinder end was experiencing a 25%–30% loss in compression work.

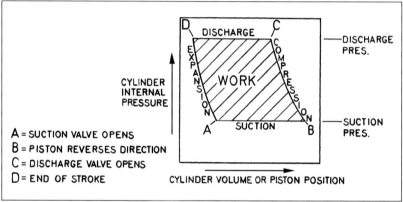

Fig. 25–7. Carnot Cycle for a reciprocating compressor.

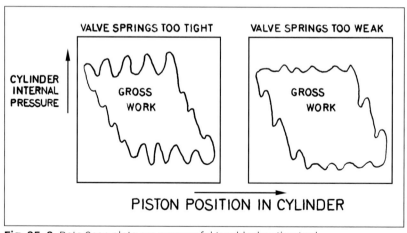

Fig. 25–8. Beta Scan plots are a powerful troubleshooting tool.

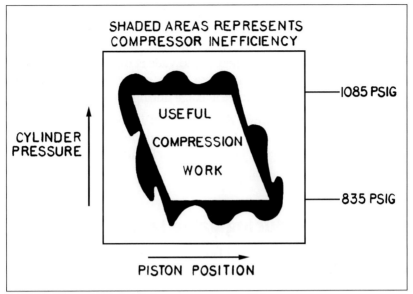

Fig. 25–9. Beta Scan plot defines compressor efficiency.

Interpreting Beta Scans

The area encompassed by the Beta Scan plot is proportional to the compression work performed by the piston. Unfortunately, not all of this work is of use in moving gas. For instance, the top horizontal line shown in figure 25–9 is the compressor discharge pressure. The area of the plot above this line is wasted compression work caused by:

- Pulsation in the discharge line
- Discharge valve opening too slowly
- Excessive resistance to flow of gas through the discharge valve

The bottom horizontal line in figure 25–9 is the compressor suction pressure. The area below this line also represents wasted compression work due to the same problems listed above, except only the suction valves are involved. The peaks and valleys indicated on the compression and expansion cycles are due to valve leakage and represent wasted work. There should be no gas flow into or out of the cylinder during the expansion or compression cycles. If both the discharge and suction valves did not leak, the lines on the Beta Scan plot representing the expansion and compression steps would resemble those of the Carnot Cycle; that is, smooth curves. Drawing a curved line tangent to the peaks and valleys of the expansion and compression steps inside the Beta Scan quantifies the extent of wasted horsepower due to valve leakage during these steps.

The shaded area shown in figure 25–9 is then the sum of the compression work wasted due to valve inefficiencies and piping pulsation problems. To this lost work must be added the detrimental effects of piston-ring leakage.

Reducing Valve Losses

One cost-effective means of reducing compression valve losses and enhancing compressor efficiency is to replace valve plates with thermoplastic plates equipped with additional flow ports (i.e., openings in the plates for gas passage). Modifying valve plates in this manner will reduce horsepower valve losses due to the frictional pressure drop.

Although modifications of this type save energy and enhance capacity, they are appropriate only in those cases where increased valve losses are related to increased gas flow. Large inefficiencies in reciprocating compressors are often related to increased compression ratios and not to gas flow rates. Compression leaks through worn piston rings and leaky valves are enhanced at higher compression ratios. Often, an inexplicable temperature rise across a compressor cylinder end, as reflected in a hot discharge valve cap, will moderate to a normal temperature rise when the compression ratio is only moderately reduced.

For example, for one machine equipped with plastic poppet valves (i.e., compressor cylinder valves designed for high capacity but low compression ratios), valve losses as measured by a Beta Scan were reduced from 25% to 10% when the compression ratio was reduced from 1.42 to 1.28. This occurred even though the gas volume moved through the compressor increased by 50%.

Reciprocating compressors may be limited by a third factor (in addition to engine horsepower availability and cylinder volumetric efficiency): rod loading.

Rod loading

One frequent cause of downtime in reciprocating compressor operation is rod breakage. Once the manufacturer's designated rod loading is exceeded, the rod will likely fail. Rod loading is calculated by:

$$\text{Rod loading} = A_p \cdot P_d - [(A_p - A_r) \cdot P_s]$$

where:

A_p = Piston area, sq. in.

A_r = Rod area, sq in.

P_d = Discharge pressure, psig

P_s = Suction pressure, psig

Thus, regardless of the horsepower load or speed, there is a maximum pressure increase that a reciprocating compressor can tolerate. The discharge pressure to be used in the above calculation is not the discharge line pressure; it is the peak pressure developed inside the cylinder (i .e., behind the discharge valve). As can be seen from the Beta Scan plot in Figure 25–9, this peak pressure may be drastically higher than the discharge line pressure.

Both pulsation problems and inadequate valve lift, or valve speed, raise the cylinder's internal peak discharge pressure. For example, a piston-rod failure on one compressor was precipitated when weak valve-plate springs were replaced with stronger springs to reduce valve leakage, thus requiring a greater valve-plate pressure differential to open.

Gas engine drives

Examining figure 25–10 would suggest that the operation of a reciprocating compressor is the essence of simplicity. Reality, however, belies this conclusion.

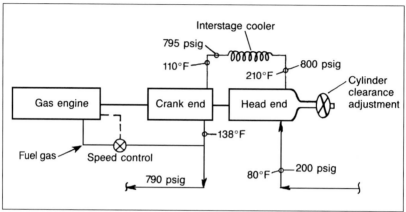

Fig. 25–10. A two-stage compressor set up for head-end only operation.

- **Engine speed versus horsepower.** An engine with a rated horsepower of 80 at 400 rpm delivers less horsepower at a slower speed. In practice this means that if the compressor discharge pressure is increased, we cannot simply slow the compressor to move less gas and maintain the overall engine horsepower output. Strange as it seems, it may be necessary to throttle the compressor suction to keep from overloading the engine when the compressor discharge pressure rises.

- **Suction pressure versus horsepower.** Figure 25–11 shows a typical relationship between the compressor suction pressure and the engine horsepower at a fixed discharge pressure. To the left of point A, the engine horsepower drops as the suction pressure decreases. To the right of point A, the engine horsepower rises as suction pressure decreases.

The explanation for this odd behavior lies in the combined but opposing effects that suction pressure exerts on the volume of gas moved and the compression work required per standard cubic foot per day (scfd) of gas compressed. Throttling on the suction reduces the scfd of gas compressed but increases the work required per scfd of gas.

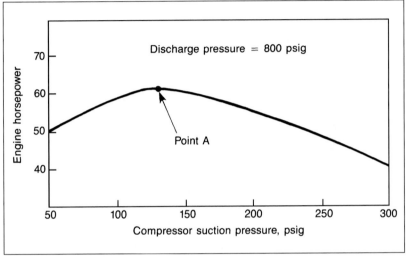

Fig. 25–11. A performance curve for a typical reciprocating compressor.

To calculate whether pinching the suction valve reduces or increases engine horsepower, use figure 25–12 and proceed as follows:

- Calculate the compression ratio, which is the absolute discharge pressure divided by the absolute suction pressure.
- Using Figure 25–12, determine the theoretical horsepower (Thp) per mole.
- Calculate:

$$hp = \frac{Thp \cdot Mscfd}{6.7}$$

where:

hp = Actual engine horsepower required to drive a typical compressor including auxiliaries

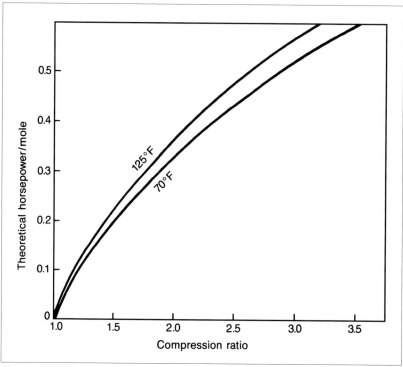

Fig. 25–12. Compressor horsepower for methane.

To confirm this calculation, watch the position of the fuel-gas control valve shown in figure 25–10. Observe if it opens or closes when the valve on the compressor suction is partially closed.

Unloading a Reciprocating Compressor

The machine shown in figure 25–10 has two compression stages. The first stage is called the *head end*; the second stage is referred to as the *crank end*. Each stage may be taken out of service by removal of the valve plate from the compressor intake valves for that stage. But why would we ever wish to take a compressor stage out of service?

Since the engine horsepower required may increase as the suction pressure rises, an engine might not be sufficiently powerful to drive both stages. If the engine is running at its maximum horsepower, it slows down as a result of the increased load. As the engine speed is reduced, it develops less horsepower and slows further. Finally, the engine stalls. One way to prevent overloading a compressor that is operating to the left of point A on figure 25–11 is to cripple one of the stages by removing the valve plates from the intake valves. Theoretically, this should cut the compressor horsepower in half. In practice, the effect is surprisingly different.

An unanticipated temperature change

One day a worker came to me with a problem.

"Why is this line hot?" he inquired. "This line is on the discharge of the crank end" (see figure 25–10), I explained, "and gas heats up as it is compressed."

"Ah yes, but the valve plates on the intake valves on the crank end have removed." The worker paused to inspect the control panel and continued, "See, there is no pressure rise across the second stage of the machine."

"It's just that the intercooler isn't working properly," I responded.

He persisted, "The second stage discharge is much hotter than the interstage cooler outlet. The gas is not being compressed, but yet it is heating up."

Back in the office, I calculated the horsepower lost in the disabled second stage:

$$\text{Whp} = \frac{\text{scfd}}{379 \text{ scfd/mole}} \cdot \frac{1}{24\text{hr/d}} \cdot \frac{10 \text{ BTU}}{\text{mole/°F}} \cdot \frac{\text{DT}}{2{,}500 \text{ BTU/hp}}$$

where:

Whp = Engine horsepower wasted in the disabled compressor stage

DT = Temperature rise across the disabled stage, °F

Next day I told the worker that 16 hp of the 80-hp engine was wasted in the disabled second stage. Also, that the disabled intake valves still permitted compression to take place, but that no net pressure rise resulted. To minimize this effect, the second-stage compression valves were entirely removed. The overall effect was to unload the engine. The temperature rise across the disabled stage dropped from 26°F to 16°F.

Adjustable clearance pockets

The best way to unload a reciprocating compressor is to increase the cylinder clearance. When the piston completes the end of its stroke, the space between the piston face and the cylinder head is called the starting volumetric clearance. Increasing the clearance reduces the scfd flow of gas to a compressor running at a constant speed and a fixed suction pressure. The effect of adjusting the cylinder clearance is proportional to the suction stroke volumetric efficiency (EVS). EVS is calculated as follows:

$$\text{EVS} = 100 - L - C \left[\frac{Z_1}{Z_2} \cdot \left(\frac{P_2}{P_1} \right)^{1/k-1} \right]$$

where:

C = Clearance based on cylinder volume, %
Z_1 = Compressibility suction
Z_2 = Compressibility discharge
P_2 = Discharge pressure, psia
P_1 = Suction pressure, psia
K = Isentropic exponent, typically 1.3 for a lean natural gas
L = Mechanical leakage correction factor, %
　(see figure 25–13)

From this equation, we can see that for small compression ratios, a large increase in the percentage of clearance has little effect; for large compression ratios a change in clearance has a drastic effect on the volume of gas compressed.

The two-stage compressor shown in figure 25–10 is restricted, due to its mechanical configuration, to adjustment of the head-end (i.e., suction) cylinder only. Therefore, if it is necessary to unload a compressor by disabling one stage, it is best to disable the crank end. If the head end is removed from service, fine-tuning the engine load by adjusting the head-end cylinder clearance is no longer possible.

Physically, the cylinder clearance is adjusted with a device similar to a rising-stem valve handle. Turning the handle clockwise pushes the adjustable cylinder head toward the piston, which increases the flow of gas through the compressor. This is the exact opposite of what normally transpires when we close a valve by turning the valve handle clockwise (see figure 25–1).

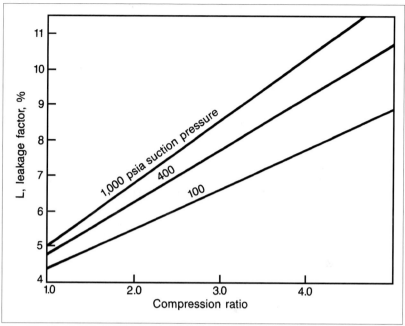

Fig. 25–13. Mechanical leakage reduces a compressor's volumetric efficiency.

Suction scrubber vessels

Liquids that enter a compressor cylinder will break the valve plates. For this reason, liquid knock-out drums called scrubbers are always placed on the suction side of reciprocating compressors. Accumulated liquid is drained under level control from the scrubbers through a dump valve.

These valves are prone to leaking through, which results in gas being lost. Such a leak is identified by comparing the scrubber temperature with the temperature of the scrubber-liquid dump line downstream of the level-control dump valve. Any noticeable temperature drop across this valve is indicative of a significant loss of gas.

Intercooler temperature

The required engine horsepower, but not the volumetric capacity, of a two-stage reciprocating compressor is affected by the intercooler outlet temperature. As a rule of thumb, each increment of 12°F increases the engine horsepower load by 1%. Frequently, all that is needed to minimize this problem is an occasional external water wash of the intercooler aerial fin-tube cooler.

Troubleshooting Checklist for Gas Compression

Centrifugal
Surge
Over-amping motor drive
Suction throttling
Fouled rotor
Gas molecular weight too low
Excessive polytropic head

Reciprocating
Running below rated speed
Horsepower increases as suction
 pressure rises
Two-stage machine not operating
 in tandem
Clearance pockets
 need adjustment
Level control in suction scrubber
Valve plates damaged
Too high spring tension
Valves fouled
Water wash dirty intercooler

Compression
Not running at rated speed
Horsepower higher at increased
 suction pressure
Two-stage machine not operating
 in tandem
Valve parts left in crank end when
 operating on head end
Clearance pockets
 need adjustment
Level control in suction scrubber
Damaged valve plates
Water wash intercooler

section **5**

The Process Engineer's Job

Thinking is good, but life is short.

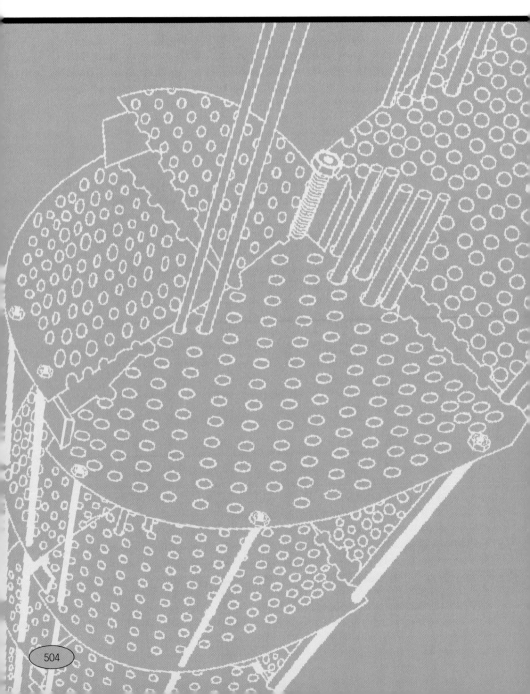

Suggestions for New Process Operating Engineers

The operations engineer's job is one of the most difficult in the refinery. With little supervision, the new engineer must quickly become an effective member of an operating team on a complex facility. The challenge is to get into a technical position from which you can effectively carry out the job. Your boss will likely inform you about what needs to be done and the objectives you might achieve. But you may not be told how to become knowledgeable in the process unit you will follow.

The program described is formulated to provide a firm, technical basis from which you can become a competent operations engineer. It will take one to two months to complete. You ought to take the time to learn the technical and operational aspects of a unit before becoming involved in business and administrative problems.

Drawing the Unit Flowsheet

Spend a day reviewing available process flowsheets, piping and instrument drawings, and operating manuals. Unless you have had an operating job before, you will not be able to understand the unit fully from these documents alone. Next, study the unit in the field. This is difficult and may easily take a week.

At the start, do not be concerned with control-room instrumentation. Take a marking crayon and trace through process lines and equipment. Mark the equipment number and function on each piece of process equipment. For instance, on a heat exchanger one would write:

Shell side	Hot diesel product
Tube side	Cold crude feed

Do not assume that any existing markings are correct. Draw your own conclusions; if it disagrees with the existing markings, check with an experienced operator. Indicate the inlet and outlet on process equipment with arrows. Mark all process pumps, drums, towers, compressors, furnaces, and exchangers.

From your field observations, draw a flowsheet showing only process equipment. This is a learning exercise, so do this even though an up-to-date process flowsheet already exists. At this point, do not include utility and process controls.

You should now know the location, function, inlets, and outlets of all process equipment. Next, retrace the lines connecting process equipment. Locate all process control valves. Mark the control valve number and function on the valve body. For instance, a typical notation on a control valve might be:

FRC-327

Debutanizer Reflux

Show the direction of flow through the control valve and the direction the control valve travels to close. After locating each process control valve, draw them on your process flowsheet. Show the appropriate control valve numbers.

Next, relocate each control valve. Determine the transmitter that is controlling each valve. For example:

- For flow controllers, find the associated orifice flow taps.
- For pressure recorder control, find the associated pressure taps.
- For temperature recorder control, find the associated thermowell.
- For level recorder control, find the associated level float chamber.

On each transmitter mark its function. For example, on a level float chamber write:

LRC-48

Depropanizer Bottoms

Note the actual physical location of thermowells. Make a list of local flow indicators (FI), temperature indicators (TI), level gauges, and pressure gauges. After finding a transmitter, enter it on the process flowsheet and show which valve it controls. Also, draw in local instrumentation.

Trace through the cooling water system. Starting at the cooling water pumps, mark the cooling water lines and control valves, as per above. Repeat this procedure for the unit's steam and condensate systems. Treat each pressure steam system as a separate system. Find and mark the valves that control the various steam pressures. Do not forget the fuel-gas and fuel-oil systems. Include items such as the fuel gas knock-out drum.

Eventually you will want to learn the other systems on the unit (plant air, instrument air, sewers, pump-outs, nitrogen, caustic). These are best left alone at the beginning. Having traced through the utility systems, sketch them in on your flowsheet. You should now know the "where and how" of the unit operation. The crayon markings will provide you with a quick refresher when you need it on a troubleshooting job.

Learning the Outside Operators Functions

The process operating engineer is dependent for information on communications from the outside shift operators. To sift the truth from fiction, one needs to know what these people do.

Work with a competent outside operator. Follow him around for a week as he rotates through the unit's jobs. Do this on the 4 P.M. to midnight shift, as this will not interrupt your regular duties. In a week, you ought to be able to learn the routine outside jobs. Find out where all of the sample points are and which vessels are routinely drained or vented. What are the critical operating parameters that the outside person follows?

The Control Board

Having studied the process flow in the field, you will find it much easier to understand the control board. However, it is difficult to get a feel for the dynamics of the unit's operation without hands-on experience with the controls. Ask the chief operator if you can take a turn running the control board, under supervision, for a few hours.

Take a look at each individual instrument. Do you know which pointer is the set point and which indicates the actual variable? Did you know that the indicated position of a control valve on the board is the air pressure to the diaphragm of the valve and not its actual position? How is a controller switched from auto to manual? Refer to chapter 28 to help answer these and similar questions.

Designate on your flowsheet the indicators that are displayed on the control panel; also show which valves are controlled from the control board. Mark all the level alarms on your flowsheet. Find the mercuroid switches or level float pots outside that activate the inside level alarms. These are located on the process vessels. With a crayon, indicate their functions.

Bottlenecks

One of the most important job functions of the process operating engineer is defining projects to increase unit capacity. To start on this, you need to determine the unit's bottlenecks. List individual pieces of process equipment. Then investigate, in the field, how the equipment limits throughput or product recoveries. Interview experienced operating personnel and get their help in completing the list. For example, if you observe a reflux control valve on the discharge of the debutanizer reflux pump wide open, you would note:

P-303, limits debutanizer fractionation, results in high vapor-pressure gasoline

Or if you find the fuel gas control valve to a furnace wide open, you would note:

B-201, inadequate burner heat release, limits crude run

The unit bottleneck list is the starting point for developing a project list. You will want to revise the bottleneck list in subsequent months as you make additional field observations at higher unit throughputs or varying operating conditions.

This list is also a good learning tool. Having completed it and the previous steps, you should have the knowledge to become a productive member of the operating team. However, before becoming involved in longer-term activities such as budget preparation, project definition, and operator training, you will want to organize and execute a performance test.

Your First Performance Test

The ability to run a complete performance test requires experience and preparation. For a first performance test, you should simply obtain a complete unit pressure profile. This is done with a single pressure gauge, which you move downstream from point to point. Obtain pressure drops across heat exchangers and control valves. Then, obtain a complete temperature survey. Where no thermowells exist, use an infrared temperature gun. Next, record all flow rates. You will probably find many flow inconsistencies. Finally, obtain a complete set of samples. Learn how and where the samples are taken (see chapter 27).

For your first performance test, do not be concerned with maximizing throughputs or optimizing operating parameters. A detailed discussion on how to carry through a complete performance test is presented in chapter 27.

On the process flowsheet you have drawn, enter the operating conditions observed during the performance test. Then calculate exchanger and furnace duties; calculate internal, unmeasured material flows; and do an overall component material balance. Enter the calculated values on your unit flowsheet. There will be plenty of inconsistencies, but they can be resolved in the months to come.

Personal initiative and technical expertise do not count for much on a refinery process unit if the engineer does not know the details of the unit operation. Trying to introduce new technical ideas without first completing this prerequisite will only alienate the shift operators and nothing will be accomplished. Process units are like people: each is unique. Get to know the individual characteristics of your unit before you try to change them.

Look for the Small Problems First

I was called to a refinery to help determine why kerosene production had declined 3,000 B/SD. The refinery manager explained that a process engineer had spent six months on the problem and had run several computer simulation analyses on 16 different crude oils, including complete tray-to-tray heat and material balances for three operating modes.

The result of the involved simulation study led to the recommendation that improved tower control, using advanced, feed-forward computer control and increased operator attention would maximize kerosene production.

I was given the computer simulation results to take back to the office to study. But before leaving, I went out to the crude unit to examine the operation directly.

I found that the instrument air tubing connection to the kerosene-draw control valve was loose. That valve controls the flow of kerosene from the crude tower to the side-stream stripper. The control valve simply wasn't getting enough air to open fully. I tightened the connection, and kerosene production rose to the normal rate.

It is essential that process engineers and operators look for the simple problems that can cause big losses in production before jumping into a massive computer simulation analysis to determine why a unit is not operating properly. Those analyses should rather be used to increase the capability of a properly operating unit.

It should be understood that computer simulations are usually based on a properly operating unit, with only some major unit problems built in. As a result, they do a remarkable job of determining new operating conditions when feedstock or product specifications change, increasing the capability of a properly operating unit, or helping to find a major unit problem. But they are just not sophisticated enough to simulate a loose tubing connection, or other simple problems. Those can only be found by close examination of the unit—not in the office poring over data sheets but out in the plant looking at all of the components, pipes, controls, valves, and tubing.

Another thing should be remembered: Advanced, computerized controls have been proven to provide substantial improvements in product yields and quality by their ability to control setpoints much tighter than manual or analog control methods. But these advantages can only be achieved when a unit has already been operating properly and producing product within specifications.

The fanciest, most sophisticated controls in the world cannot make a poorly operating unit operate well.

How to Learn a Process Unit

Drawing a Process
Flowsheet
Review operating manual
  and drawings
Trace process lines
Locate pumps, exchangers,
  and vessels
Label lines and equipment
Draw a process flowsheet
  from observations
Locate control valves
Locate instrument transmitters
Trace the utility and fuel system
Find level alarm pots

Learn Outside Operators' Duties
Follow experienced operators
Cover all the outside jobs
Work on the 4 P.M. to midnight shift

Control Room
Instruments that indicate
Instruments that control
Alarm points and trip levels
Try for hands-on experience

Completing Unit Flowsheet
Run an abbreviated
  performance test
Enter observed data on flowsheet
Calculate thermal duties
  and unmeasured flows
Overall component
  material balance

Planning a Performance Test

"**H**ow do you know operations have deteriorated so badly," queried the engineering manager, "if you don't have data showing when they were better?" The manager was referring to a report just presented by the unit process engineer claiming energy use and product fractionation at the refinery's crude unit had badly deteriorated. It appeared as if no one had bothered to run a performance test on the crude unit in the past 10 years.

A well-documented performance test report is an excellent starting point for almost every troubleshooting assignment. The report documents the capabilities of the unit at a particular point in time for future process engineers. Perhaps more significantly to the current process operating engineer, a well-run performance test highlights unit limitations and operating deficiencies.

In many refineries, a person is assigned as the process operating engineer to follow a particular unit for two years. One of the job goals during this period should be to run a thorough unit performance test. The data gathered during the performance test should be sufficient to calculate the mass, energy, and hydraulic characteristics of the unit as follows:

- An overall unit material balance. The pounds of feed to a unit ought to equal the total pounds of product.
- A measurement of the degree of fractionation between products.
- An overall unit energy balance. The energy content of the steam, electricity, and fuel used should equal the total BTUs rejected to air and water coolers plus the enthalpy difference between feeds and products.
- A measurement of the heat-transfer coefficient of the heat exchangers.
- A pressure drop survey. The total head developed by the pumps and compressors should equal the pressure loss in the unit plus the pressure difference between feed and products.

- A measurement of the hydraulic efficiency of pumps and compressors.

In summary, the unit performance test is a strategy used to follow the changes in pounds, BTUs, and hydraulic head.

How to Start

The success of a performance test is directly proportional to the amount of effort put into its planning. As illustrated by the following list, this planning will be a many-faceted undertaking:

Laboratory

The refinery lab must be ready to handle the large number of samples that will be taken during the performance test. The lab should be apprised of the number and type of samples to be run. Most importantly, the lab supervisor must be committed to completing the sample analysis by some reasonable date. Many performance tests are compromised by samples being lost or improperly handled in the lab. The longer the samples sit on a shelf, the greater the chances are of this happening. Do not forget that the lab supervisor will need to schedule technician overtime to support the test.

Instruments

Flow indicators should be zeroed and calibrated. The orifice taps for flow meters should be blown out and re-packed with glycol, and the process engineer must make sure that every flow meter needed for the test actually exists in the field as an operable entity. A preliminary unit material balance should be made. This will delineate if any of the feed or product meters are badly out of calibration.

Pressure and levels indicated in the control room should be verified by checking the outside readings. Do not bother to change pressure gauges in the field; a single, calibrated test gauge should be used for all pressure measurements. Board-indicated thermocouple readings are usually quite reliable. However, all dial thermometers should be removed and checked for accuracy. Be careful when unscrewing dial thermometers. The thermowell may be leaking. If oil is leaking around the threads, screw the thermometer back in place.

A few hours spent with the instrument mechanic, explaining the purpose of the test, will be a worthwhile investment.

Interfacing with other units

Units sending and receiving products from the facility being tested should be advised of the test date. A meaningful performance test cannot be completed when feed and product flows vary.

Shift operators

The chief operators and panel operator on shift during the performance test are the keys to success. These people must be personally advised and consulted before the test commences. Simply handing the chief operator a set of written instructions during the test is insufficient.

Mechanical department

A pipe fitter will be required during the test to move the calibrated pressure gauge for the hydraulic survey. The refinery's mechanical department should be lined up to provide the services of this craftsman.

Flag sheet

Recording data will be one of the principal activities of the process engineer during the test. The best way to do this is to enter the data directly on a unit flag sheet, such as is shown in figure 27–1. Except for sample analysis, the flag sheet shows all the test data. The advantage of entering the data on the flag sheet (which is a modified process flow diagram) is that it highlights data inconsistencies that can be resolved during the test. Recording data in only a tabular format makes it more difficult to locate potential errors.

On-stream analyzers

In general, on-stream analyzers are not as reliable as lab analysis and should not be used. However, for measuring the oxygen content of furnace flue gas or the H_2S and SO_2 content of sulfur-bearing gas streams, portable analyzers are best. These portable instruments need to be obtained and calibrated in advance.

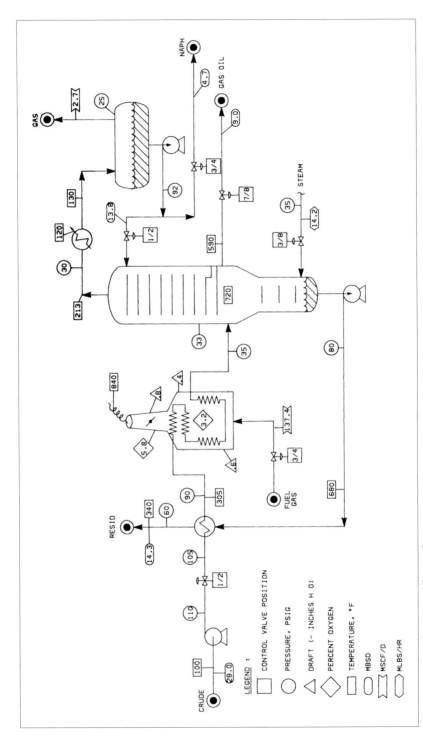

Fig. 27–1. Flag sheet used for performance test.

The Day Before the Test

This is the critical day. The roadblocks encountered the day before the performance test can be resolved through hard work. If these same problems crop up the next day, a meaningless test can result. Some mundane but critical details to attend to are:

- **Sample containers.** The required number of bottles and cans should be obtained. This is especially important in regard to gas chromatograph pressure sample bombs, which are usually in limited supply in most refineries. Label each container with a weatherproof tag, as it usually rains on performance test days. Differentiate the sample tag from those tags used for routine check samples.

- **Sample points.** Verify that all sample points are unplugged. Are sample coolers operable? Do the points where a gas chromatograph bomb sample is to be obtained have the necessary connections? Ask the chief operator if he feels that all the samples designated to be taken can be obtained safely.

- **Pressure points.** Tag and number all locations where a pressure reading will be taken. Verify that each point is unplugged. Then have a suitable connection installed to accommodate the single pressure gauge that will be used during the test. These connections must be assembled so that the pressure gauge can be read in an upright position. A gauge will read differently when installed upside down or at an angle. Make sure that all draft gauges used on furnaces are unplugged and zeroed. If the draft gauge reading changes when the flexible tubing is slid up and down on the gauge connection, the gauge connection is plugged.

Optimize unit

The performance test should serve as a bench mark for operations. Hence, unit operations during the test should be the best obtainable with the resources at hand; for example:

- Optimize furnace excess air.
- Close the furnace stack damper consistent with obtaining a slight draft at the inlet to the convective section (0.1 inches of water maximum).
- Minimize the tower top reflux rates consistent with meeting product specs.
- Optimize the tower pumparound rates, consistent with operating targets.
- Run the turbine drives at the minimum speed consistent with head requirements.
- Line out the unit at a constant feed and product rate.

Written instructions

Prepare a one-page typed note addressed to the shift operators. Explain the objectives of the performance test, the type of data to be collected, and the duration of the test. Have the unit superintendent sign the note.

The Day of the Test

If the process operating engineer has prepared correctly, the actual performance test will go smoothly. Hand every person who will participate in the test a copy of the written instructions and the flag sheet. As the data are assembled, enter them on the master flag sheet. When the unit seems pretty well lined out, have the pipefitter begin the pressure survey. Typically, this activity will take the longest to complete.

Then begin taking the unit samples. Some experienced engineers take duplicate samples. Send one to the lab and save the other, just in case of a mix-up. It is best to personally observe the shift operators when they catch the samples. The performance test samples will likely be obtained at unfamiliar spots, and erroneous samples may thus be submitted. More frequently, the person catching the sample will not allow sufficient time to thoroughly flush out the sample line.

Measure the furnace oxygen content both ahead of and downstream of the convective section. Record the temperatures and flows on the flag sheet. Note the operating position of flow control valves and liquid levels. If an exchanger or compressor bypass is open, record the valve position. Try to estimate the volume of products lost to slop or sewers.

With all the information (except the lab data) assembled on a single flag sheet, perform a quick engineering analysis of the test. Are there any glaring inconsistencies in the pressure survey? Does the total of the recorded product flows equal 150% of the feed rate? Are the temperatures shown in violation of the laws of thermodynamics?

The best time to correct these errors is during the performance test rather than two weeks later when the data are being analyzed. An example of a completed flag sheet is shown in figure 27–1.

Correlating the Data

It is an interesting and instructive exercise to take a copy of the original unit design process flowsheet and pencil in the corresponding performance test data. Beyond this, the operating engineer should compare the unit design parameters against actual operating conditions and ask himself, "Why are they different?"

Material balance

If a reasonable overall comparison of pounds of feeds and products (±5%) cannot be calculated, the validity of the test is compromised. A table showing a component balance should be made. Present these data in mol/hr and B/D. For example, on an alkylation unit it is important to determine if the moles of isobutane consumed equal the moles of alkylate produced. On a crude unit one might hope that all the barrels of pentane in the crude charge can be accounted for in the products. A sample component material balance is shown in table 27–1.

Summarize the degree of fractionation between products. A table or curves comparing current operations with past or design valves is instructive. Figure 27–2 shows how to present such data for a crude unit.

Table 27–1. Sample of a Component Material Balance for a Gas Pant, mol/hr.

	Methane	Ethane	Propane	Isobutane	n-Butane	Pentanes	Hexanes
Feeds							
Wet gas	104	47	39	18	17	7	2
Unstabalized naptha	3	11	42	28	32	47	77
Total	107	58	81	46	49	54	79
Products							
Dry gas	104	53	9	3	2	3	1
LPG	0	3	70	3	1	0	0
Alky butane	0	0	3	34	10	1	0
Blending butane	0	0	0	2	30	8	3
Gasoline	0	0	0	0	7	49	67
Total	104	56	82	42	50	61	71

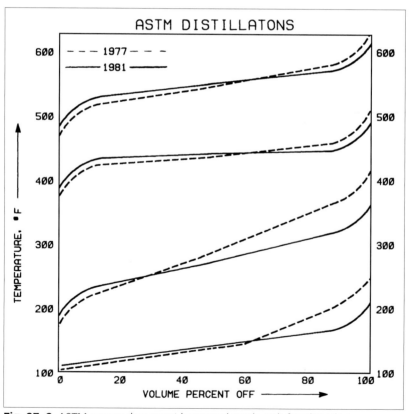

Fig. 27–2. ASTM curves document improved crude unit fractionation.

Energy balance

The performance test energy balance is a wonderful moneymaking tool. Table 27–2 shows such a balance for a typical process unit. The sum of energy inputs (steam, fuel, electricity, and feed enthalpy) should equal the sum of energy losses (flue gas, water and air coolers, ambient heat loss, and product enthalpy). This approach of tracking all of the BTUs through the process facility reveals many energy-saving opportunities.

The efficiency of each furnace should be calculated. This is done in two ways: calculating the heat absorbed by the process or the heat content of the flue gases.

Table 27–2. Energy Balance Results from a Crude-Unit Performance Test.

Furnace	Overall Unit Energy Balance			MM BTU/hr	
	Excess Air	Stack Temperature	Efficiency	Heat Release	Duty
B-1	23%	760°F	78%	244	190
B-2	33%	778°F	76%	237	181
Totals				481	371
Steam to towers 16.0 M lb/hr					17
Total heat input					388
Heat losses accounted (excluding flue gas)					
Tower-top condenser duty					95
Product heat to coolers					47
Enthalpy of products above crude					173
Loss in desalter brine					12
Pumparound trim cooler duties					18
Total					345
Unaccounted heat losses					43

The heat-transfer coefficient (U) for each exchanger should be calculated. This is an especially valuable exercise since it highlights which exchangers need to be cleaned. The thermal efficiency data for furnaces and the exchanger heat transfer coefficients can be tabulated and compared with historical data or design values.

Calculating missing flows

Reflux pumparound rates can usually be calculated from a heat balance around individual towers. This is the best way to account for unrecorded flows. It is not uncommon for a reflux or pumparound duty, calculated from the tower heat balance, to be 10% to 20% higher than the measured flow would indicate. This is probably due to ambient heat losses. For the sake of consistency of the total test report, it is best to stick with the duties from the heat balance calculations. If the difference between a duty calculated in the two ways described above is much more than 20%, there is a significant error in the test data.

Hydraulic losses

Knowledge of the pressure drop through individual heat exchangers is an excellent weapon in the battle against energy inefficiency. Available head should be utilized to maximize velocities through exchangers and not dissipated across control valves. Do not forget that:

$$U \sim M^{0.7} \text{ (tube-side flow)}$$

$$U \sim M^{0.55} \text{ (shell-side flow)}$$

where:

U = Heat-transfer coefficient

M = Mass velocity (lb/hr/ft^2)

For example, modification of the tube-side pass partitions can double an exchanger's tube-side velocity (see chapter 9). Again, pressure drops should be tabulated and compared against previous data to identify partially plugged equipment.

Pumps

Calculate the flow (gpm) and head (feet of liquid) for each pump. Compare the observed head to the predicted head obtained from the pump curve at the corresponding flow. This is a measure of the loss in pump efficiency caused by wear to the impeller, pump case, or wear rings (see chapter 11). A tabulation of pump and heat-exchanger efficiency results for one performance test is shown in table 27–3.

Table 27–3. Performance Test—Pump and Exchanger Efficiency Summaries.

	Pumps			
Number	Observed Flow	Observed Head, ft	Design Head at Observed Flow	Pump Needs Overhaul
P-1	32	38	60	Yes
P-2	105	202	200	No
P-3	77	40	48	No
P-4	20	380	470	Yes
P-5	28	70	76	No
P-6	203	300	305	No
P-7	17	180	170	No
	Heat Exchangers			
Number	Observed U*	Design Clean U	Design Service U	Needs Cleaning
E-1	77	104	63	No
E-2	11	98	61	Yes
E-3	40	69	48	No
E-4	47	140	105	Yes
E-5	33	85	70	Yes
E-6	62	73	55	No

*U = BTU/hr/ft²/°F

Furnace-draft gauge readings can pinpoint restrictions to flow on the flue-gas side of the convective section. Excessive draft at the inlet to the convective section causes cold air to be sucked in and wastes energy (see chapter 15).

Capital Projects

It is a rare unit that has been designed in an optimum fashion for the current mode of operation. A good performance test will almost always define capital projects to save energy or improve yields. A few examples taken from actual refinery experiences are:

- Upgrading piping insulation
- Additional furnace convective tubes
- New turbine governor speed controller
- Trimming pump impellers

- New exchanger bundles
- Additional feed product heat exchangers
- New side-stream steam stripper
- Larger pumparound circulating pump

Performance Text Checklist

Objectives
Material balance
Prepare sample points
Unplug pressure taps
Optimize unit operation
Issue written instructions

Preparation
Ensure laboratory manpower
Instrument calibration
Interface with other units
Communications with shift operators
Obtain services of pipefitter
Flag sheet
Locate portable analyzers

The Day Before the Test
Sample containers
Prepare sample points
Unplug pressure taps
Optimize unit operation
Issue written instructions

Correlating the Data
Component balance
Fractionation efficiency
Overall unit energy balance
Furnace efficiency
Heat exchanger Us
Calculate missing flows
Hydraulic losses
Pump efficiency
Identify capital projects
Normalizing data

Understanding Control Board Instruments

The panel board in a refinery control room consists of instruments to indicate, record, and control process variables. The usual variables are:

- Flow
- Temperature
- Pressure
- Liquid Level
- Composition

The great variety in instrument manufacturers and models makes the operating engineer's job that much more difficult. Every control room seems to contain at least some unique instrumentation. This is somewhat superficial in that the fundamental components of instruments are the same.

Some of the instruments only indicate or record the process variable and have no control over function whatsoever. Other instruments, called controllers, are the nerve center of the process unit. The operating engineer troubleshooting a process problem should locate all of the controllers on the panel board. Those controllers will have the following features.

Set point

The set-point dial determines the liquid level, flow rate, pressure, and so on, that the controller will attempt to maintain. The controller will try to fulfill this function by varying the signal to a control valve. Figure 28–1 shows a typical loop.

The set-point is only functional when the controller is set on auto. In this mode, the controller shown in figure 28–1 compares the desired set-point flow to the flow measured by the orifice flow meter. The controller then moves the control valve to return the measured flow to the set-point flow. This is called closed loop control.

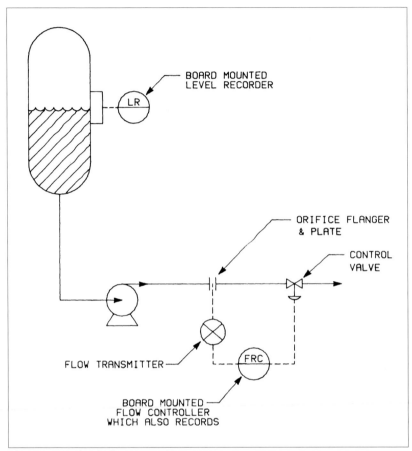

Fig. 28–1. A simple flow control loop.

Controllers may be switched from the auto to the manual mode. When a controller is in manual, the set point can be moved without affecting the control valve. Operating personnel must then adjust a different controller to provide a steady signal to the control valve. This is called open loop control, since to have control the operator must move manually to bring the measurement to the desired value.

Cascade control

Some controllers will have a switch that can be set on L (for local) and R (for remote). These are cascade controllers. If the cascade switch is set on remote, a second controller is moving the set point of the first controller.

Figure 28–2 explains the operation of a cascade controller. In this sketch, the flow controller is being reset by the vessel's level controller. The purpose of this arrangement is to help maintain a steady flow from the vessel while still controlling the vessel's level within an acceptable range.

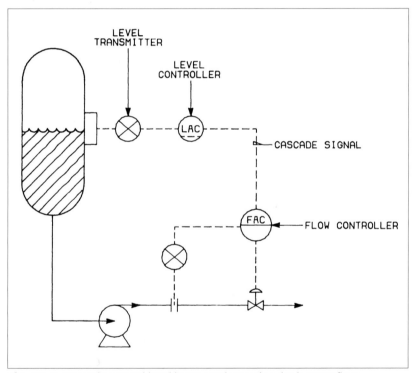

Fig. 28–2. Cascade control level in a vessel resetting the bottom flow.

As an example, with a sudden rise in the vessel pressure, the cascade flow controller would immediately close the control valve to keep the bottom flow constant. With normal level control the flow would increase, and the control valve would not begin closing until the level controller sensed a drop in the bottom level.

When the cascade switch is turned to local, the cascade feature is turned off. The controller is now simply a flow controller in auto. If the controller is set on manual, it doesn't matter whether the cascade switch is set on local or remote since the controller is on open-loop control.

A typical cascade controller-recorder instrument is shown in figure 28–3. The particular instrument shown is set for remote automatic control. The cascade control set-point temperature is 155°F. The actual temperature being recorded is 145°F. The output signal to the control valve is 132°F. This output signal will eventually increase to match the 155°F set point.

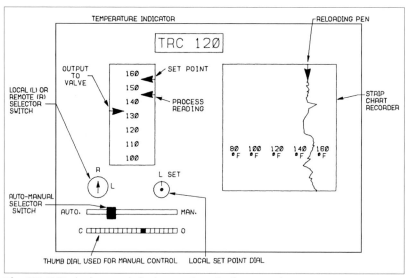

Fig. 28–3. Typical cascade instrument controller-recorder.

Control valve position

The control room is not the place to determine the true position of a control valve. One needs to go outside and look at the valve itself. The indicator on the controller that normally indicates the valve position may be totally wrong. Operating personnel miss this crucial point so often that it is worth emphasizing.

For most pneumatic controllers, the valve position shown in the control room is only an indication of the air signal that the controller is transmitting to the control valve. The control room instrument does not normally receive any feedback from the control valve. So, what factors influence a control loop to cause an incorrect valve position to be shown on the control panel?

Most often, the control valve gets stuck. If a steel nut or welding rod is jammed into the valve seat, maximum air pressure (assuming an air-to-close valve) will certainly not close it. However, the control room instrument, having signaled maximum air pressure, will show the control valve to be shut.

Not infrequently, someone will accidently turn off the air supply to a control valve's signal box. (This is that small box with three pressure gauges attached to the control valve.) This type of failure is usually found reasonably fast by the operators as the control valve becomes totally inoperative.

A more insidious problem results from improper use of a control valve hand jack. The hand jack is a wheel attached to a control valve that can be manually used to move the valve. There is an indicator on the hand jack showing whether it is being used to open or close the control valve. Normally, the hand jack should be in the neutral position. In this position, it is disengaged from the control valve. Luckily, few control valves have such a hand jack.

An operator may decide to close a control valve partially with the hand jack. He sets the hand jack in the one-half closed position and then goes home. Now, this control valve can still be closed from the control panel, but it cannot be opened more than halfway. How can anyone sitting in the control room deduce that the valve's hand jack is engaged? The control room instrument will, of course, show the valve to be wide open, when in reality it is only half open.

This problem is really not changed by the use of a valve positioner. If a control valve is stuck, the valve positioner will not retrieve the situation. However, a position transmitter can be installed. This device indicates the actual valve position in the control room to the operators.

The infamous Three Mile Island nuclear power plant incident occurred partly because the operators became confused between the actual position of a relief valve on the containment building and the indicated position of the relief valve on their control panel.

In summary, the valve position shown in the control room may or may not correspond to the real control valve position.

Computer Control Consoles

In the 30-year span between the publication of the first and fourth editions of this text, most major refiners have switched from pneumatic controllers to computer control with CRT displays. While the benefits for computer control are certainly real, 99% of the benefits of computer control can be obtained with relatively inexpensive programmable controllers.

Regardless, the real value of computer control has nothing to do with control at all. It is the data-monitoring capability of the system that justifies the conversion from conventional pneumatic to computer control.

For each control point (usually corresponding to a control valve in the field) there is a display showing:

- **SP.** This is the set point entered by the operator. For example, if he wants a heater outlet temperature to be 705°F, the SP valve shows on the console as 705°F.

- **PV.** Before the set point, the process variable valve (PV) is displayed. This is the actual heater outlet temperature, for example 702°F.

- **The relevant valve position** is shown below the PV display. In the preceding example, the fuel-gas control valve position (0% closed; 100% open) would be displayed. If the SP valve is 705°F, and the PV valve is 702°F, the valve position (assuming the control loop was on auto) would be opening. If the control loop was on manual, the SP valve is meaningless and the control valve will not move.

Caution: Some variables on certain CRTs can be manipulated by touching the screen with your finger.

Data Monitoring on Computer Consoles

On most control systems, it is a simple matter to display a trend of any process variable versus time. For example, the computer could plot heater outlet temperature on the vertical axis and time on the horizontal axis. Moreover, the span of either axis can be changed. You could set the vertical axis to read from 600°F to 800°F. If you wish to study span temperature fluctuations, however, this span could be changed to read from 695°F to 705°F.

The horizontal axis could be switched to span one hour, one shift, one day, or one week. Many computer systems only preserve hourly average readings after several days. Therefore, the data from a brief process excursion cannot be retrieved. Most unfortunately, some systems do not save data after 7 to 14 days.

In some computer systems, it is possible to cross-plot a wide variety of variables. For example, one could study regenerator pressure versus the flue-gas slide value position. Or a variable calculated continuously by the computer (FCCU catalyst circulation) could be plotted as a function of gasoline production.

All the information displayed on the computer console can be printed. Usually, you print out the process flowsheet of a portion of the unit showing temperatures, flows, pressures, and control-valve positions at some point in time. Also, a particular plot or process variable may be printed. However, you could print every process variable on one-minute intervals that the computer is monitoring in the entire refinery. I have, in my home, 300 pounds of computer output representing a four-day operating period for a refinery in Chicago. The ability to accumulate data with a computer, therefore, can be potentially overwhelming. Unfortunately, the ability of engineers to analyze problems can be drowned in this sea of data.

Zeroing an Instrument

Calibrating pressure or flow indicators is a job for the instrument mechanic. The operating engineer will, however, frequently need to zero instruments, especially flow recorders. This is a necessary first step in running a performance test or troubleshooting a material balance problem, and is done as follows:

1. Find the orifice plate that is measuring the flow to be checked.

2. Trace the small lines from the orifice taps back to the flow transmitter. This is done to ensure that you will be working with the correct transmitter.

3. Make sure the unit's chief operator knows what is going on. Tell him to put the controller connected to the flow transmitter on manual. Check that trips associated with the flow transmitter are bypassed. (I once forgot to do this on a sulfur plant and achieved instant and lasting notoriety.)

4. Unscrew the transmitter cover. At the base of the transmitter is a screw marked zero. Some transmitters will have the zero screw exposed without removing the cover.

5. Close the orifice tap valves to the transmitter and open the meter bypass valve (see figure 17–5).

6. Turn the zero screw until the local flow indicator or the flow indicator at the control panel reads zero.

7. Return the flow transmitter to service. Be sure the bypass valve is tightly closed. Reactivate any trips that have been bypassed.

After zeroing a flow recorder, you may note that with the meter bypass valve still open, the displayed flow is still 200 B/SD. If the range of the flow meter is 10,000 B/SD and you will be running at 7,000 B/SD, the meter will only read about 30 B/SD high. If the flow is going to run between 0 and 1,000 B/SD, however, the indicated flow will be about 150 B/SD high. Even on a computer digital display, the flow indication is still a function of the square root of the pressure drop through the orifice plate. Hence, the error introduced in an incorrect setting of the zero point diminishes rapidly as the flow is increased. While this is rather obvious on a strip-chart flow recorder, it is just as true but more subtle on a computer digital-display CRT.

Tuning Instruments

Once, a noted expert on advanced process control techniques lectured our engineering staff. Having understood nothing of the lecture, I discussed the subject with my colleagues. They too had failed to absorb this advanced technology. Moreover, it became apparent that almost none of us had a grasp of the fundamentals of process control.

The process operating engineer does not need to become a control expert. He ought, though, to have a rudimentary idea of how control-room instruments are tuned. He may find he is the only one in the refinery who can actually tune instruments.

The obvious function of a controller is to hold a set point. There is an inherent contradiction in achieving this goal: Stability versus a fast recovery. Normally you would like a control valve to close off instantly to reduce a flow back to the set-point valve. However, there is no practical way of doing this without overshooting the target and winding up with too low a flow. Hence, a too-rapid return to the set point initiates prolonged cycling. The science of tuning controllers is to effect the best compromise between the competing goals of fast recovery and stability.

The optimum settings on a controller will change with equipment wear and throughput rates. Also, during smooth operations it is difficult to tell if one has good or poor control. This makes instrument tuning a recurring and challenging task.

Controller dials

When one pulls a controller out of the instrument panel, several small screws or dials come into view. These are usually labeled proportional band or gain, reset, or rate (or derivative). Turning these screws profoundly affects the performance of the control loop. This is a job best left to the instrument mechanic if he really knows how to tune instruments. Quite often, the operations engineer will have to try his own hand at instrument tuning. This is how to go about it.

Proportional band

This formidable term has a surprisingly simple definition. Suppose the proportional band is set on 100%. Now the set-point dial on the front of the controller is moved full scale. The corresponding control valve outside will then also move full scale.

For example, a flow controller can be set from 0 to 10,000 B/SD. The proportional band is set at 100%. The flow control set point is increased from 0 to 10,000 B/SD. The flow control valve, which had been closed, opens all the way.

If the proportional band had been set at 20%, increasing the set point by 10% would open the control valve by 50%. That is, the control valve would respond five times (100% ÷ 20%) as fast to a change in the set point. Another way of saying this is that the control loop has a gain of five.

Alternately, the proportional band could be set at 400%. Now, increasing the set point by 40% results in only a 10% movement of the control valve. The control valve is responding at one-quarter the speed of the set point. The gain for the control loop is now one-quarter.

What does this mean in the field? If the controller is responding too slowly (that is, the process is changing too quickly for the control valve to keep up), the proportional band should be dialed down. This will speed up the reaction time of the control valve.

Gain

Gain is the reciprocal of proportional band. The intrinsic or total gain of a control loop is the process gain times the controller gain.

If the intrinsic gain of a system exceeds 100%, control instability can result. In the field, this means that one cannot arbitrarily keep speeding up a control valve's response time by dialing down the proportional band. A flow control loop with a gain of over 100% would likely have a flow recorder chart similar to the one in figure 28–4.

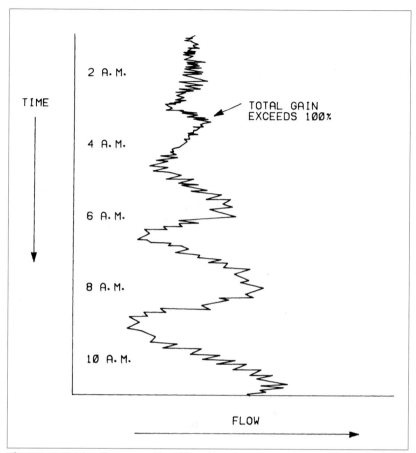

Fig. 28–4. Setting the proportional band too low causes increasing oscillations.

The optimum proportional band setting results in oscillations of diminishing amplitudes such that each succeeding oscillation is about one-quarter the preceding one. That is, set the proportional band as low as possible consistent with diminishing swings in control valve movement. This will give the fastest control valve response time and still have a reasonably stable control loop.

Offset

Offset is the difference between the controller set point and the actual controller reading when the control valve is at rest. The operating engineer would naturally expect any decent controller to have an almost zero offset. The characteristic of a controller that mainly reduces offset is termed reset.

Reset

Treat the reset screw with respect; a little can go a long way. Turning the reset screw down forces the control valve to return the variable being controlled closer to the set point. For example, a flow controller is set at 5,000 B/SD. After a period of diminishing oscillations, the flow lines out at 5,600 B/SD. When the reset screw is calibrated in minutes per repeat, one would turn the screw setting down to reduce the flow toward 5,000 B/SD. However, too much reset will tend to make the control loop unstable.

Some reset screws are calibrated in repeats per minute. For these controllers, one turns up the screw to obtain more reset action. Level control loops are inherently sensitive to reset and can become unstable with small amounts of reset.

Rate (derivative)

Certain process parameters respond slowly to changes in control-valve position. A furnace outlet temperature controller is an example. Even when the fuel-gas control valve swings wide open, the furnace tubes themselves take awhile to heat. Such controllers (often on temperature) contain a rate dial. As a process variable moves away from the set point, rate increases the controller gain (reduced proportional band) as a function of the speed with which the process is moving away from the set point. This will help bring the variable back to the set point faster and with less chance of instability than simply reducing the proportional band.

Tuning Summary

The best proportional band setting provides a damping factor that is satisfactory for process control. Too low a proportional band causes excessive cycling. Too large a proportional band causes prolonged and large deviations from the set point.

The optimum reset setting (minutes per repeat) enables the measured parameter, after an initial deviation, to return and cross the set point by a reasonable amount. Too low a reset time permits the measured parameter to cross the set point excessively and results in prolonged cycling. Too high a reset results in extensive deviation from the set point (i.e., offset) before the measured parameter crosses the set point. The best rate setting is the one beyond which a further increase does not minimize—or even increase—cycling.

Ziegler-Nichols tuning method

For simplicity, a detailed tuning method is presented for controllers having proportional band and reset only.[1]

1. Set reset to the maximum practical setting (minimum reset action).
2. Narrow the proportional band by a small step.
3. Raise the set point briefly and then return it to the desired valve.
4. Repeat the above two steps until the process has sustained the oscillations.
5. Note the proportional band and time between peaks for the final setting for step 4.
6. Set the new proportional band at twice the proportional band found above.
7. Set the new reset valve at 80% of the reset value in step 1.

The preceding discussion is for orientation. Tuning instruments is an art form best learned by watching a professional performance. Practice on simple control loops is a worthwhile exercise, but be sure to record the initial control settings in case you get into trouble.

Author's note: I've updated and expanded this chapter in a new book published in 2008.[2]

Control-Room Instrument Checklist

Controller Terminology
Set point
Manual—open loop
Automatic—closed loop
Cascade control
Local
Remote
Recorder

Control Valve Position
Air signal is not valve position
Valve sticks
Air supply to diaphragm shutoff
Hand jack engaged
Control valve positioner

Zeroing Instruments
Put instrument on manual
Bypass associated trips
Open transmitter bypass valve
Percent error at low flow

Tuning Instruments
Response time versus stability
Proportional band
Gain
Reset time
Rate or derivative
Ziegler-Nichols tuning method

References

1. "Control Mode Adjustments," The Foxboro Company, January 1965.

2. Lieberman, N.P. *Troubleshooting Process Plant Control.* Wiley Publications, 2008.

How to Make Field Measurements

Decisions are no better than the quality of the data analyzed. Often, the troubleshooter is faced with data that are so obviously contradictory that he despairs of reaching any conclusions. Resolving refinery process problems is hard enough without having to worry if the facts one is working with are really correct.

Shift operators cannot be consistently relied upon to make nonroutine field measurements. They sometimes do not know how or do not care to do the job with accuracy. Even taking a pressure measurement requires motivation, intelligence, and knowledge. The following sections will amplify the need to oversee the collection of field data personally.

Pressure Surveys

There is more to taking a pressure reading than just screwing a gauge onto a connection and reading the pressure. The following procedure should be followed:

1. Obtain a new pressure gauge with the smallest usable range. Employ this single gauge for all measurements.

2. Slowly remove the plug or gauge from the connection where the pressure will be checked. Never try to pull the plug from a connection that will not depressure.

3. When working with H_2S or other toxic substances, fresh-air breathing equipment is in order.

4. Crack open the valve to the connection to ensure the point is not plugged. The safe way to unplug connections is discussed in a subsequent section. Unplugging drains, vents, and bleeders haphazardly can, and has, injured the perpetrator.

5. Always install the pressure gauge in an upright, vertical position. The gauge will read differently when it is held at an angle or upside down.

6. Open the valve one or two turns and wait a moment before taking the reading. The pressure should come up quickly. If not, the pressure reading will be low and wrong due to a partly plugged connection.

7. Unscrew the gauge slowly. Let the pressure bleed off around the threads. Watch the indicated pressure fall to zero before removing the gauge.

8. Now crack open the valve and make sure the connection is still open. If the connection has plugged during the previous manipulations, the preceding procedure will have to be repeated.

Correction for elevation

Pressure surveys are taken to find pressure drops (ΔP) due to friction and not to determine an absolute pressure. Therefore, consistency is more important than accuracy. To obtain consistency in liquid-filled systems, the measured pressures must be corrected for differences in elevation. This is done as follows:

$$P_c = P_m \pm \frac{D \cdot (\text{sp gr})}{2.31}$$

where:

P_c = Pressure corrected for elevation, psig

P_m = Measured pressure, psig

D = Vertical height between the base level and the level at which Pm was recorded in feet. When the level at which the pressure measurement is taken is below the base level, subtract $[D \cdot (\text{sp gr})]/2.31$ from P_m.

Manometers

When measuring small pressure drops, a water-filled manometer or a ΔP cell is needed. The manometer is connected to the two points across which the pressure drop is to be found. For pressures much above atmospheric, a ΔP cell is used. For pressure drops below a few psi or when a trend in ΔP is being followed, a manometer or ΔP cell must be used. Taking two successive measurements with a single-pressure gauge will not yield sufficient accuracy.

A mercury manometer is required for checking pressures in vacuum systems. You can purchase compact mercury manometers that have been especially built for low-pressure (0 to 5 psia) vacuum service. They are much handier when climbing up a vacuum tower than a standard 30-inch manometer.

Differential Pressure Measurement in Distillation Towers

To properly measure the ΔP across a trayed distillation tower, proceed as follows:

- Obtain two ordinary pressure gauges. Use the smallest appropriate range gauge to minimize the loss of accuracy.
- Gauges should always be installed in a straight-up position because they will read differently at different orientations.
- Do not overtighten gauges by hand. This will change the gauge calibration by a small amount. Use a wrench on the stem of the gauge to tighten.
- If the gauge reading increases slowly, the connection is partly plugged and the reading will be too low. Remove the gauge and clean the gauge port. If this does not correct the problem use the angle-worm drill out device shown later in Figure 29–2 to clear the pressure tap.
- At the base of the tower place one pressure gauge at the top of the level glass. Check that the tower liquid level is below this point. We do not want to measure a head of liquid plus the real tower pressure. Label this first gauge as the *Test* gauge.
- Repeat the previous step with the second gauge. Label this gauge as the *Reference* gauge. Compare the two readings.
- Simultaneously with each reading taken with the Test gauge, have someone verify any changes to the Reference gauge reading and make appropriate manual adjustments to the Test gauge valve.
- At the top of the tower, connect the Test gauge either on the tower top head or below the relief valve but not on the vapor line.

- As many intermediate pressure points as feasible should be checked on the side of the tower but not on feed or draw nozzle piping.
- If the tower top pressure must be checked on the overhead vapor line then the calculated nozzle exit loss pressure must be added to the observed vapor line pressure. This nozzle exit loss is calculated as follows:

$$\Delta P = 0.34 \bullet \frac{DV}{62} \bullet \frac{(VG)^2}{28}$$

where:

ΔP = Nozzle exit loss, PSI

0.34 = Typical nozzle exit loss coefficient

DV = Vapor density, lbs. per cubic ft.

VG = Nozzle velocity, ft. per second

62 = Density of water, lbs. per cubic ft.

28 = Inches of H_2O per PSI.

If the pressures are swinging too badly to take accurate readings, then that in itself indicates a severe problem. If the tower bottom pressure is changing faster than the tower top pressure, this is a sign of tray flooding. If the tower top pressure is swinging more than the tower bottom pressure, this is an indication of erratic tower pressure control.

For dirty services, an oil filled gauge with a diaphragm is best. Digital gauges are not too good. They are expensive, subject to damage, and the reading can be altered by slight electrical currents originating from nearby radio transmissions or even the minor current generated by your hand. Calibrating the Test gauge is a waste of time. The absolute pressure readings will not be used. Only the pressure differences. Also, all the pressure readings will be referred back to the Reference gauge for correction, regardless of their calibrated reading.

Finally, for towers with vapor densities above one pound per cubic foot, the pressure readings should always be corrected for elevations as explained in the previous section. For tall depropanizers and de-ethanizers, this correction is extremely critical.

Temperature

Temperatures indicated in the control room are normally very reliable. If necessary, they can be checked at the thermowell with a portable temperature indicator. Dial thermometers are not to be trusted. Simply unscrew the dial thermometer and insert a glass thermometer into the thermowell. Note that the thermowell is sealed off from the process fluid.

It is often necessary to determine a process temperature at a point where there is no thermowell. The refinery inspection department usually has a kit for measuring metal surface temperatures. A temperature-sensing disk connected to the end of a rod is pressed against the bare metal surface through a hole cut in the insulation. The temperature is then read from a dial.

The right way to get the job done is with an ordinary glass thermometer. Figure 29–1 shows how to insert a thermometer under a piece of insulation to measure a process temperature. Note how the thermometer is packed into the hole with loose insulation. Wait 10 to 15 minutes before reading the thermometer. To convert this surface temperature reading to the approximate fluid temperature inside the pipe, use the following rule of thumb:

$$T_f = T_m + 6\% \, (T_m - T_a)$$

where:

T_f = Fluid temperature

T_m = Temperature measured with thermometer

T_a = Ambient temperature

The 6% correction factor was developed by comparing temperatures measured as per above against thermowell readings made with a thermocouple. The field troubleshooter should re-determine the correction factor for individual services. The correction factor also varies with pipe condition, wind speed, and the techniques of the observer. For low-pressure steam lines, a correction factor of 10% has been observed. For asphalt lines a correction factor of 30% has been measured. If the top of the line is cooler than the bottom of the pipe then the line is not running full. Lines containing vacuum tower off-gas conduct heat so poorly that quite different temperatures may be observed on the top and the bottom of the line.

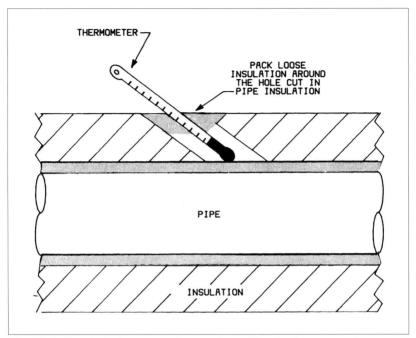

THERMOMETER

PACK LOOSE
INSULATION AROUND
THE HOLE CUT IN
PIPE INSULATION

PIPE

INSULATION

Fig. 29–1. How to find a process temperature without a thermocouple.

You cannot hold your hand on a pipe hotter than 140°F. A line that feels cool to the touch is cooler than 80°F. The experienced troubleshooter always spits on bare piping before touching it.

For measuring hot (1,000+°F) furnace flue-gas temperatures, a velocity thermocouple should be used (see chapter 15). Using an ordinary static thermocouple will indicate a lower-than-actual temperature due to the effects of reradiation from the thermocouple. Furnace tube metal temperatures are approximated visually:

> 1,300°F—dull red
>
> 1,500°F—bright glowing red
>
> 1,700°F—white

An optical pyrometer can successfully be used to follow firebox tube metal temperatures. This instrument will read the color of the heater's tubes. In practice, this method will permit you to follow the increase in tube metal temperature as the heater's run length progresses and the tubes get hotter. Reading the optical pyrometer is somewhat subjective, so it is best to have the same individual take all the readings.

Infrared surface temperature indication

An extremely handy device to track surface temperatures is an infrared thermometer gun. The main advantages of this device are speed (responding within seconds to the surface temperature) and the elimination of the need to touch the probe to the surface.

The main disadvantage is that the nature of the surface (color, roughness) influences the surface's emissivity. Because infrared radiation is a function of the surface's emissivity as well as its temperature, the surface appearance will alter the measured surface temperature.

I have used my infrared pyrometer in many applications such as:

- Finding the condensate level in the channel head of a steam reboiler.
- Determining the amine liquid level in the bottom of a fuel gas scrubber.
- Determining the pressure (assuming saturated steam) in a steam line.
- Finding maldistribution of vapors in the header box of an air-cooled fin-fan exchanger.

Flows

When there is no meter installation to measure a process flow, the operating engineer will have to use his imagination to devise a method to obtain the necessary data. Portable external Doppler flow meters are very tricky to use and their successful application seems quite limited. A few mundane but sure-fire methods to get the needed flows are outlined here.

Measuring steam flow to a reboiler

You can often calculate unmetered internal unit flows, such as reflux rates, from an enthalpy balance if only a reboiler or preheater duty is known (see chapter 27). If the heat exchanger uses condensing steam as the energy source and it too is unmetered, a simple, if primitive, method is available.

On the downstream side of the steam condensate control valve or steam trap, divert the condensate into an empty 55-gallon drum. Note the time it takes for the drum to fill. Add 5% (for low-pressure steam) to 15% (for high-pressure steam) to the observed pounds per hour of flow. This accounts for condensate flashing as it drops to ambient pressure. For instance, a reboiler having a duty of 25 MM BTU/hr will fill a 55-gallon drum in about one minute.

Measuring flow from a vessel

Lower the liquid level in the drum to minimum. Next, block in the outlet from the drum and note the time for the drum to fill to its maximum safe level. From the vessel geometry, calculate the normal flow rate from the drum. This type of flow interruption can be tolerated in many distillation tower reflux drum services or when a product is being run down to the field.

Gas flows

The meter factors provided for flow indicators in gas service are for a particular density gas. If the pressure, temperature, or molecular weight of the gas is significantly different from design, the meter factor will have to be corrected for the new gas density. Indicated flow will vary with the square root of the gas density.

Drilling Out Bleeders

Opening plugged drains, vents, bleeders, and other small connections is a serious business. The troubleshooter, who must frequently utilize little-used sample points and pressure connections, should be cognizant of the hazards involved. Here is a typical story that you will do well to avoid repeating.

The operating engineer decides he will check the drain from the bottom of a depropanizer reflux drum for water. The reflux drum is full of liquid propane under 300-psi pressure. He goes out alone with neither gloves nor goggles. The engineer opens the drain valve a few turns without effect. He opens the valve all the way, but it is plugged and nothing comes out. He then takes an old welding rod and shoves it up through the valve body. Suddenly, without the slightest warning, the propane blows through the valve with tremendous force.

The engineer is enveloped in a white cloud of freezing propane vapor. He cannot get back to the valve to close it (liquid butane and propane cause frostbite upon skin contact). Disoriented and shaken, he goes off for help, leaving the unit and the rest of the refinery in jeopardy from a possible ignition or detonation of the vapor cloud.

Opening plugged connections with a bent wire is okay for water or nontoxic vapor services. For liquid light hydrocarbons or hot-oil streams, there is a far safer way to get the job done. Figure 29–2 shows a device for drilling through small connections. A similar apparatus is commercially available, or the refinery machine shop can easily fabricate one.

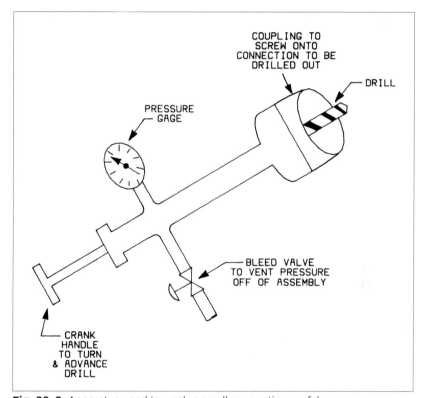

Fig. 29–2. Apparatus used to unplug small connections safely.

These bleeder unplugging devices are manufactured in a variety of shapes and sizes. For bleeders close to grade, a device is available with a 90° turn at the business end. All that is needed is a clearance above grade of 4 to 6 inches. For sample points that are used routinely, permanently mounted unplugging assemblies are available. (Check your "Mine Safety Equipment" catalog for the Angle Worm, or search online for "Angle Worm bleeder cleaner.")

Sampling Techniques

Many troubleshooting problems are easily resolved once the proper sampling technique is used. A few examples point this out.

Gasoline vapor pressure

A debutanized light naphtha was stored in a floating roof tank. Hydrocarbon vapor emissions from this tank were obviously excessive; yet, repeated samples of naphtha, drawn from the debutanizer bottoms, showed a very low butane content. The paradox was resolved when the process engineer observed that the light naphtha sample was taken in a plastic bottle. By the time the lab ran the sample, most of the butane had evaporated. A sample obtained in a pressure bomb revealed the true butane content of the naphtha.

Note that when a liquid bomb sample is taken, a small amount of liquid is always drawn off the bomb. This is to leave room for the thermal expansion of the bomb contents. If this precaution is not taken, the bomb could rupture and possibly explode on warming. This happened in the Amoco refinery in Whiting, Indiana.

Weathering

A good way to gain lasting enmity in the lab is to submit an unweathered, unstabilized naphtha sample for ASTM distillation. To weather off such a sample, one should obtain a full quart sample of the unstabilized naphtha and drop a few small pebbles into the bottle. These will act as boiling stones. Place the bottle in a warm, well-ventilated area for a few hours. Leave a thermometer in the bottle. When no more bubbles evolve and the bottle has warmed to room temperature, measure the volume percent evaporated and submit the sample to the lab. Also check the temperature to make sure the sample has not autorefrigerated. If the bubbles have stopped but the sample is cold, gently stir the contents of the bottle. When the ASTM analysis is received from the lab, do not forget to correct for the volume weathered off.

Hot oil

When obtaining a sample of hot oil, one is often interested in the composition of the front end. If the sample is taken close to its process temperature, the lightest components will simply flash off and not be accounted for. A well-designed sample cooler will solve this problem. For resid streams, tempered cooling water will probably be needed to prevent plugging the cooling coil.

If you are only interested in front-end composition, ask the lab to run an atmospheric distillation rather than a vacuum distillation. This will yield a more accurate measurement of front-end (–680°F) composition, and be far less expensive to run.

Gas samples

A sample that contains H_2S and O_2 in concentrations of interest to us should never be obtained in a steel sample bomb. The H_2S and O_2 will react to form water and solid sulfur. The metal walls of the steel bomb will catalyze the reaction. I use a mountain-bike tire inner tube which is pumped up with a small rubber handpump to catch such samples. The sample must still be analyzed promptly to minimize the disappearance of H_2S and O_2.

Levels

A liquid level may easily be found in a vessel, even without a gauge glass or level indicator. Locate two connections on the side of the vessel that span the level in question. Unplug the taps and then connect a ΔP cell across them. After correcting for the liquid density, the indicated differential pressure will show the height of liquid above the bottom tap.

When all else fails, a liquid level in a vessel can be found with a portable gamma-ray source and a Geiger counter-type radiation receiver. Most refineries have personnel licensed to handle radioactive materials in connection with x-ray inspection of process piping. Their equipment can be used, along with the proper radiation receiver, to locate a liquid level in a vessel accurately. Obviously, this is a procedure that will not be used routinely. Note that this can create a potential safety hazard to operators unfamiliar with the dangers of exposure to radioactive materials.

Troubleshooting Checklist
for Field Measurements

Pressure Surveys

Using a single gauge
Depressure connection
Install gauge in upright position
Correct for elevation
Use a manometer for small ΔPs
Digital ΔP cell

Temperature Measurements

Dial thermometers are unreliable
Insert glass thermometer
  under pipe insulation
Use velocity thermocouple
  for hot flue gas
Use optical pyrometers
  for furnace tubes
Infared thermometer

Flows

Portable external flow meters
  are unreliable
Collect steam condensate in a barrel
Observe increase in vessel liquid level
Correct gas meters for fluid density

Plugged Bleeders

Unplugging small connections
  is dangerous
Use a safe drilling assembly apparatus

Sampling Techniques

Light hydrocarbon liquid
  bomb samples
Weathering naphtha
Hot oil

Liquid Levels

ΔP cell
Gamma rays

Turbines, Expanders, and Variable Speed Motors

Conventional motor-driven pumps and compressors potentially waste energy either by throttling on the discharge flow or recirculating the process fluid. Variable speed drivers such as steam turbines, gas-fired turbines and power recovery expanders do not suffer from this deficiency. Their speed can be altered to adjust flows and pressures as process conditions require.

Steam Turbines

The most common variable-speed driver you will encounter in a refinery or process plant is the topping steam turbine. For example, 400-psig motive steam is exhausted to 100-psig exhaust steam. The steam flow path is illustrated in figure 30–1:

- The motive steam passes through a spring loaded valve. Should the turbine overspeed or vibrate excessively, this valve will slam shut. It's called the overspeed trip. It's reset manually with a lever that latches back into place. If the turbine operates at 3550 rpm to 3600 rpm (to match the parallel motor-driven pump), then the turbine would be set by the unit machinist to trip off at about 3750 rpm.

- The steam then flows through the governor speed control valve. Actually, both the trip and the governor valves are an integral part of the turbine. The turbine speed opens or closes this steam valve to maintain the required turbine speed. The desired speed is set by a speed control knob. One doesn't actually set a particular speed. It's just a knob that is turned to make the turbine run faster or slower.

- The steam enters the steam chest. Note that the pressure drop between the motive steam and the steam chest pressure is 100 psig. This pressure loss is called an *iso-enthalpic expansion*. That means the heat content of the steam is preserved. In simple terms, the velocity of the steam does not increase. The steam cools as it expands.

The reduction in the sensible heat content of the steam is converted into latent heat of the steam. Also, the lost pressure of the steam is converted into latent heat. I call this a parasitic expansion. It's bad. Bad in the sense that the expansion energy of the steam is no longer available to do work. Yes, the energy is still all there. But now more of it is residing in latent heat and thus unable to do work.

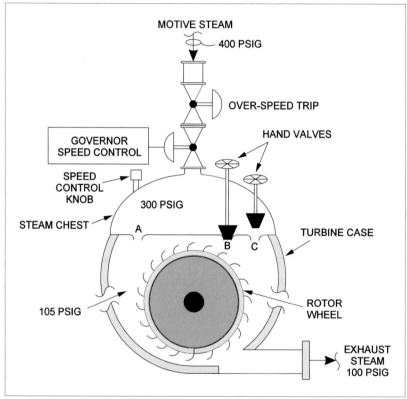

Fig. 30–1. A single stage topping steam turbine

- To extract work from steam, the steam needs to be moving fast. The turbine is just like a windmill. The wind, or the velocity of air striking the windmill's sail cause it to spin. The steam striking the turbine blades (which surround the rotor wheel shown in Figure 30–1), also causes the turbine wheel to spin. The faster the velocity of the steam hitting the turbine wheel, the more work we can extract from each pound of motive stream.

- The steam passes into the steam chest and flows out of the nozzle ports. I've labeled the three nozzle ports A, B, and C. The nozzle ports connect the steam chest to the turbine case. The steam enters the steam chest at a pressure of 105 psig, slightly higher than the exhaust steam pressure. The 300-psig pressure energy of the steam is converted into kinetic energy or speed. As the speed flows through the nozzles it accelerates. The steam expands from 300 psig in the steam chest to 105 psig steam in the turbine case. As the steam accelerates, its energy content is unchanged. But, the steam's heat content has gone down. It cools and usually partly condenses. What has happened to all this heat in the steam? It has been converted to momentum, or kinetic energy, or velocity, or speed. This is called an *iso-entropic expansion.*

- The high velocity steam strikes the blades on the turbine wheel. It's the velocity, not the pressure of the steam that causes the turbine wheel to spin. The faster the steam impacting the turbine wheel, the more work can be extracted from each pound of steam.

- In the steam chest, I have shown three ports or openings in the following three conditions:

  1. The *A* port is open. It's called the main nozzle port and can neither be opened nor closed. About 60% to 80% of the motive steam is designed to flow through the A port.

  2. The *B* port is closed. Turning its hand valve, shown in Figure 30–1, pushes a plug into the port and closes it. No steam is flowing through the B port.

  3. The *C* port is open. About 10% to 20% of the motive steam is designed to flow through the C port.

- The port valves B and C can be used to save steam. When I closed the B hand valve, the flow of steam hitting the turbine wheel was reduced. The turbine slowed. This caused the governor speed control valve to open. The pressure drop across the governor was reduced as it opened. This raised the pressure of the steam inside the steam chest. The flow rate of steam through nozzles A and C increased. Thus the velocity of steam escaping from nozzles A and C also increased. The steam then impacted the turbine wheel blades with greater velocity and momentum. As a result, I could extract more work from each pound of steam.

- On the other hand, let's say I wanted to increase the turbine's speed. Perhaps I need to drive a pump faster to get more flow and more pump discharge pressure. But, I cannot do this because the governor speed control valve is 100% open. Well then, I would open hand valve B. which would reduce the pressure in the steam chest, but would also permit more steam to flow into the turbine case from the governor.

Thermodynamics in Action

Opening hand valve B would increase the total amount of steam entering the turbine case to generate greater turbine horsepower. Closing hand valve C would reduce the total amount of steam entering the turbine case, but this would save steam. In summary, the purpose of the hand valves is to allow the turbine to run more efficiently at different horsepower loads.

The iso-enthalpic expansion of the steam through the governor valve reduces the ability of the steam to do work. The steam's velocity does not increase. It just expands to no purpose. However, the heat or the enthalpy content of the steam is not diminished.

The iso-entropic expansion of the steam through the port nozzles accelerates the steam. The steam loses a lot of heat, which is converted into kinetic energy. But this is good. It's the speed or velocity of the steam impacting the turbine wheel that provides shaft horsepower to the turbine.

In summary, the evil iso-enthalpic expansion preserves the heat content of the steam as well as its total energy. The good iso-entropic expansion converts some of the steam's heat to velocity but also preserves the total energy of the steam. The energy content of the steam is not reduced until the steam strikes the turbine wheel's blades. Then, the kinetic energy or the momentum of the steam causes the turbine to spin.

Windage Losses

We saw that the steam pressure drops from 105 psig to 100 psig as it flows through the single stage turbine shown in figure 30–1. This minor efficiency loss is called a "windage loss." I always ignore it in my calculations. However, if you do have a way to reduce the turbine exhaust steam pressure, then this will certainly result in an energy savings. At the lower turbine case pressure, the steam will hit the turbine blades with greater velocity because the steam is expanding into a lower pressure region. In most refineries however, the exhaust steam from a topping turbine flows into a lower pressure steam header for reuse in re-boiling distillation towers or stripping hydrocarbon liquid products.

Adjusting Turbine Speed

As I write this in July 2008, the CNN news has just announced that crude oil is selling for $147.28 per barrel. In November 1965, the day I started work for Amoco Oil, crude was selling for about $3.50. Please guys, will you all go out today and slow down your turbines. It's really quite easy. Look at the process control valve downstream of the turbine driven pump. If this control valve is less than 60%–70% open, adjust the turbine's speed control knob (located on the front of the steam chest) to reduce the turbine's speed. Continue to reduce the turbine's speed until the process control valve downstream of the pump is in a mostly wide open, but still controllable position.

This will cause the governor speed control valve to close. Next, close one of the turbine's nozzle port hand valves to partly open the governor. The objective is to minimize the wasteful iso-enthalpic expansion of the steam by maximizing the steam chest pressure.

What's the purpose of these adjustments anyway? The dual objectives are to:

- Save steam. Work varies with speed cubed. If you can reduce the speed of a turbine by 10%, then about 27% of the motive steam will be saved, assuming a constant steam chest pressure.
- Have fun. I love to play with process equipment. Especially if the game involves saving energy and reducing greenhouse CO_2 emissions.

On larger steam turbines, perhaps those over several thousand BHP (brake horsepower), there are no hand valves. There is instead a steam *rack*. The steam rack essentially does automatically what the hand valves (also called the port valves or horsepower valves) can do manually. That is they close automatically to keep the governor in a mostly open position. Or, they open automatically to prevent the turbine from slowing because the governor is wide open.

Malfunctions

If you find that the hand valves can be opened or closed without affecting the governor, then the nozzles are eroded. This is caused by someone partly opening a hand valve. These valves should always be 100% open or shut. Any position in-between causes rapid erosion to the valve seats. Sometimes the governor does not open 100%. Then there is a large ΔP between the motive steam supply pressure and the steam chest. For 400-psig motive steam line pressure I would like to see the steam chest pressure over 375 psig. If not, then get the unit machinist to alter the governor setting so that it opens further. If he says the governor is wide open, see if you cannot open it further by pushing against the governor's spring tension with your foot. I've used this trick several times to debottleneck steam turbine's driving large centrifugal pumps.

Back pressure from the exhaust steam header will also reduce turbine horsepower. Salt deposits from moisture in the motive steam reduces turbine efficiency. If the deposits get too thick, they break off. This unbalances the rotor and causes excessive vibrations and shutdowns.

Calculating Turbine Horsepower

How much work can be extracted from steam in a turbine? Reference the Mollier Diagram in back of your steam tables. Then follow this calculation procedure:

1. Plot the motive steam temperature and pressure as measured in the steam chest. Not the steam supply header pressure.
2. Drop straight down along the lines of constant entropy until you intersect the exhaust steam pressure measured at the turbine steam outlet.

3. Note the enthalpy difference between the motive and exhaust steam conditions. Read enthalpy on the slanted vertical axis in BTUs per pound of steam.

4. Multiply the BTUs per pound by the steam flow and divide by 2550 BTU per horsepower. I would then multiply by about 95% for turbine efficiency to arrive at the available horsepower at the coupling.

Overspeed Trip

There is a widespread misconception in the process industry that the trip is a backup to the governor. This is not so. A turbine cannot be run safely unless both the trip and governor are working. I'll explain.

Let's refer to figure 30–2. If the control valve on the discharge of the pump closes, the work required to spin the pump is reduced. That's because the flow goes down faster than the increase in the pump discharge pressure. This means the speed control governor would start to close.

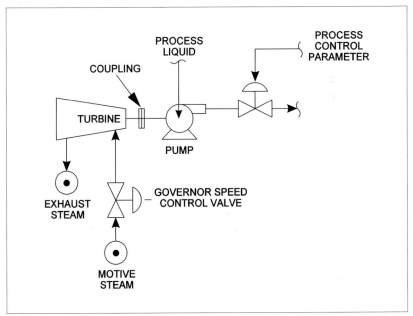

Fig. 30–2. Turbine driven process jump.

But let's assume the governor valve is stuck. Then the turbine and pump would run faster and faster until the turbine tripped off. This is fine. However, the turbine will trip off repeatedly whenever the process discharge valve closes too much. During the 1980 strike in Texas City, we dealt with this problem with a piece of wire. That is, Joe Petracelli tied a piece of No. 9 wire around the trip.

"Norm," said old Joe, "You can't run a turbine with the governor stuck unless you disable the overspeed trip valve. Every time the pump load drops off, the turbine will trip off because of excess speed. I know it's dangerous to wire up the trip. But that's what we'll have to do until those asses in maintenance get the governor speed control valve on this amine pump working again."

Condensing Steam Turbines

The lower the exhaust steam pressure, the more work that can be extracted from each pound of steam. Exhausting steam under vacuum conditions can increase the amount of work extracted from each pound of steam by a very large amount. You do not even need your Mollier Diagram or steam tables to approximate this benefit. Just use the following relationship:

$$\Delta W = \frac{(T_3 - T_2)}{(T_3 - T_1)}$$

where:

ΔW = Increased percent of work extracted from the steam

T_3 – Motive steam temperature

T_2 – New exhaust steam temperature

T_1 – Original exhaust steam temperature

For example, exhausting 100 psig saturated steam to the atmosphere generates about 0.060 horsepower per pound. Exhausting 100 psig steam to a surface condenser (see figure 30–3) generates about 0.12 horsepower per pound of motive steam, when the condenser is operating at 100°F and 28 inches of mercury vacuum. (Two inches of mercury pressure on an absolute basis).

Why not double check my calculations right now with your Mollier Diagram? If you then suddenly rush out of your office to back-flush or acid clean the water side of your surface condenser, you deserve an A-plus in your understanding of thermodynamics.

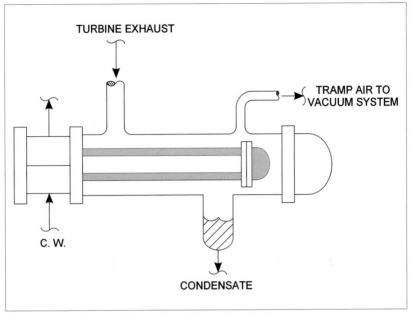

Fig. 30–3. Surface condenser for turbine exhaust steam.

Variable Speed AC Motors

An ordinary alternating current electric motor is a fixed sped device. At home they are single phase motors that operate at 1800 rpm. At work, they are mainly three phase motors that operate at 3600 rpm. (In Europe it's 1500 rpm and 3000 rpm). To vary the speed of an AC motor, one must vary the frequency of the power supply from the standard 60 Hertz (50 Hertz in Europe). Varying frequency involves an additional capital investment; an additional loss of overall efficiency of power transmission; and another operating complication. Why bother?

The big benefit is that the downstream process control valve shown in figure 30–2 is not needed. To maintain the required pump discharge flow or pressure, the pump speed is altered automatically by altering the driver's electric supply frequency.

Any variable speed driver has this potential advantage, including turbine driven pumps. Switching to a variable speed electric motor will save energy in accordance with the following calculations:

$$\text{Pumping Energy Saved} = \frac{(\Delta H) \cdot W}{775}$$

where:

Pumping energy saved = BTUs

ΔH = Head loss across the control valve, feet

W = Weight of fluid pumped, pounds

The pumping energy saved should be multiplied by a factor of three to four. It takes that much heat to generate a single unit of pumping energy or work. For example, it takes about 10,000 BTU per hour to generate one KWH (3,000 BTU/Hr.) of electrical energy.

Variable speed motors are quite practical. I've worked with one for many years at the Chevron Delayed Coker in El Segundo, California, that was serving on the Heavy Coker Gas Oil Pumparound circuit.

On the other hand, why bother? Most of these benefits could be obtained by optimizing a centrifugal pump's impeller size to minimize parasitic losses through the downstream process control valve. If the process itself requires a variable flow and head, then the frequency adjusted electric motor becomes more economic, as changing out a pump impeller is an all day job.

My logic may be archaic as energy costs have doubled in the past twelve months (July 2007 to July 2008). What will the future bring us?

Roto-Flow Turbo-Expander

One of my first projects as a young Amoco Oil engineer was to design a new system to recover butane form refinery fuel gas at VRU 300, in Whiting, Indiana. An ancient engineer, Joe Belansky, suggested that I increase the lean oil flow to the absorber. But I was far more creative than Old Joe. So I designed the Roto-Flow Turbo Expander Project (shown in Figure 30–4) that became famous throughout Amoco Oil in the 1960s. It worked like this:

- Gas was compressed in a variable speed centrifugal compressor from 300 psig to 500 psig.

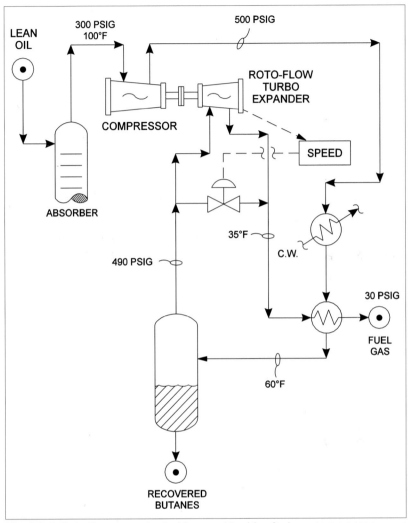

Fig. 30–4. Roto-flow expander turbine used to drive fuel gas compressor.

- The gas was chilled to 60°F. At 60°F and 490 psig, most of the butanes were recovered as liquid hydrocarbons.
- The gas was depressured through my new Roto-Flow Turbine Expander down to the fuel gas system pressure of about 30 psig. The rotational energy from the expander turbine drove the new fuel gas compressor.
- The fuel gas auto-refrigerated as its expansion energy was transmitted to the centrifugal compressor. The cold gas was then used as a refrigerant to chill the fuel gas to 60°F.

The energy to drive this complex but creative design was derived indirectly from expanding the 300 psig absorber off-gas to the 30 psig fuel gas header. The unit was commissioned and it worked fine. Except butane recovery was almost nil. What had gone wrong?

Unfortunately, shortly before the expander startup, old Joe Belansky had the operators open a hand valve on the turbine driven lean oil pump. The pump, running faster, doubled the lean oil flow as well as the butane recovery. Thus, there was very little butane left over for my new, creative design to recover from fuel gas.

The Roto-Flow Turbo Expander ran for a few more months and then it was "temporarily" shut down for repair. But, Joe Belansky wrote a big red number three priority on the maintenance repair work order slip. And there the expander sits to this day. Rusting back into the sandy soil of Whiting, Indiana. A monument dedicated to the creativity of a young process engineer. (The parameters shown in Figure 30–4 are for illustration only, and do not reflect any calculation or actual design).

Gas-Fired Turbine

A gas-fired turbine is pretty much the same as a jet engine. The technology was developed by Frank Whittle, a junior naval officer in the British Royal Navy prior to World War II. The British weren't all that interested at first. But the Nazis in Germany adopted a more positive attitude to jet propulsion. Eventually, the British government gave Mr. Whittle's technology to General Electric (GE), in 1943.

The main part of a gas turbine drive was originally a centrifugal compressor. More currently it's an axial air compressor. The compressor discharge pressure is typically 100 psig. The high pressure compressed air then mixes with the fuel gas in a combustion chamber. The hot products of combustion are directed at a high velocity against a multi-wheeled power recovery turbine. About two-thirds of the turbine's power output is used to drive the air compressor. About one-third is used for such functions as:

- Thrusting an airplane forward
- Spinning a gas compressor
- Driving very large process pumps

Effect of Rotor Fouling

Let's further assume we are compressing natural gas containing entrained brine. As the gas is compressed, the gas gets hotter. Moisture evaporates and salt deposits accumulate inside the compressor wheels.

Let's assume that the fuel gas rate to the turbine is fixed. That is, the horsepower or torque developed by the turbine is constant. As the salt deposits accumulate on the gas compressor end, efficiency and capacity are both reduced. However, the reduction in capacity predominates. This unloads the turbine. The turbine and the air compressor will now run faster. I used to track fouling rates of our natural gas compressors in Laredo, Texas, by monitoring the speed increase of the gas-fired turbines.

The maximum fuel gas rate to a gas-fired turbine is determined by the exhaust outlet temperature. The material of construction of the combustion chambers and turbine blades sets this maximum. A typical maximum outlet is 1,100°F. The best way to overcome this limitation is to increase the combustion air flow from the front end air compressor. Cleaning the inlet air filters always helps. Detergent washing the compressor internals is also often very beneficial. I recall in Laredo, we obtained about 5% extra air flow when ambient air temperatures dropped by about 10°F.

Our three giant 10,000 horsepower natural gas compressors in Laredo were Solar-Centaur Split Shaft designs. A process type sketch is shown in figure 30–5. The interesting feature of this design was that there were two independent shafts. Each shaft spun at a different speed. The gas turbine drive shaft was split into dual sections. Two of the turbine wheels drove the front end air blower. One turbine wheel drove the back-end 10,000 horsepower natural gas compressor.

Loading or unloading the natural gas compressor did not affect the speed or the horsepower output of the air compressor side of the gas-fired turbine. But the reverse was not true. Increasing the horsepower output from the air compressor side or the air compressor speed did increase both the speed as well as the capacity of the natural gas compressor side.

The big problem on the gas compressor side was rotor fouling. As the fouling progressed, the speed of the gas compressor would increase relative to the air compressor side. I tracked gas compressor fouling with this idea. When the fouling reached a predetermined point using this method, I would shut down the gas compressor for off-line nut blasting. As I have explained in this reference,[1] if fouling progressed too far, the deposits would break off the rotor. Then the machine would become unbalanced and trip off on high vibrations, which meant that the gas compressor would have to be disassembled and cleaned on a shop bench in Dallas, Texas.

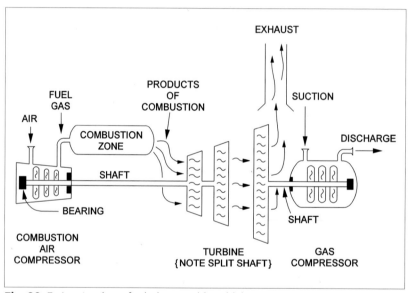

Fig. 30–5. A natural gas fueled gas turbine driving a gas compressor of the split shaft design.

Raising the gas compressor suction pressure would cause the gas compressor to slow down relative to the air compressor. I also found this confusing until I better understood the nature of centrifugal compression. Our big problem on the air side of the Solar-Centaur units were fouling of the suction air intake filters with moths.

Power Recovery Turbine

The Roto-Flow Turbo Expander shown earlier in figure 30–4 is a variable speed turbine driving a variable speed compressor. However, most of power recovery turbines are fixed speed machines. In refineries, I have worked with power recovery turbines of two sorts.

Fluid catalytic cracker units (FCU)

The 25 psig and 1,200°F flue gas from the regenerator (see figure 30–6) expands to atmospheric pressure. The expander turbine drives the FCU main air blower, which burns off the coke on the circulating catalyst. Just like the gas-fired turbine, the resulting flue gas is so hot that the blower's effluent contains more recoverable energy than is needed to drive the blower. The excess energy is absorbed by the motor shown in figure 30–6.

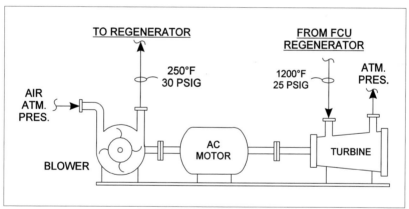

Fig. 30–6. FCU power recovery turbine used for electric power generation and driving air blower.

The motor was used to start the regenerator's combustion air blower. Once the expander turbine is producing its full power, the motor is no longer required. However, an ordinary alternating current, three-phase electric motor will work like an electric generator. No modifications are required. The motor exports power into the electric grid that powered the motor during the unit start-up.

As the temperature of the flue gas increases, more power is produced from the expander turbine. But, this extra power cannot make the blower run faster. The extra power turns the motor into an electric power generator, which then exports this power into the grid. Or, the motor acts like a brake on the expander turbine.

Hydrocracker reactor effluent

The cool, 2,000-psig liquid from the low temperature, high pressure separator vessel shown in figure 30–7 drops to 100 psig through the power recovery turbine designed for flashing liquids. The energy of the expanding evolved vapors, plus the energy released from the pressure reduction of the liquid stream, is used to drive the pump that initially pressurized the liquid. Because of the flashing vapors, more energy is available from the expander than is needed to drive the centrifugal pump. The excess energy is not available to make the pump run faster. This excess energy is absorbed by the motor, which then generates electric power. The excess power is then exported back into the utilities' power grid.

The motor acts like a brake on the turbine and thus prevents the pump from running faster and pumping more feed. The speed of the entire assembly shown in figure 30–7 is a function of the 60 Hertz frequency of the electric grid and the phases of the motor.

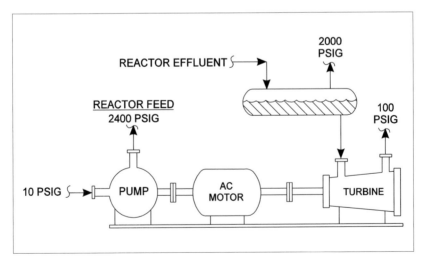

Fig. 30–7. Expander turbine with power recovery motor is not a variable speed turbine.

Troubleshooting Checklist for Turbines, Expanders, and Variable Speed Motors

Steam turbines speed adjustment
Use of horsepower valves
Optimizing steam chest pressure
High-speed control valve ΔP
Optimizing governor position
Back pressure on exhaust steam
Moisture in motive steam
The steam rack
Troubleshooting using
 thermodynamics
Use of Mollier Diagram
Windage losses in turbines
Eliminating control valve
 on pump discharge flow

Erroded-out hand valves
Calculating turbine
 horsepower output
Trip valve malfunctions
Condensing steam
 turbine problems
Variable-speed AC motors—use
 in flow control
Expander power recovery turbines
Gas-fired turbines—Rotor fouling
Gas-fired turbines—Compressor
 fouling
Expander turbines in cat cracker
 regenerator flue gas

References

1. Norman Lieberman, *Troubleshooting Natural Gas Operations,*
    Second Printing, 2008. Lieberman Publications.
    (ISBN 978-0-98166542-2-21)

2. Norman Lieberman, *A Working Guide to Process Equipment,*
    3rd edition, 2008. McGraw-Hill Publications, New York.
    (ISBN 978-0-07-149674-2)

3. Everett Woodruff and Herbert Lammers, *Steam-Plant Operations,*
    4th edition, McGraw-Hill, 1977. (ISBN-07-071731-1)

Forced Draft Air Coolers

As a young engineer working for the Amoco Oil Refinery in Whiting, Indiana, I had never seen an air cooler. It was cheaper for Amoco to use Lake Michigan as a giant cooling water pond. Hydrocarbon contamination of the circulating lake water was not something that concerned my employer. Air coolers are far more expensive then water coolers. Perhaps double the cost, unless one has to supply the cooling water to the exchanger from an expensive cooling tower.

Most air coolers have the fan located beneath the finned tubes. Such coolers are called *forced draft* fin-fan coolers. If the fan is located above the tube bundle, it's called an *induced draft* fin-fan cooler. The reason most air coolers are forced draft is not because of process considerations. It's to provide easier access to the fan bearing (shown in figure 31–1), of the electric motor and drive belt.

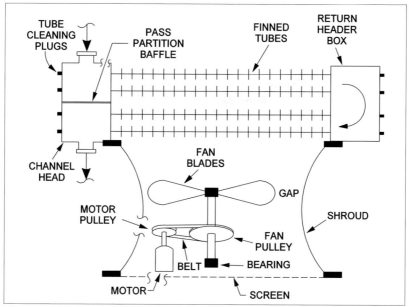

Fig. 31–1. Forced draft air coolers. Note the gap between fan blade and shroud.

The screen shown in figure 31–1 does not influence the air flow. The screen's function is to protect your head from being sliced off. However, you can observe an interesting air flow pattern through this screen. I use a strip of cloth. The observations I have made are summarized in figure 31–2. Note that about half the air flow that is passing through the screen is recirculated air. This air is not recirculated from the air cooler outlet but from fan discharge, as indicated by the temperatures noted on this figure. This means that half the air flow generated by the fan is not being used for cooling. This air recirculation is the major problem with fin-fan forced draft air coolers.

The recirculation problem is a function of two factors:

- The gap between the fan blade and the inner diameter of the shroud. (per figure 31–1)
- Restrictions to the air flow on the underside of the finned tubes

The following statements are not exactly correct, but apply reasonably well for fixed speed motors.

- Air flow produced by the fan is constant, regardless of finned tube exterior fouling.
- Fan discharge pressure is constant, regardless of finned tube air-side fouling.
- The amperage or power from the motor is constant, regardless of finned tube fouling.

Effect of Air-Side Fouling

Fouling or crushing of the fins on the top of the tubes is not important. Straightening the fins or water washing the top of the bundle to promote more air flow is a waste of time. I've tried.

The fin tubes need to be washed from underneath the tubes with a narrow jet of carefully focused water. Be really careful. First, do not use too much water pressure. The fins are made of aluminum and are easily bent. Bent fins underneath the tubes will positively restrict the air flow. Secondly, this is done with the fan shut off. After you remove the screen, it is not sufficient to electrically lock out the fan motor at the breaker. Tie the fan off with a strong rope. Suppose a gust of wind starts the fan spinning? A 40-foot-by-20-foot bundle can take four hours to clean properly. Soaking the fins with detergent first will help if the deposits are greasy.

Low air outlet temperature is a sign of lack of hot process side flow, if the process side outlet is also cool. If the air outlet temperature is low, and the process side outlet is hot, this is a sign of tube-side film resistance or tube-side fouling, as discussed in the next section of this chapter. If the air outlet temperature is high, and the process outlet temperature is cool, this indicates that the cooler is doing a good job of removing heat.

Excessive Shroud-to-Fan Tip Gap

As shown in figure 31-1, there is a gap between the tip of the fan blade and the shroud wall on the inside of the shroud. This gap ought to be a uniform ¼ inch. But, as time passes the shroud, which may be 20 feet in diameter, will distort. The blade tip is slowly worn down. If you spin the fan by hand on any older air cooler, you will see that the fan blades are almost touching the shroud in certain areas. In other areas there could be a 4-inch gap between the blade tips and the shroud. This gap promotes the air recirculation shown in figure 31-2.

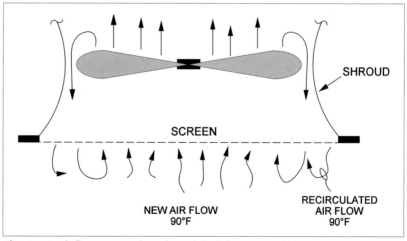

Fig. 31–2. Air flow pattern for a forced draft fan. Some air is always recirculated.

This problem is easy to correct. A long mesh strip is bolted inside the shroud. The strip is called a *fan tip seal*. It's sold by the air cooler manufacturer. It comes in a large roll. The mesh is about 9 inches wide and 2 or 3 inches thick. A tool fastened to the fan cuts a shallow grove in the mesh, thus eliminating the gap. I've only used this fin tip seal retrofit once. At the hydrogen plant effluent cooler in Aruba. The results indicated a nice 10% to 20% increase in air flow for no real cost.

Other Methods to Increase Air Flow

One somewhat novel method to partly restore air flow through an air cooler is to reverse the polarity of the motor leads. This will reverse the direction of air flow. If the fouling deposits are loosely adhering to the fins, the dirt will be blown off the underside surface of the finned tubes.

Increasing the fan blade pitch or angle form 10° to 15°, up to a maximum of 20° to 25°, only increases the air flow by about 5%. At least that's all I achieved the one time I tried it. But, on the other hand, it only required an hour for two maintenance people to make the adjustment.

Increasing the speed of the spinning fan is a far more aggressive way of increasing air flow by perhaps 20% to 30%. This is done quite simply by increasing the size of the motor pulley shown in figure 31–1. If you increase the diameter of this pulley by 10%, then the fan will run 10% faster. But, the amp load on the motor driver circuit will increase with speed cubed. Or in this case, by 34%. You should make sure that the full limit amp load on the motor driver circuit breaker is rated for this extra electric power requirement. Also, check that the fan blades are rated for the extra torque at the greater speed.

Water Spray

In areas of low ambient relative humidity, spraying water in a finely divided mist just below the screen can reduce process outlet temperatures by about 10°F to 15°F. The idea is to use the water spray on hot summer afternoons. I would not want to use such a spray continuously throughout the year. I saw the water spray used sparingly in Lithuania and it worked really nicely on a vacuum tower overhead pre-condenser.

On the other hand, don't spray water on top of the tube bundle. The fins will corrode and inhibit rather than promote heat transfer. Perhaps the worst example of this practice I observed was in Aruba. Sea water from the fire monitors was cascaded on the delayed coker fin-fan coolers and salt accumulated between the fins.

Slipping Belts

The drive belt shown in figure 31–1 cannot slip if it is a modern serrated type belt. Most new air coolers are so equipped. Older units may have smooth belts and in these units drive belt slippage was a constant problem. A slipping belt will cause a visual reduction in fan rotational speed and decrease in the amperage load on the motor drive. Increasing the belt tension by realigning the motor corrects this simple problem.

Incidentally, don't forget the fan bearings need to be re-greased every few months. Most of the maintenance problems with air coolers are associated with these fan bearings drying out.

Air Leaks

The tube bundle ought to sit firmly atop the shroud. Usually it does. But sometimes there are glaring gaps underneath the tube bundle. A "relatively" large amount of fan's discharge air flow can escape through these gaps. By relatively large, I have in mind just a few percent of the air flow. On the other hand these gaps are easy to fix.

Incidentally, an air cooler without its fan running typically will have perhaps 30% of its normal capacity. I like to take advantage of this capacity in my process designs. During a power failure, when a cooling water exchanger duty drops to zero, the air cooler can still be used as a heat sink.

Process-Side Problems

A common problem with air coolers subject to tube-side fouling is a leaking pass partition baffle. Should the tube side pressure drop become excessive, then the channel head pass partition baffle shown earlier in figure 31–1 may pull away from the channel head tube sheet. Hot fluid could then partly bypass the tubes. I identify this problem by checking the temperature with my hand at the return header box. If the return header box is cooler than the channel head outlet nozzle, then the pass partition baffle is definitely leaking. This becomes far more complicated with an exchanger with more than two passes. It's best to check the exchanger header box drawings and sketch the pass partition baffle configuration with a marker right on the channel head and the header box covers.

Tube-Side Fouling

In a refinery air coolers are extensively used in several dozen parallel banks of bundles in fractionator overhead condensers. When one of these banks begins to foul, the process side flow markedly diminishes, but the pressure drop only slightly increases. The fouling mechanisms for the following kinds of condensers are typically ammonia chloride sublimation reactions:

- FCU fractionator overhead condensers
- Delayed coker fractionator condensers
- Crude distillation unit overhead condensers
- Hydrocracker fractionator condensers

The word *to sublime* means to change directly from a vapor to a solid salt.

I am personally most familiar with ammonia chloride salts. Both ammonia bisulfide salts and dry amine chloride salt will form a white deposit that smells quite strongly of ammonia. The chlorides originate from either hydrolyzed $MgCL_2$ and $CaCL_2$ in crude, or from chlorides added to the naphtha reformer which are inadvertently stripped off the catalyst. The ammonia may come from organic nitrogen compounds in FCU or coker feed. On the crude unit, ammonia is absorbed into the crude from poorly stripped sour water used as the wash water in the crude unit's desalter.

The ammonia chloride salts sublime inside the air cooler tubes and thus reduce the process flow. My practice and experience, rather than any definitive technical rationale, is to use the following methods:

- On fluid catalytic cracking units (FCU) I simply shut off the air fan for twenty or thirty minutes and allow the tubes to heat up. This allows the salts to revaporize. This method would not work too well in cold ambient conditions.
- On coker fractionators, I have used slug water washing with success. Taking one air cooler bundle at a time, just inject as much water as possible, without shutting off the fan, for about fifteen minutes. Watch the water boot level in the reflux drum, as refluxing water saturated with ammonia chloride back to the column is a really bad idea. Slug wash a single tube bundle at a time to avoid overloading the boot.

For crude distillation towers, the results of either method will probably be disappointing. The reason is the predominate salt deposit is typically iron sulfide in the crude unit overhead condensers. The iron sulfide deposits $Fe(HS)_2$ are neither water soluble nor can they be revaporized. The exchanger must be cleaned off-line by pulling the cleaning plugs (shown in figure 31–1) and running a water lance through each tube separately! A typical air cooler bundle may have several hundred such cleanout plugs. Some of which always get cross-threaded and leak after each cleaning.

The result of cleaning one air-cooled tube bundle in the bank of two dozen may not be what you have anticipated. The process outlet temperature will go up by a lot, rather than down. The reason is that you have restored the full process flow through this single bundle. However, to prove that all is well, note that the air outlet temperature has also increased because more heat is being removed at the higher tube-side flow rates. Also, the reflux drum temperature will have dropped by a tiny amount. Of course, if you cleaned only one bundle out of twenty-four, the reflux drum temperature change may be too small to observe. Once, at a Chevron plant, the operators noted in disappointment that the very first air cooler we cleaned on the delayed coker by slug washings had a 20°F outlet temperature increase. But, as we continued cleaning the bundles, the reflux drum temperature slipped down, and they could see the accumulated benefits of their work.

Air Louvers

As I noted previously, aerial coolers maintain an appreciable cooling capacity even when the fans are shut off. Thus, many air coolers are equipped with air louvers that can be adjusted by an air-operated control valve. These louvers are intended to be manipulated to keep the tubes from over-cooling. They look like Venetian blinds set atop the tube bundle. After a year or two of use, these louvers often stop working. At best a few of them can be partly closed by hand. Mechanically speaking they are not reliable and I have rarely seen them used with any long-term success. If your unit is equipped with such louvers make sure all the louver's slats are actually open if you are short of cooling capacity. During cold weather the tube bundle should simply be covered, if shutting off fans is not sufficient to reduce cooling capacity.

To make adjustments on the air flow, two-speed motors driving the fans work quite well. This is a reasonable retrofit option to control cooling capacity, as compared to retrofitting with louvers. With two-speed motors the following approximate capacity ranges can be expected:

- Full speed—100%
- Half speed—70%
- Fan off—30%

In really frigid climates, steam coils are needed underneath the tube bundles. These steam coils work extremely well but are quite expensive and are a huge waste of energy. Better just to pull a fire retardant tarp over the top of the forced draft air cooler bundles.

Troubleshooting Checklist
for Forced Draft Air Coolers

Protective screen removal
Air recirculation
Blade tip to shroud clearance
Air-side fouling
Vane tip seals
Reversing air flow
Larger motor pulley to increase
  fan speed
Slipping belts
Adjusting fan blade angle
Cool air with water sprays

Fin corrosion
Air leaks around bundle
Tube-side fouling
Leaking process side pass
  partition baffles
Salt sublimation inside tubes
Iron sulfide deposits inside tubes
Slug water wash tubes
Air louvers
Two-speed fans
Steam coils below tube bundles

References

1. Henry Z. Kister, *Distillation Troubleshooting* AICHE, Wiley-Interscience, 2005.

2. E.T. and N. P. Lieberman, *A Working Guide to Process Equipment,* 3rd edition, 2008. McGraw-Hill Publications, New York. (ISBN 978-0-07-149674-2)

Advances in Heat Exchanger Design

This chapter covers new proprietary methods available to enhance heat transfer efficiency and some older techniques to increase heat exchanger coefficients using the existing shell, floating head cover, and the channel head but modifying the tube bundle. Figure 23–2 (chapter 23) shows the components of a conventional shell and tube exchanger.

The simplest method to increase heat transfer is to increase the number of tube-side passes, if the controlling resistance to heat transfer is on the tube side and current tube-side velocities are low. Low is less than 3 feet per second. (See chapter 9, figure 9–2 for details).

Stainless Steel Tubes

Other than high velocities, the next best method to control fouling on the both shell and tube sides is to prevent the formation of rough surfaces due to corrosion. A smooth, mirror-finished surface will retard the accumulation of fouling deposits. At an FCU in Whiting, Indiana, stainless tube bundles replaced carbon steel tubes in the preheat exchanger's train, and resulted in suppressed fouling.

Two notes of caution regarding re-tubing with stainless:

- Do not put stainless tubes in direct physical contact with carbon steel tube support baffles or carbon steel tube sheets. The results will be galvanic corrosion of the carbon steel components. On the other hand, nine chrome tubes are consistent with carbon steel components.
- 304, 316 and 317 stainless should not be used in crude preheat. At least, not upstream of the desalter. The problem is chloride stress corrosion cracking.

Low Fin Tubing

Serrations are cut into the outside surface of tubes to form fins about ¼-inch high and about a dozen fins to the inch of tube length. The increased surface area is about 250%. But this only makes sense if the controlling resistance to heat transfer is on the shell side. If the controlling resistance to heat transfer is due to shell-side fouling, low fins will lead to disaster. Accelerated rates of fouling between the fins will ruin heat transfer efficiency. In one refinery, I retrofitted the propane refrigerant condensers with low fins. It worked great until the cooling water (tube side) fouling increased due to lack of proper water treatment (see chapter 16).

Sintered Metal Tubes

Rough surfaces are bad for sensible heat transfer due to fouling. But in clean services, rough surfaces are critically important to boil water and hydrocarbons. Rough surfaces provide nucleation sites for bubbles to form. For one butane reboiler, an old reboiler bundle with pitted carbon steel tubes had double the heat transfer capacity of a new bundle.

There are two ways to prevent loss of heat transfer on reboilers when a newly retubed bundle is commissioned. A sintered metal coating can be applied to the tubes, or the tubes can be lightly sand blasted to roughen their surface.

You can see the effect at home. Place a new, unscratched cup of water in your microwave. Apply lots of heat but the water will not boil. Drop a tea bag in the cup and the water will explode out of the cup.

Unequal Seal Strips

Seal strips are used to prevent internal shell side bypassing. Most exchangers in the United States have vertically cut tube support baffles (see figure 32–1). This means the impingement plate is on top of the tube bundle right beneath the shell-side inlet. To make room for the impingement plate, several rows of tubes are omitted from the top of the bundle. This creates a gap for the shell-side inlet flow to bypass the tubes by flowing across the empty space at the top of the bundle.

It has been common practice to block this bypass area with a pair of flat strips called *seal* strips. The strips are typically 3 inches high and ¼ inch thick. They extend (assuming the shell-side nozzle is a top entry) from the tube sheet opposite the shell inlet nozzle to the furthest tube support baffle. If the pair of seal strips both extended all the way to the other tube sheet, the feed inlet flow would be trapped. The flow would be boxed in by the impingement plate, the first tube support baffle, the top of the shell and the two seal strips.

But, if both seal strips end at the first tube support baffle, then the shell-side fluid may bypass part of the bundle. The section of the tube bundle between the tube sheet and the first tube support baffle will be bypassed as the shell-side flow slips through the top of the shell. Figure 32-1 illustrates both the problem and the solution. As shown, the seal strip along the edge of the first tube support baffle is extended to the tube sheet. This forces the fluid to flow across the tubes below the impingement plate.

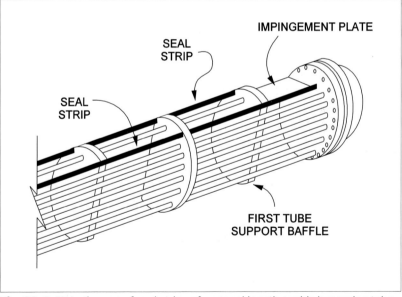

Fig. 32–1. Note the use of seal strips of unequal length avoids bypassing tube area to the right of the first tube support baffle.

This is difficult to follow. But, if the controlling resistance to heat transfer is shell-side film resistance and thus seal strips are used, this idea will improve the heat transfer coefficient by 5%–10%.

Advanced Techniques

Some, "newer," ideas are:

- Helical baffles
- Tube inserts
- Twisted tubes

I use the term "newer" advisedly because some of these techniques are several decades old. All of these concepts, in addition to being available for some time, have several other things in common:

- All are proprietary and a licensing fee is required.
- None are in widespread use.
- I have never had any client claim to use any of the new methods with consistent success.
- All have potential cleaning and/or mechanical problems that do not exist in conventional shell and tube exchangers.

Tube Inserts

These are springs that are inserted into the tubes. Some types are fixed and some spin with the flow. The objective is to create turbulence that reduces the tube-side film resistance, and the rate of tube-side fouling.

My first industrial experience with these inserts was bad. At the Texaco Lube Oil Reprocessing Plant, the vacuum tower top pumparound had an air cooler. The cooler tubes were badly plugged and could not be cleaned. The spiral inserts were stuck inside the tubes. I dealt with this problem by purchasing a new tube bundle.

Broken springs are often found in the channel head. Obviously the tube-side pressure drop will increase in a clean service. However, if the inserts work as intended, the service pressure drop might actually be reduced by the inserts. Total Fina Elf markets various tube inserts for use in crude preheat exchanger trains when crude charge is flowing through the tubes.

Twisted Tubes

I have shown a sketch of a twisted tube in figure 32–2. It looks like a 1-inch hollow drill bit. The idea is that the twisted surface generates more turbulence on both the tube side and the shell side than a straight tube. The twisted tube does not require any tube support baffles. The tubes touch and thus are self-supporting. However, without any tube support baffles, an external sleeve or shroud is needed to keep the tubes in place. Care must be taken in handling the shroud so as not to alter the alignment of the twisted tubes.

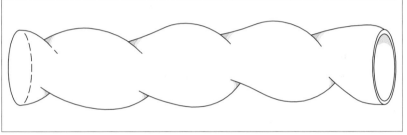

Fig. 32–2. A twisted tube. Bundle does not have tube support baffles.

The proper point of contact between adjacent twisted tubes is essential to maintain the design cleaning channels. It is designing and maintaining these cleaning channels that appear to be the main issues with twisted tube bundles. An exchanger that cannot be cleaned on the shell side ought not to be employed in refinery applications.

One of the purported benefits of the twisted tube exchanger is that about 30%–40% more tube surface area can be contained in a given size shell because the average spacing between the tubes has been reduced. Also, added turbulence induced by the twisted tube reduces heat transfer film resistance and thus increases the overall heat transfer coefficient.

Helical Tube Support Baffles

The flow through the tube side of this type of exchanger is conventional. However, the shell-side flow is unique. It is neither perpendicular to the tubes nor parallel. Rather, the shell-side flow follows a screw-type pathway across the tubes. It is the angled slope of the baffles that induce this sort of helical or screw-type flow to the shell-side liquid (figure 32–3).

The advantage of this sort of flow is that dead zones are eliminated in the exchanger areas where ordinarily the flow changes directions along the edge of the tube support baffles. For this strategy to work, the controlling resistance to heat transfer must be on the shell side.

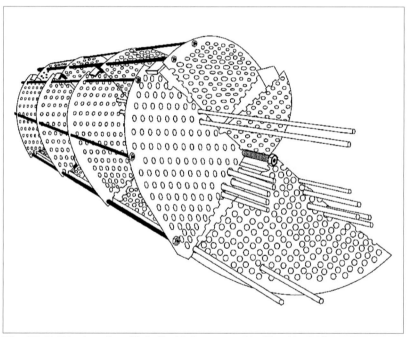

Fig. 32–3. Tube bundle with helical tube support baffles. Liquid flows in a screw type path.

While the theory of the helical baffles makes good process sense, the results have been mixed. I'll report three representative examples of many reports by my clients:

- **BP—Shell, South Africa—Visbreaker bottoms cooling.** Results were a failure due to fouling. I have spent 20 years trying to design exchangers to cool visbreaker bottoms with no success. I would not think helical baffles would solve this intrinsically fouling service.

- **Exxon, Texas—Coker feed preheat.** The vacuum resid preheating against gas oil pumparound does appear to work well. Fouling rates appear less than a conventional shell and tube exchanger. The exchangers do appear to be working as per their design.

- **Citgo, Texas—Resid preheat.** Exchangers have fouled more rapidly than anticipated. The problem has been attributed to the angle of the helical baffles being wrong for this particular service. This angle is critical. It must be carefully specified for each individual service by the exchanger design engineer.

Spiral Heat Exchangers

In 1963, I did my unit ops lab with a spiral heat exchanger, which was not new then. It is a practical heat exchanger used to cool relatively small flows. As shown in figure 32–4, there is only one flow path each for the hot and the cold fluids. Thus, no channeling or bypassing is possible. It's intended for use with fluids containing solids, such as FCU slurry oil product (not the much larger slurry pumparound).

I have heard some good reports from my clients who use spiral heat exchangers to cool cat (FCU) slurry. The one reported problem concerns gasketing the top plate which is hard to seal. This problem becomes progressively more difficult as the exchanger diameter increases.

A spiral heat exchanger is a true counter-current flow device. Thus, no correction factor is needed, as in shell and tube exchangers for non-true counter-current flow.

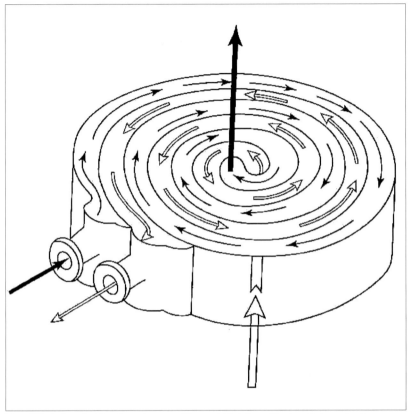

Fig. 32–4. A spiral heat exchanger. The cold inlet and hot outlet are in the center of the spiral.

Practical Ideas to Enhance Heat Transfer Efficiency

Several more conventional methods to enhance heat transfer are:

- Minimize clearance between tube support baffles and shell ID, as per TEMA specifications.
- Use ½-inch rather than the standard ¼-inch space between tubes. This will reduce dirt bridging between tubes on the shell side. This bridging problem creates dead zones with no flow and no heat transfer.

- Specify the highest possible allowable pressure drops on your exchanger data sheets. This permits vendors to design for a high velocity, which suppresses fouling rates.
- Include block valves and bypasses to stop flow for brief periods. The result is thermal spalling and/or melting of deposits from tube surfaces.
- Use floating head, not U tube exchangers. U tube exchangers cannot be visually inspected after cleaning the tube side of the bundle.
- Do not rerun cracked recovered slops through heat exchangers after the slops have been exposed to air. Polymerization will result at about 300°F to 350°F. The polymers form gums which promote fouling. Virgin materials do not polymerize.
- Use vertically cut baffles in boiling or condensation services for the shell-side flow.
- Charge from tanks with floating suctions. Dirt will settle out of the bottom of the tank. Wait until just before a unit turnaround to start the internal tank mixers. Exchangers will then foul rapidly.
- Watch for hydrocarbon leaks in cooling water system. Hydrocarbons promote biological fouling inside the tubes.
- Elevate condensers above reflux drums for drainage to avoid condensate backup.
- Finally, the most important point. Place the fluid with the lowest Reynolds number (i.e., the high viscosity fluid) on the shell side. The resulting vortex shedding as the liquid flows perpendicularly across the tubes will avoid laminar flow and the resulting high heat transfer film resistance. As long as the tube pitch is rotated square, the shell will still be able to be cleaned even though the shell-side flow is dirty.

I once switched sides on a water cooled exchanger. Originally the tube-side service was a high viscosity industrial fuel oil. The shell-side service was cooling water. After reversing the sides, the overall heat transfer coefficient increased by a factor of three.[3]

Troubleshooting Checklist for Advances in Heat Exchanger Design

Spiral heat exchanger
Helical baffles
Tube inserts
Twisted tubes
Stainless steel tubes
Galvanic corrosion
Cleaning U tube exchangers
Low fin tube exchangers
High viscosity fluids

Use of seal strips
Excessive clearance
Unequal seal strip length
Leaking pass partition baffle
Low velocity
Sintered metal tubes
Smooth tubes in reboilers
Dangers of inadequate cooling

References

1. E.T. and N.P. Lieberman, *A Working Guide to Process Equipment,* 3rd edition, 2008. McGraw-Hill Publications, New York. (ISBN 978-0-07-149674-2)

2. *Tubular Exchanger Manufacturers Association Standards Reference Manual.*

3. N.P. Lieberman, *Process Design for Reliable Operations,* 3rd Edition. Lieberman Publications.

Positive Feedback Loops

My ex-girlfriend once told me that we had a good relationship because she provided me with lots of positive feedback. From her comments, I concluded that positive feedback is desirable. And in your relationship with your partner this is so. But positive feedback in process control loops is to be avoided. I've encountered many such loops in my work. For instance, let's consider a propane-butane splitter.

The reboiler is on automatic temperature control to maintain a fixed tower bottoms temperatures. The objective being to keep the propane content of the butane bottoms product constant. Reflux flow is also being controlled to keep the tower top temperature constant. The objective being to keep the butane content of the propane overhead product constant. (See chapter 12, figure 12–1).

As the reflux rate automatically increases, so does the reboiler duty. If we had only increased the reflux rate then the tower bottoms temperature would fall. As the reboiler duty increases, so too does the vapor flow up the tower. All this is fine and that's how distillation towers work. That is, the reflux and reboiler rates increase together.

But, the higher vapor rates can promote excessive entrainment. This is called *jet flood*. The excessive entrainment will cause droplets of butane-rich liquid to be carried up the tower. The entrainment of the liquid droplets will slightly increase the butane content of the vapor leaving the top tray. The vapor leaving the top tray is saturated vapor at its dew point temperature, so the increased butane content slightly raises the tower top temperature. The small increase in tower top temperature raises the reflux rate a bit more. The extra reflux cools the entire column. The reboiler duty automatically increases to restore the tower bottoms temperature. The extra vapor flow promotes the entrainment of more droplets of butane-rich liquid. This increases the tower top temperature which calls for more reflux which generates more vapor flow. The problem feeds upon itself until the operator intervenes. The intervention takes the form of switching from automatic reflux temperature control to manual reflux flow control.

Should the operator not intervene by switching from auto to manual, the propane product would never come back to the required specification. The propane would forever be contaminated with butane. Such is the nature of a positive feedback process control loop.

Surging Centrifugal Compressor

In 1974, I was working as the outside operator on a sulfuric acid alkylation plant at the Amoco Refinery in Texas City. Normally I was the operating superintendent of this unit. However, the Oil, Chemical and Atomic Workers Union had been on strike for two months.

One night, due to the lack of olefin feed stock availability, we had reduced the alkylation unit feed in half. Gradually the iso-butane refrigerant flow to the centrifugal compressor decreased. (See chapter 6). As shown in figure 33–1, this compressor was equipped with a minimum flow recirculation valve. As the refrigerant vapor flow fell, the spill-back valve opened automatically. The compressor suction temperature increased, which of course increased the compressor discharge temperature, which increased the compressor suction temperature. This is not a positive feedback loop yet, because the cold refrigerant vapor will restore an equilibrium suction temperature at very roughly 150°F.

As the suction temperature to the compressor rises, the vapor density drops in proportion to the percent temperature rise in degrees Rankine (i.e., 150°F + 460°R = 610°R). The reduction in gas density caused the centrifugal compressor to pump fewer pounds of gas. Thus, the compressor suction pressure began to rise. At a higher compressor suction pressure the flow of refrigerant iso-butane vapors from the alkylation reactors was suppressed. This caused the compressor suction pressure to rise further, which increased the discharge temperature, thus increasing the suction temperature and pressure, which suppressed the refrigerant flow. (See chapter 25).

As the refrigerant flow dropped, the spill-back valve shown in figure 33–1 opened further. This accelerated the rate of suction temperature increase and accelerated the reduction in the flow of the cool refrigerant. A true positive feedback loop had now been created.

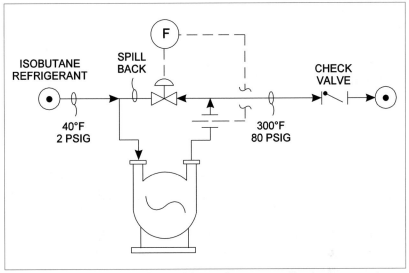

Fig. 33–1. Spill-back maintains a minimum flow to centrifugal compressors.

It is a characteristic of centrifugal compressors that reduced gas density pushes the machine to its surge point. When a compressor surges, the vapor flow stops and reverses. The reversal of the gas flow pushes the rotor backwards. The shaft slams up against the thrust bearing. Surge will tear up both the rotor and stationary elements of the compressor sooner rather than later.

When the compressor flow stopped due to the surging, the recirculation spill-back valve opened all the way and the flow of cool iso-butane refrigerant stopped completely. This just made the problem worse.

This happened really fast. The analysis I offer was done in retrospect. However, what impressed me beyond all else at the time was the banging of the discharge check valve shown in figure 33–1. With each gas flow reversal this 12-inch check valve slammed shut with an ear shattering clang. Added to the surging, and the animal-like roar repeated regularly every 10 seconds, well . . . the clanging and the roaring completely unnerved me.

In 1974, the compressor controls were all local. It was 1:00 AM and I was all alone. I had been transferred from the Chicago Engineering Headquarters just four months previously. I was confused, frightened, and soaked with sweat. Overcome by panic, I pushed the big red compressor Emergency Stop button. A serene and quiet peace then descended over No. 2 alky unit at the Amoco Oil Texas City Refinery.

Exploding De-Aerator Atmospheric Vent

Most operators that have attended my Troubleshooting Seminar have observed the following startling upset. Water and steam suddenly blow out of the de-aerator vent shown in figure 33–2. This is an example of a positive feedback loop created by a level measurement. It happens when the steam supply valve that maintains the vessel pressure opens 100%.

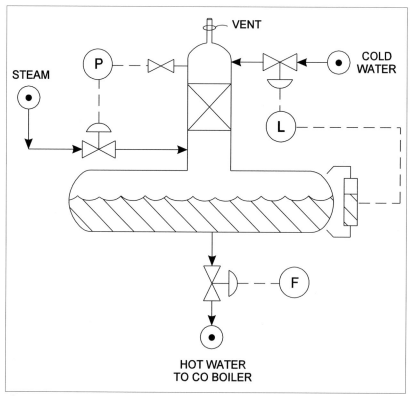

Fig. 33–2. Deaerator level surges and blows water out the vent.

The steam mixes and condenses in the cold water supply that feeds the de-aerator. The cold water flow is reset by the vessel level. The condensing steam heats the cold water to its boiling point temperature at the set-point pressure. For example, if I increase the set-point pressure by one psi, then the steam valve would open to heat the cold water by 2°F. But what happens once the steam supply valve is 100% open and more cold water enters the de-aerator vessel?

Obviously, the de-aerator pressure starts to drop. As soon as this happens the water stored in the lower portion of the de-aerator begins to boil. The volume of expanding water quickly swells as the density of the boiling water is radically reduced. The expanding water volume rises and covers the top tap of the level control indicator shown in figure 33–2. (See chapter 35—i.e., being, "tapped-out").

The boiling water between the level taps is much less dense than hot water. The level indicator really just measures the pressure difference between the top and bottom taps. As the boiling water is less dense than the hot water, the measured level drops, even though the real level in the vessel is expanding.

The reduction in the measured level opens the cold water valve. The surge in cold water flow drops the de-aerator temperature and pressure. As the de-aerator pressures falls, the boiling water volume expands. The density in the de-aerator drops, which opens the cold water level control valve, which drops the temperature and pressure in the de-aerator, which lowers the de-aerator pressure. A positive feedback loop has now been created.

The surging level quickly blows out of the atmospheric vent. The water lost from the vent must be compensated for by more cold water flow. But, without any more steam to heat the incremental cold water, the de-aerator pressure drops further and faster. This magnifies the effect of the positive feedback loop.

I was working at the Exxon (then Tenneco) Refinery near New Orleans in 1985 on a level instability problem for their new CO boiler de-aerator when this happened. It was a truly terrifying experience. To stop this eruption the panel operator switched the cold water level control to manual and closed the cold water makeup valve to 50%. Of course, the flow of hot water to the CO boiler had to subsequently also be reduced.

Inherent Instability

Note that all process systems are inherently unstable. The system will restore its own equilibrium provided that a certain stress point is not exceeded. This stress point is called the *tipping point*. Below the tipping point, negative feedback predominates. The system will not be able to restore its own equilibrium. Beyond the tipping point, instability feeds upon itself. Positive feedback rules. Short of extreme outside intervention, the system can never again achieve equilibrium.

Think about global warming and CO_2 accumulation in that context. What sort of extreme intervention will be required to restore equilibrium to our planet?

Fired Heaters Stalling

Operators call this *flooding the fire box with fuel*. A very common problem. Your stack damper and air registers are wide open. No more combustion air is available. The heater fuel is regulated on automatic temperature control to achieve the required process outlet temperature set-point of 700°F. (See chapter 15). You decide to increase the set-point to 702°F to recover more diesel oil from crude.

The fuel gas valve opens from 49% to 50%. The fuel gas flow increases from 6.0 MMSCFD to 6.1 MMSCFD. But the extra fuel doesn't burn, because you cannot increase the air flow. Then what?

Since the fire box is at 1,400°F and the unburnt fuel is at 60°F, the extra fuel gas cools the fire box to 1,399°F. The process outlet temperature drops by 0.1°F. The fuel gas flow increases to restore the process temperature set-point. But without more air, the cold extra fuel gas just makes the problem worse.

A positive feedback loop has been created. Black smoke will pour out of your stack. Your heater outlet temperature will continue to drop. If there are sizable tramp air leaks in the convective section, the heater may blow up and kill you!

Quickly! Switch from auto to manual on the fuel gas regulator. Manipulate the fuel gas valve to a safer, partly closed position. Running above the tipping point on automatic control on a heater is dangerous. Only manual intervention will work. Positive feedback loops can never correct themselves without outside help.

Stalling a Thermosiphon Reboiler

Kenny Kitchner didn't like me. Kenny was the Chief Operator on B shift on my alky unit in Texas City. Most people when they first met me think they won't like me. But, after people really get to know me, they become quite convinced they don't like me. Kenny was one of those people.

For example, one day Kenny said, "Mr. Lieberman, let's reduce the debutanizer top reflux rate a 1,000 BSD. This will reduce the reboiler duty and we'll save 2,000 pounds per hour of reboiler steam. We're overrefluxing the tower. Mr. Lieberman, this will improve your energy index. Let's try. It can't hurt to try."

"Good idea. Go ahead Kenny," I said somewhat surprised by his friendliness and positive attitude, "It can't hurt to try."

When the panel operator cut back the reflux rate, here's what happened:

- The reboiler outlet temperature went up by 1°F, as less cold reflux was running down the tower.
- The reboiler steam temperature control valve (see figure 33–3) cut back from 50% to 49%.
- The flow of vapor into the tower dropped off a bit.
- The pressure drop of the vapor flowing through the bottom tray dropped by a bit.
- The bottom valve tray (or any type of perforated tray deck), leaked a few more drops of liquid.
- The flow of liquid into the reboiler was reduced.
- The reboiler outlet went up 1°F.
- The steam flow control valve to the reboiler closed from 49% to 48%, to restore the set-point temperature.
- The reboiler duty went down. The flow of vapor through the bottom tray went down. The pressure drop of the vapor flow through the bottom tray went down, and the bottom tray began to leak more.

This reduced liquid flow to the reboiler and caused the steam flow to the reboiler to automatically cut back.

A positive feedback loop had been created. The reboiler outlet temperature was fine. But the tower bottoms temperature was dropping like a rock. My gasoline product to the storage tank was becoming progressively contaminated with excessive butane. A few hours of this and the tank roof vents would blow. Maybe I would even sink the roof of the 250,000 barrel floating roof storage tank.

In a panic, I switched the reboiler steam flow from automatic to manual, and opened it 100%. Nothing happened. The reboiler had stalled out. This means the reboiler duty was no longer limited by the flow of steam to the tube side of the reboiler. The duty was limited by the flow of hydrocarbon liquid from the draw-off sump beneath the bottom tray, into the shell side of the reboiler.

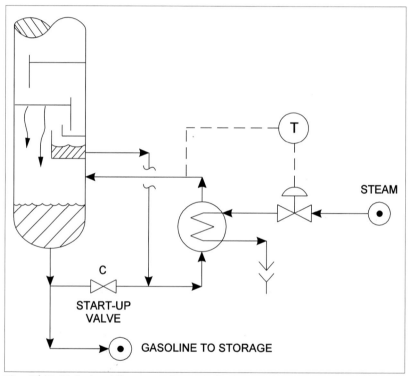

Fig. 33–3. Stalled-out thermosyphon reboiler.

Unfortunately most of this liquid was bypassing the sump and dropping into the bottom of the tower. Now what?

I looked over at Kenny. He was smiling and sipping hot coffee from his old cracked mug.

"Yes, Mr. Lieberman. Any instructions?" he said happily.

Finally. An hour later, Mr. Kenney Kitchner opened the local start-up gate valve (shown in figure 33–3) beneath the debutanizer. This gravitated liquid directly into the shell side of the reboiler and restored the vapor flow through the bottom tray. This broke the positive feedback loop. Thus, I learned, how a stalled-out reboiler, which has lost thermosiphon circulation, is re-started.

Of course, Kenny was well aware that I did not know that the start-up valve existed. So, he maliciously created the positive feedback loop by reducing the top reflux rate too much and stalling out the reboiler. Incidentally, Mr. Kitchner later became an important public official in Galveston County.

Our Little Planet

In 1965, when I went to work for Amoco Oil, it was assumed that incremental CO_2 emissions would be absorbed in the ocean.

Currently about 40% of CO_2 released from hydrocarbon combustion is absorbed by the ocean. The rest is accumulating in the atmosphere at a rate of about 2 ppm per year. However, the ocean surface waters are warming and their pH is dropping. Both factors reduce the solubility of CO_2 in water. Sometimes in the next few decades the ocean will become a net emitter of CO_2 rather than an absorber. By that time CO_2 accumulation will rise to about 5 ppm per year. My examination of the data indicates this will occur about 2030.[1]

The tipping point will occur when methane trapped in the frozen tundra and the offshore hydrate deposits is released. Methane is 23 times per mole more powerful a greenhouse gas than is carbon dioxide.

Or the tipping point will occur when Syncrude and Suncor and Petro Canada really escalate their production of tar sands in Ft. McMurray, in Northern Alberta (as planned by 2013).

Or the tipping point will occur when Exxon reactivates their billion dollar Rio Blanco oil shale project outside of Fruita, Colorado (shelved in the 1980s).

Or the tipping point will occur when the United States, China, and India construct all the coal-fired power plants now being planned and designed (about one a week).

Or the tipping point will occur when my insane clients continue to build plants to convert natural gas to gasoline and distillates at a net energy loss of 40%.

Keep in mind that the actual reduction in the measured atmospheric accumulation rate of CO_2 since 1965 has been zero. Our one success story is Freon-type molecules. Their accumulation rate has been arrested. But such substances only account for 1% of global warming. Still, it shows that progress is possible if people work together to a common goal.

What's to be done. First the emphasis ought to be not on consumption, but production. All barrels of hydrocarbons extracted will be consumed. But which production should be restricted? A good candidate is the Alberta tar sands in-situ production.

Steam is used to melt out the tar sands in their underground formation. Speaking to the onsite operators, it takes 4 pounds of steam (not the advertised 2 pounds) to recover a single pound of black tar sands oil. On a heating valve basis, that is a ratio of 25% of fuel consumed to make steam for 100% of recovered tar sands heavy oil.

Then the tar sands oil is charged to a fluid coker. There about 20% of the heating value of the tar sands oil is converted to fuel.

The only finished product produced in the refinery (called an upgrader) is diesel oil, which is consumed internally in the Ft. McMurray operation. Mostly, synthetic crude is exported, all of which must be re-refined in United States and Canadian refineries.

I would guess then that for each BTU worth of energy appearing in distillate products, up to another BTU is consumed. It's this sort of production that ought to be eliminated first. Venezuela and South Africa have somewhat different processes to produce hydrocarbon products, that also consume huge amounts of fossil fuels.

Troubleshooting Checklist
for Positive Feedback Loops

Negative versus positive feedback

Excess reflux rates

Incipient flood versus tower top temperature control

Centrifugal compressor surge

Surge and gas density reduction

De-aerator sudden water eruption

Loss in de-aerator pressure

Boiling water density reduction

The *tipping point* concept

Stalling a fired heater

Lack of combustion air

Stalling thermosiphon reboiler

Use of the start-up line

Global positive feedback loop

Reference

1. N.P. Lieberman, *Process Engineering for a Small Planet.* Lieberman Publications, 2009.

Troubleshooting Control Loops

In this chapter, I have drawn five examples from my refinery experience illustrating troubleshooting process control loops for the following variables:

- Condensate liquid level
- Tower pressure
- Steam flow to a stripper
- Temperature in a vacuum tower

None of these examples involves instrument malfunctions. The examples are intended to illustrate how control loops cause process equipment malfunctions when the process control engineer and panel board operator do not analyze the interaction of the controlled variable with the process. As I discuss in chapter 35, this lack of analysis can lead to horrific unit failures, as occurred at the British Petroleum Texas City Refinery, in 2005.

Troubleshooting Condenser Level

I advanced my career more in my first week at the Good Hope Refinery than in the previous sixteen years at Amoco Oil. Here's my story.

The Good Hope Refinery had a very large 120,000 BSD fluid catalytic cracking unit (FCCU). The catalyst was regenerated by burning off the coke. The combustion chamber was served by four large air blowers. The largest blower was driven by a condensing steam turbine. The steam turbine exhausted into the surface condenser shown in figure 34–1. The steam turbine exhaust pressure was 18 inches of mercury (or 300 mm of mercury on an absolute basis). Historically, at the current cooling water temperature of 90°F, a vacuum of 24 inches of mercury (or 150 mm of mercury on an absolute basis) was obtained. I then made the following field observations:

- The air blower was surging.

- The surge was caused by the air flow produced by the blower dropping essentially to zero.
- The very low air flow was a consequence of the blower discharge pressure being less than the regenerator pressure.
- The low blower discharge pressure was caused by the blower running at 4680 rpm as compared to its normal speed of 5000 rpm (revolutions per minute).

To stop the surging, the plant operators had started to vent the air blower discharge to the atmosphere. The lack of combustion air flow had unfortunately reduced the FCCU feed rate from 120,000 BSD to 90,000 BSD. The owner of the Good Hope Refinery, Jack Stanley, suggested that I fix the problem before I went home that evening.

"Take your time Norm. But just don't leave the plant until K-501 is lined up to the FCCU regenerator," Mr. Stanley directed.

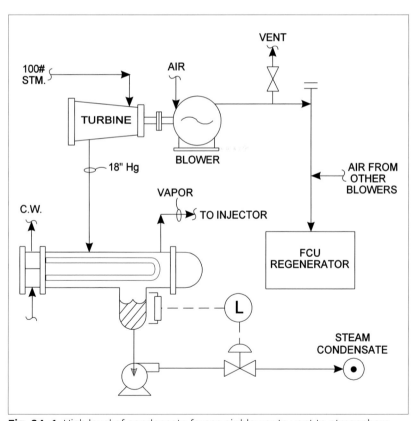

Fig. 34–1. High level of condensate forces air blower to vent to atmosphere.

Mr. Stanley never cared how we dressed, or what we said, or how well our reports were organized. The only thing Mr. Stanley cared about were results.

I now made the following additional field observations (referring to figure 34–1):

- The condensate level in the surface condenser boot was high, but still visible in the water boot.
- The local boot LIC (level indicator control) was set at 60% (see figure 34–2).
- The black pointer was indicating 60%. (The black pointer is inside the little box which contains the control used to change the level set point). That is, the boot level control valve was holding the desired 60% set point.
- The water being pumped out of the boot felt much colder than the vapor flowing out of the surface condenser. (There were no temperature indicators on either the vapor or the liquid outlets from the surface condenser).

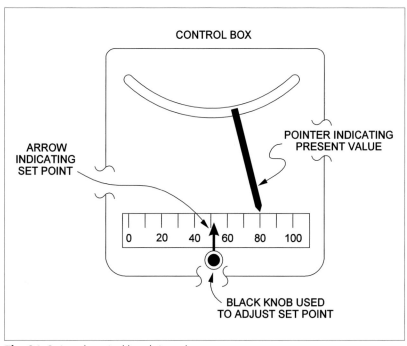

Fig. 34–2. Local control box internals.

In 1980, I did not know all that much about troubleshooting surface condenser controls or vacuum systems (see chapter 13). However, I considered that the condensate from the boot was sub-cooled to be a possible indication of condensate backup in the surface condenser. I thought that the indicated level in the boot gauge glass was reading too low (i.e., lower than the real level). One of the reasons I loved working in the Good Hope Refinery is that the operators understood that engineers had to be allowed to do their job without undue interference. So I proceeded as follows:

- Referring to figure 34–3, I opened valve A and closed valve B. I then opened valve C slightly to allow a small amount of air to be drawn through the top tap of the gauge glass.
- I closed valve A: and opened valve B. I then opened valve C slightly to allow a small amount of air to be drawn into the bottom tap of the gauge glass.

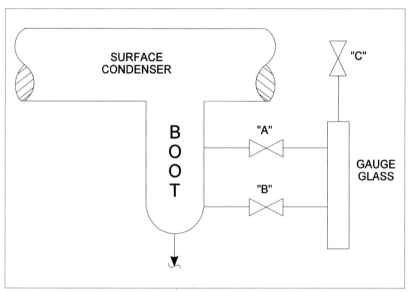

Fig. 34–3. Surface condenser boot level indication.

I had now proven that the gauge glass taps were not plugged. Also that the visible level in the glass was correct. This indicated, from the process engineer's perspective, that the water level in the boot was below the bottom row of the condenser tubes and that a further reduction in the condensate water level in the boot should not affect the condenser's capacity. But, on the other hand, why was the water condensate draining from the boot sub-cooled?

I opened the door to the control box (shown in figure 34–2). The little red arrow indicated the level set point was at 60%. The black pointer indicated that the condensate level control valve on the discharge of the condensate boot pump was holding the 60% level set point. The plant operators were busy making gumbo and frying catfish in the control center.

I turned the little black knob clockwise half a turn to move the little red arrow from 60% to 40%. The control valve on the condensate pump discharge opened a bit. The water level in the gauge glass dropped by about 10 inches.

I could feel the condensate pump discharge line warming beneath my fingers. Concurrently, the surface condenser vapor outlet to the first stage steam ejector (i.e., jet) cooled. Now, both the vapor outlet and the liquid outlet temperatures from the surface condenser felt quite similar to my touch.

The vacuum gauge started to creep up from 18 inches of mercury to 24 inches of mercury. Or, on an absolute basis, the surface condenser pressure dropped from 300 mm of mercury to 150 mm of mercury. I could hear the steam turbine and blower winding up. The blower speed increased from 4680 rpm to 5000 rpm. (*Note:* turbine horsepower increases with speed cubed. The extra 320 rpm was therefore worth an extra 22% of turbine brake horsepower).

The blower discharge pressure began to rise. Two hours later (I had to wait on the gumbo and catfish), the K-501 air blower discharge was lined up to the FCCU regenerator. I still recall now how the full midnight moon flitted between the swamp cypress trees as I drove back home.

Next morning, the owner asked for a quick summary of the previous night's events.

"Mr. Stanley," I responded, "It's just an example of *process control technology in action.*"

I still do not quite understand why the physical level in the condenser was higher than the observed level in the gauge glass. Perhaps at very low pressures, the water foams? (See chapter 18, figure 18–4 and figure 18–5). Or perhaps there was a layer of lighter oil floating on top of the water condensate in the boot? (See chapter 4, figure 4–4). Maybe there was an air leak on the top of the glass which pushed the water level in the glass down? It really did not matter because the objective of good process control engineering is to achieve the desired results, which I also learned from Jack Stanley.

Override Control—Colorimeter to Tower Pressure

This next example occurred at the Coastal Refinery in Eagle Point, New Jersey and also at the Exxon Refinery in Baton Rouge, Louisiana. I have combined the technical aspects of both incidents into one example for instructional purposes.[1]

Because of extraneous chemicals, certain crudes have a tendency to foam when heated above the boiling point of water. If there is a crude preflash tower, as shown in figure 34–4, I have observed that the following series of events will transpire:

- The foam level rises in the bottom of the tower.
- The low density of foam causes the level-trol output to drop (see chapter 18).
- The reduced level-trol output causes the tower bottoms level control valve to close, even though the foam or froth level in the tower is actually too high.
- The foam level rises and covers the flashed crude inlet nozzle.
- Black crude oil is entrained up the tower. That is, the tower floods.
- The flooding causes the tower top temperature to increase.
- As the reflux rate is on the tower top temperature control (TRC), the reflux rate increases.
- The higher reflux rate increases the tendency to flood and promotes even worse entrainment due to crude droplets containing asphaltines and resins, which turns the naphtha brownish-black.
- The catalyst in downstream hydrotreating units is quickly coked-up by the asphaltines and resins.

There are two methods to control the entrainment problem. You can lower the temperature or increase the pressure. Both methods will suppress the foaming of flashed crude oil.

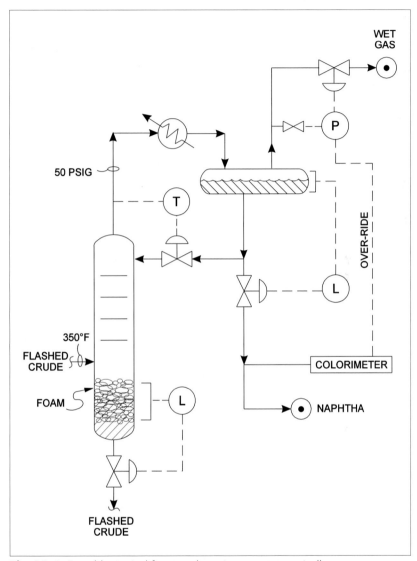

Fig. 34–4. Override control from analyzer to pressure controller.

From the *process control* perspective, there is no way to rapidly reduce the flashed crude temperature by the requisite 20°F to 30°F. The correct way to suppress the foaming is to raise the tower pressure by 5 or 10 psig. The operators were already aware of this technique prior to my involvement in this project. Every few hours (at least in theory) they drew a sample of the overhead naphtha product for visual inspection. If the naphtha was not "water white," they would raise the tower pressure (see figure 34–4) by a few psig. But, reliance on this sort of manual intervention is certain to lead to black naphtha escaping from the crude unit.

My contribution was to have an online, continuous colorimeter installed on the overhead naphtha produced. Normally the tower pressure was maintained by the PRC set point at 50 psig. However, if the colorimeter output rose above a certain point, then the tower pressure set point would automatically be increased by some 5 or 10 psig to suppress entrainment. This is a rather simple example of override pressure control by an on-stream analyzer.[1]

Other options I've tried with overall negative results are:

- Lowering the tower bottoms level. This is good, provided we do not blow foam out of the bottom of the tower. The resulting light naphtha vapors cause some nasty downstream problems.

- Adding silicon defoaming chemicals. A really bad practice. The silicon is distilled overhead with the naphtha product. It accumulates over a period of months on the hydrotreating catalyst. The catalyst pores plug up with the silicon. Catalyst activity is severely degraded and naphtha desulferization becomes progressively worse.

These sorts of projects and operational changes are typical of my work. My consulting business is built upon successful results. The reason I'm successful is that I try out my new ideas when no one is watching, or I automate an existing manual operating practice, or copy a process control concept I've seen working in another plant. I learned this from Jack Stanley. The only thing that matters is results. The next story amplifies this principle.

Flow Control Gone Wild

Ray Hanson was the operating engineer at the Chevron Refinery in El Paso, Texas. Ray's problem was black HVGO (Heavy Vacuum Gas Oil), as shown in figure 34–5. One possibility was that the vacuum tower bottoms stripping steam rate was excessive. This could certainly cause black asphaltines to be entrained into the HVGO product, due to the high vapor velocity.

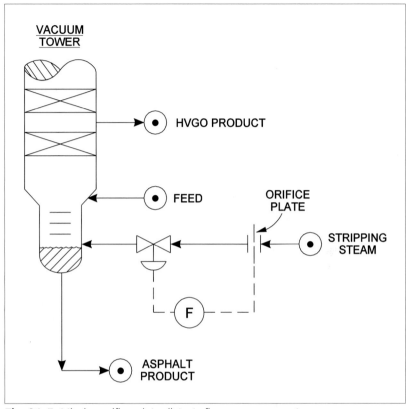

Fig. 34–5. Missing orifice plate distorts flow measurement.

Ray assured me that the stripping steam rate was a reasonable 2,000 pounds per hour. Ten pounds of tower bottoms stripping steam per barrel of asphalt is indeed a proper steam rate for the production of paving asphalt. Ray assured me that the steam flow meter was correct, as the instrument department had verified its calibration recently. But, Ray is not the sort of person that I trust. There is only one person that I do trust. And that person is myself.

So I decided to verify the measured flow of steam myself. The first step to verify a flow is to check the size of the orifice plate. The instrumentation department had recorded the orifice plate size at one inch, and the orifice flanges at three inches. The size of the orifice is stamped on the handle of the orifice plate, facing the flow.

However, my attempt to read the orifice plate was not successful, because there was no orifice plate between the 3-inch orifice flanges!

Instead of the opening between the orifice flanges being 1 inch in diameter, the opening was 3 inches in diameter. That is, the area between the orifice flanges was nine times bigger than it was supposed to be. (Area varies with diameter squared). Thus, I reasoned that the flow of asphalt stripping steam was not 2,000 pounds per hour but roughly 18,000 pounds per hour!

I was able to obtain an approximate check of the 18,000 pounds per hour of stripping steam by measuring the flow of overhead condensate from the vacuum tower. Ray reduced the flow of stripping steam to the minimum needed to control his asphalt product flash point specification. The heavy vacuum gas oil product turned a pleasing, transparent golden color.

Misleading Differential Temperature

My favorite method of controlling wash oil flow rate in a crude, coker or vacuum tower is shown in figure 34–6. The wash oil is largely evaporated by the up-flowing flash zone vapors. The heat needed to evaporate the cooler wash oil comes from the enthalpy of the rising flash zone vapors. The loss of enthalpy from the flash zone vapors cools and partially condenses the rising vapors.

The idea is to prevent the packing in the wash oil bed from coking. The bed must be kept wet to retard coke formation. Typically in a vacuum tower, I would use enough wash oil to reduce the vapor temperature between T-1 and T-2 shown in figure 34–6 by 35°F to 60°F.

Usually we think of a temperature indicator as a quite reliable instrument. And this is so. But in a vacuum tower we may be measuring a localized temperature representing a few square inches, while the cross sectional area of the vessel may be a thousand square feet. As shown in figure 34–6, there was a leak on the gas oil product chimney tray. Cool product reduced the indicated temperature at T-2. The process control engineer had designed the wash oil flow control valve to be reset by the differential pressure between T-1 versus T-2.

Unfortunately the leaking gas oil pan increased the measured differential temperature between T-1 and T-2. This caused the wash oil flow control valve (FRC) to close off. The reduced wash oil flow rate caused the packing in the wash oil bed to coke. The resulting excessive wash oil bed pressure drop precipitated a unit shutdown, at which time the leak in the gas oil product chimney tray was identified.

Fundamentally, the ΔT control resetting the wash oil flow rate was a good process control design. But the control engineer must also anticipate the possibility of malfunctions such as a non-representative localized temperature. The error here was the total reliance on a closed-loop control scheme that was accepted, without examination, by both the panel board operators and the unit engineers.

Note that in this costly failure none of the instrumentation had malfunctioned and that the control logic was entirely correct. However, in the final analysis we are still dependent on the unit engineer and console operator understanding how the equipment works. In this case, the wash oil rate was only sufficient to cool the rising flash zone vapors by a few degrees, as indicated by a relatively simple heat balance calculation.

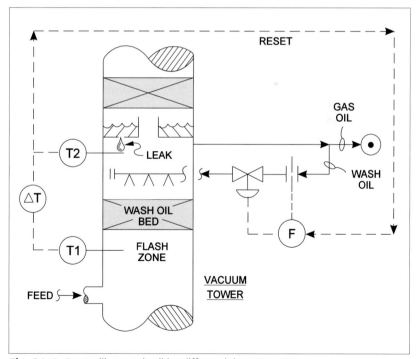

Fig. 34–6. Controlling wash oil by differential temperature.

I suppose that the preceding calculation could have been programmed into the computer control as a backup to the differential temperature measurement, that was used to reset the wash oil flow rate.

Coke Drum Overpressure Incident

Two of my clients had essentially the same malfunction in 1989. The problem was a pressure-sensing failure on a coke drum, as shown in figure 34–7. A common problem in delayed cokers is carryover of coke, which on occasion partially plugs the overhead vapor line. This will cause an increase in the coke drum pressure, unless the coke drum feed rate is reduced.

Unfortunately, the panel pressure recorder did not show a high coke drum pressure. The pressure transmitter was located not on the coke drum but on the vapor line, 50 feet from the coke drum vapor outlet nozzle. Also, the high pressure alarm (HPA) shown in figure 34–7 did not sound. It received its signal from the same pressure transmitter, which was downstream of the restriction in the vapor line.

A local field observation from a pressure gauge located on the coke drum itself indicated that the vessel pressure exceeded 100 psig. The drum normally operated at 30 psig top pressure. The relief valve was set to open at 50 psig, but it too failed to open. The pressure relief valve also was located 50 feet from the vapor outlet nozzle on the vapor line itself, and not on the vessel. The only factor that prevented a catastrophic vessel failure was the inherent ductility of the carbon steel coke drum. Had this been a high chrome vessel, which is far less ductile, this story might have had a very terrible ending.

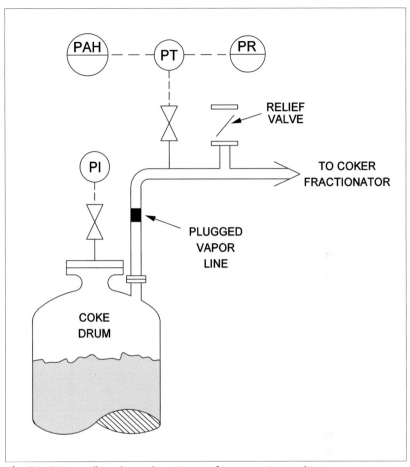

Fig. 34–7. Vapor line plugged upstream of pressure transmitter.

Summary

In the five stories related in this chapter there were no actual instrument failures or any control loop malfunctions. Yet the control loops either failed to prevent a process upset or in some instances created a positive feedback loop which made the malfunction worse. In hindsight, adequate safeguards could have been provided in the unit control logic to override the erroneous control loops. But, in a more realistic sense, we depend on the panel operator's training, experience, and intelligence to intervene as required. Automatic control can never be a replacement for competent operators who understand how the process variables interact with the control logic.

Troubleshooting Checklist for Troubleshooting Control Loops

Condensate liquid level control
Steam condensate flow
Surface condenser boot level
Sub-cooling steam condensate
Plugged gauge glass taps
Override control
Use of colorimeter to control flow
Foam effect on level control
Pressure control to suppress foam
Silicon use to suppress foam

Missing orifice plate effects flow
Orifice plate installed backwards
Mislocated TI point
Liquid leaking onto thermowell
Plugged pressure-sensing tap
Sensing versus alarm
  connection redundancy
Overpressure of vessels
Control logic and positive feedback

Reference

1. N.P. Lieberman, *Troubleshooting Process Plant Control.* Wiley, 2008.

Texas City—B.P. Refinery Fire

Several years ago a terrible fire killed 15 people and injured hundreds at the British Petroleum (formerly Amoco) Refinery in Texas City. It was the most famous accident in the history of the refining industry. I still see documentaries on the TV about the horrible incident. The chief executives of B.P. resigned as a result. Litigation costs exceed $2,000,000,000 and still counting.

What happened? I don't know exactly. I've guessed at the parts I'm not sure of. But, I had worked at the refinery for several years and have supplied the process design for nine distillation towers and many of its other facilities. So, this chapter is an educated guess. I've also taken the liberty to include some general engineering safety principles in the story drawn from other accidents.

Safety in process operations is largely in the hands of the operators. Operators always want to run the plant safely. But, if they don't understand how their instrumentation works, they can't run the plant safely. This story illustrates this idea.

I was patiently waiting for the elevator with Gary. As usual we were both late for work at our jobs as process design engineers at the Amoco headquarters in Chicago.

"Norm, can you help me? I'm late with the new naphtha splitter project," complained Gary.

"Gary, I've got my own problems," I answered.

"Just fill out the heat exchanger and centrifugal pump data sheets for me. The naphtha splitter tower design is finished, but the data sheets are due next week," whined Gary.

So, in a minor way, I participated in the tower design that thirty years latter caused such havoc. I've shown a simplified process sketch of Gary's design in figure 35–1. The light naphtha overhead product was charged to a pentane-hexane isomerization unit for octane enhancement. The heavy naphtha bottoms product was charged to a reformer to produce aromatics (benzene and xylene) and high octane toluene for gasoline blending.

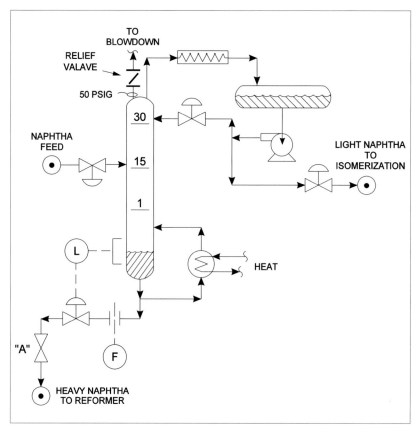

Fig. 35–1. Texas City naphtha splitter.

Note the relief valve located at the top of the tower was set to relieve at 50 psig. Also, that the relief valve vented to "blowdown" and not to a flare header. This was not part of Gary's process design. At Amoco, in 1977, the process design engineer did not usually specify where the relief valves vented. This aspect of the project was handled by a mechanical or civil engineer, not a chemical or process engineer. In retrospect, this was a mistake.

As an operating superintendent in Texas City from 1974 through 1976, I was surprised and dismayed to discover that my chief operators and panel operators did not understand how their control loops worked. Basically, Amoco never provided any realistic operator training. The two instrumentation issues that precipitated this accident 30 years later were among the many control concepts that my alkylation unit operators also did not comprehend. The instrumentation issues were *flow* and *level*.

- **Level Indication:** The bottoms level indicator shown in figure 35–1 had been calibrated for heavy naphtha. Let's assume the density of the heavy naphtha at normal operating temperatures was 0.80 sp. gr. Let's assume the density of the naphtha feed was 0.70 S.G. In a refinery, level indication is inferred from a pressure difference. This pressure difference is the pressure measured at the lower level tap, minus the pressure measured at the upper level tap. To convert from ΔP to level then:

$$\frac{(\Delta P)\,(2.31)}{(S.G.)} = \text{Feet of liquid}$$

If a level instrument is calibrated with an assumed 0.80 S.G., and the ΔP measured is 2 psi, then:

$$\frac{(2\,\text{PSI})\,(2.31)}{0.8} = 5.7\,\text{feet}$$

If we calculated the actual level in the tower, assuming a 0.70 S.G., then the tower's bottom level would be:

$$\frac{(2\,\text{PSI})\,(2.31)}{0.7} = 6.6\,\text{feet}$$

Let's make a chart of the level generated by an instrument calibrated with a 0.8 S.G fluid while the real liquid in the tower is 0.7 S.G.:

Indicated Level	Real Tower Level
5.7 feet	6.6 feet
8.7 feet	9.9 feet
11.5 feet	13.2 feet
12.0 feet	14.0 feet
12.0 feet	17.0 feet
12.0 feet	28.0 feet
12.0 feet	200.0 feet
12.0 feet	10,000.0 feet
12.0 feet	100,000.0 feet

The problem is, when the real level in the tower rises above 14 feet, the liquid has reached the top level tap. Henceforth, the head pressure exerted at the top pressure tap rises at the same rate as the pressure on the lower tap. The difference between the two pressure readings remains constant, regardless of the height of the liquid level. I call this being *tapped-out*.

- **Flow Indication:** We infer flows by measuring the pressure drop through an orifice plate:

$$Flow = K \cdot \sqrt{\Delta P}$$

Let's suppose that a meter is off-zero, meaning it's reading 4 gpm when there is really no flow. What then is the actual flow, if the flow measured by the flow meter is 8 gpm? The measured pressure drop is proportional to the square root of the ΔP:

$$Actual\ Flow\ Squared = (8^2 - 4^2)$$

$$Actual\ Flow = Square\ Root\ (64 - 16) = 6.93$$

Using this calculation procedure, I've prepared this chart:

Panel Reading Measured Flow (gpm)	Meter Error Meter Off-Zero (gpm)	Real or Actual Flow (gpm)
11	4	10.25
10	4	9.16
9	4	8.66
8	4	6.93
7	4	5.75
6	4	4.47
5	4	3.00
4	4	0.00

Note that at 11 gpm the measured flow that the meter is reading is only 0.75 gpm high. At 5 gpm the meter is reading 2 gpm high. At no flow the meter is reading 4 gpm high.

This suggests that a flow meter working at 11 gpm will have a small error of 7%. The same meter working at 5 gpm will have a large error of 40%. Thus, on start-up, when flows are low, a meter with an acceptable calibration at normal operating rates, will be very misleading at the low startup flow rates. I call this *flow meter off-zero calibration concept.*

Apparently the operators at the Texas City naphtha splitter did not understand the concepts of being:

- Level tapped-out
- Flow meter off-zero calibration

What then was the result of this lack of instrumentation understanding by the splitter operators?

The Start-up

The splitter had been shut down for some minor problem and was being restarted. The way I normally re-streamed distillation towers when I worked at Amoco was to:

- Initiate feed flow slowly.
- Keep the tower on total reflux to establish a good level in the reflux drum and the tower bottoms.
- Increase reboiler heat slowly.
- Check low point drains for water (tower bottoms and reflux drum).

Now I'll start to guess. But, I've been told by people quite knowledgeable about the incident that my guesses capture the essence of the accident.

First, the operators noticed that the tower bottom level was not particularly high. However, they failed to note that the bottom product was less dense than normal. Since, if the tower was on total reflux, the bottoms flow was mainly feed, which is lighter than heavy naphtha. Thus, the actual bottoms level was *higher* than the level indicated on the panel screen. This is not an instrument malfunction. It's just that the level indicator was calibrated for heavy naphtha and not the lighter feed composition in the bottom during start-up operations.

Once the real or actual level in the tower rose to cover the top level tap, the indicated level on the panel screen ceased increasing. The level indication was *tapped-out.*

This the panel operator failed to grasp, because he (or she) did not understand how the level instrumentation worked. Fact! The temperature profile later indicated (according to the B.P. investigation) that the tower was essentially full of liquid up to the top tray.

The panel operator now made a fatal observation. He observed that the bottoms flow rate was a substantial portion of the feed flow rate. Likely, the real bottoms flow rate was nothing. But, the flow meter was off-zero. As I just tabulated, a meter slightly off-zero calibration introduces a slight error at normal flows. But as the real flow diminishes, the off-zero error becomes progressively more significant.

Not one out of twenty engineers who attend my seminars knows how to correct an observed flow for the flow meter being off-zero. So we should not be too critical that the Texas City operators thought the flow meter (which they routinely used) was lying to them. Probably it was reasonably accurate during normal operations.

The fatal observation of the nonexistent bottoms flow rate, and a less than optimum high level indication in the splitter suggested to the operators that the splitter bottoms LRC valve was leaking through. So they closed isolation valve A (figure 35–1). I'm guessing. But what is not a guess is that valve A (per the B.P. investigation report) was found after the fire to be blocked in the field.

At this point the shift changed over. In 1974, my operators in Texas City would "relieve at the gate." That is, four operators would walk off the unit and leave one man behind. As the four punched out, their five relief operators punched in. I tried to stop this inane practice but discovered it was a longstanding refinery-wide custom. Maybe three decades later it still was?

I cannot say if the relief shift knew that both the bottoms product and overhead product lines were closed. With the tower on total reflux and the isolation valve A closed, logic would have dictated that no more feed be brought into the unit. But, I'll guess that the new shift noted that the tower bottoms level indication was showing substantially less than 100%. Actually by this time "their cup runneth over." So they brought more feed into the tower, which had no place to go.

Ladies and gentlemen, if you want to over-pressure a tower really fast, try filling the tower with liquid above the top tray and then cranking up the heat duty. Apparently that's what the relief shift did. I'm still guessing. But what is not a guess is the 50 psig relief valves located on the top of the column popped open at their set point pressure of 50 psig (see figure 35–1).

Note that there is really no equipment malfunction in this whole story. The operators were neither negligent or careless. They simply lacked a basic understanding as to how their flow and liquid measurement instrumentation functioned. I can reach back in time and hear Mundo Lira, my wonderful instrument tech in 1974 on my alky unit in Texas City.

"Mr. Lieberman, these guys are dangerous. They don't understand what they are doing. What kind of a supervisor are you? Even Paul Stelly wouldn't let these people run the board. I'm tired of running back to the control room to straighten out their screw-ups. Why don't you train these guys before you put them on the panel? Can't Amoco find some intelligent engineers to supervise these units?"

Engineering Aspect of Failure

Let's now look at the engineering aspects of this catastrophe by referring to figure 35–2. First, note that the 50 psig relief valve vented through the blowdown stack to the atmosphere.

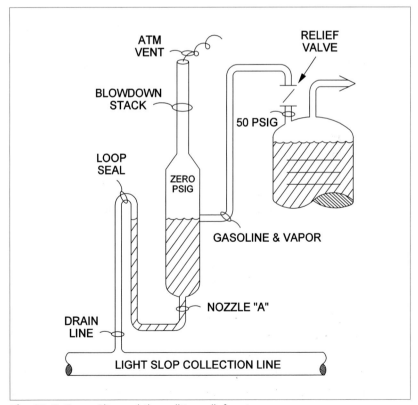

Fig. 35–2. Texas City naphtha splitter relief system.

I was 31 years old when I assumed the position of operating superintendent of the sulfuric acid regeneration, butane treating, and alkylation complex at the Amoco (later B.P.) Refinery in Texas City. My predecessor Paul Stelly trained me for my new job in just 10 minutes.

"Norm. Don't let this job ruin your life. Family first. Take care of your teeth. Take your kids out to Galveston. Listen to Mundo Lira. He's a full blooded Indian—maybe Navajo?"

All five of my distillation towers relieved to the atmosphere. I found this out the second day at work. Before each thunderstorm the operators would put steam into the relief valve vents. The relief valve vents, located at the top of my tall towers, might otherwise catch fire from lightning. After all, I was informed by Old Zip, my ancient lead shift foreman, "Norm, all our relief valves leak. Once they pop open, they never reseat properly. We pop relief valves here all the time."

Across the refinery fence adjacent to my alky unit in 1974 (and still today) is a playground. I never stopped to think what would happen if a relief valve opened when a tower was in fully developed flood. Then liquid hydrocarbons would dump down into the dilapidated playground. And suppose someone's mom was lighting a cigarette. Then what?

Death and disaster! That's what.

But in 1974, I didn't worry about my relief valves not being connected to the plant's flare header system. I didn't worry about the relief valves that vented to the atmosphere 100 feet above my head because:

- No one else was worried.
- It was that way when I got there on April 1, 1974.
- It must be okay because it was built per Amoco standards.
- I couldn't worry about everything, and I had lots of family problems.
- Anyway, only vapor should be at the top of the tower, not liquid hydrocarbons.

The naphtha splitter relief valve shown in figure 35–2 did not vent directly to the atmosphere as on my alkylation unit towers. The relief valves vented to the *blowdown stack*, which was then vented to the atmosphere.

I guess the theory was that if a relief valve opened, only the hot vapor would enter the blowdown stack. Condensed light hydrocarbon liquids could then drain to the light slop collection system as shown (we called it the *light slop sewer* in 1974).

I suppose the loop seal was there to prevent hydrocarbon vapors from backing up into the blowdown stack, rather like the loop on your bathroom hand basin drain line. I know a loop existed, but I'm guessing at its top elevation relative to the inlet nozzle shown in my sketch (figure 35–2).

As the splitter over-pressured and the 50 psig relief valves opened, boiling gasoline or naphtha feed flowed into the zero pressure blowdown stack. I mean, if you de-pressure hot naphtha from 50 psig to zero psig, it ought to flash. The saturated liquid gasoline then drained out of the blowdown stack through nozzle A. The saturated gasoline then lost pressure because of:

- Nozzle exit loss due to the acceleration of the flow as it entered the smaller drain line.
- Frictional loss in the drain line upstream of the light slop sewer.
- Finally, the bigger factor. That is the increase in elevation and consequent reduction in pressure as the liquid flowed uphill in the loop seal.

Vapor Lock

If at the top of the loop seal the saturated liquid pressure fell below atmospheric pressure, then the gasoline would try to vaporize. But, the vapor so generated would choke off the flow in the relatively restricted drain line. I call this *line cavitation limit*. The rest of the process plant world calls this *vapor lock*.

The evolved vapor chokes off the flow of liquid draining from the fat blowdown stack vessel. The liquid level in the blowdown stack rises to cover the inlet nozzle. Just like in any other tower or vessel, once the liquid level rises above the vapor-liquid mixed phase inlet nozzle, the vessel will start to flood. Liquid gasoline will be pushed out of the top vent by the flashing vapor entering the blowdown stack. The hot naphtha or liquid gasoline will then drop down to the ground on top of a contractor's office trailer, where 15 people were meeting.

Not such a good idea to locate a contractor trailer park around a hydrocarbon vent blowdown stack. But, that's another story.

Why didn't Gary and I worry about connecting the new naphtha splitter we were designing to an atmospheric vent. Engineering logic and common sense suggests that pressure relief valves should have been vented to the flare header. Let me tabulate our reason:

- No one else in the Process Engineering Division worried about relief valve dispositions. So why should we?
- Only vapors were supposed to be vented. The possibility of a tower over-pressuring because it flooded did not occur to us.
- It was consistent with Amoco standards.
- Gary and I couldn't worry about everything. Gary had dental problems and I had family problems.

Summary

This tragic accident was not due to operator fatigue or even an error in judgment. The real causes of this fire were due to:

- The loop seal and vapor lock.
- Not connecting a hydrocarbon safety relief valve to the flare.
- Operators who did not understand how their instrumentation worked (levels and flows).

I forgot to mention something. Hydrocarbon liquids would often drop out of the East Plant flare stack in Texas City in the 1970s. Mr. Durland, the Refinery Manager, was so concerned with the resulting frequent grass fires that the grass around the stack was kept very short. The grass was mowed by a team of mules to guard against any source of ignition from an engine during the grass cutting. The source of the hydrocarbon liquids were the pressure relief systems that suddenly vented into the flare header.

I used to worry about the old farmer and the worn-out mules. Suppose a slug of burning alkylate from my unit dropped down on them. Then what? Now I know.

What Should Have Been Done?

Often, on start-up, I'm never quite sure of the tower's bottom level. Is it full or empty? It's often really hard to see an interface in a dirty gauge glass especially if the fluid is water white. So, when I'm not too confident as to the approximate liquid level, I'll proceed as follows: (See figure 35–3).

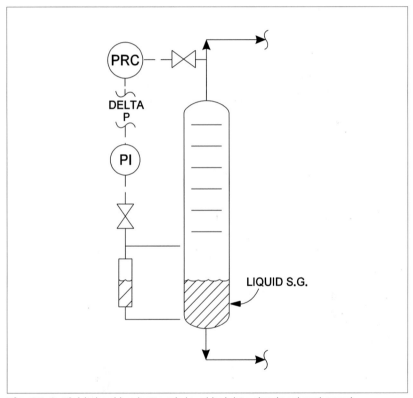

Fig. 35–3. Field checking bottom's level height using local and panel pressure indications.

1. Place a pressure gauge on the top of the gauge glass at the bottom of the tower.
2. Read the pressure at the top of the tower from the panel PRC. I'm assuming the pressure tap is properly located at the top of the column and not improperly located on the reflux drum or downstream of the overhead condenser.

3. Subtract the top pressure (PRC) from the bottom pressure (on the top of the gauge glass).
4. Multiply the resulting ΔP by 2.3 (for water). Divide by the liquid specific gravity to obtain the approximate feet of liquid above the top of the gauge glass.

Some typical specific gravities to use are:

- Water: 1.00 S.G.
- Hot Water: 0.98 S.G.
- Cold Gasoline: 0.70 S.G.
- Hot Gasoline: 0.60 S.G.
- Propane: 0.50 S.G.
- Butane: 0.55 S.G.
- Hot Butane: 0.50 S.G.
- Hot Kerosene: 0.65 S.G.
- Hot Diesel: 0.70 S.G.
- Hot Gas Oil: 0.75 S.G.
- Hot Asphalt: 0.82 S.G.

The above calculation ignores the pressure drop across the trays. That's one reason why it's just a very approximate method. One thing for sure, if the operators at the B.P. Refinery in Texas City had made this simple check, the entire sorry incident would have been avoided. I remember Mondo Lira, my instrument technician in Texas City in the 1970s, verifying a liquid level once on start-up this way. I still recall his wise words, "Mr. Lieberman, you people are worse then stupid; you're dangerous."

Too bad British Petroleum management didn't check with Mondo before they acquired Amoco Oil. By the way, you all have any relief valves in hydrocarbon service venting to the atmosphere on your process units?

Comments on the Actual Incident

The previous description of the incident was modified for simplicity and combined with several principles learned from similar accidents. The actual accident was in reality worse than my description:

- The explosion took place in 2004.
- The source of ignition was a diesel truck.
- A high liquid level alarm failed to function.
- The relief valves did not initially open at their set pressure.
- The source of heat that finally caused the blowdown stack vent to erupt liquid was an increase in feed preheat.
- A chain valve was opened to partly bypass the relief valves.
- There is a video available on the Internet that gives a somewhat biased view blaming the operators for most of the errors.
- Management, design engineers (such as your author), operating supervision and the unit operators all share in the blame for this horrendous accident.

Troubleshooting Checklist
for Texas City-BP Refinery Fire

Level indication effected
  by density

Flow indication effected by meter
  not zeroed

Nonlinear flow indication error

Concept of being *tapped-out*
  on levels

Field checking levels
  using pressures

Vapor lock in loop seals

Dangers of loop seal drains

Blowdown stack hazards

Relief valve vent disposition

Location of temporary trailers

Sources of ignition

Process design of relief
  valve locations

Failures in training

Failures in supervision

Failures in design

Environmentally Friendly Process Concepts

L *ast night I dreamed that I died. God called me up to headquarters for the final accounting.*

"Norman," the Lord God commanded, "Tell me about your life."

"Creator of the Universe, it's rather a long story," I responded.

"Not a problem," said God. "We have all eternity."

"Master of Creation, I was born in Brooklyn in 1942. I was married in 1965 and again in 1988. I've had three children, Lisa, Joe, and Irene. Then I went to . . ."

"Norman, I know all that already," God thundered. "Thou shalt inscribe in stone a full score of Process Engineering Concepts to help save the earth from environmental destruction. Thus shall ye atone for thy sins."

"Lord of the Universe, an e-mail would be faster than a stone inscription," I suggested.

"Time is relative. And you have all of eternity," commanded the Lord God. "Go forth and atone, sinner that you are."

I dreamed on. I was standing on a granite mountain with a chisel and a hammer in my hands. And thus I inscribed in the stone as God hath commanded.

Wasting Seal Flush

If you want to see Hell on Earth, visit the Suncor and Syncrude operation in Ft. McMurray in Northern Alberta. By 2015, about 4% of the earth's crude oil will be produced from Alberta Tar Sands. So much destruction for a little bit of gasoline and diesel. It's a sin to waste any of this synthetic crude. But that's what Syncrude is doing with a sinful waste of seal flush.

Bitumen (residual tar sand extract) is pumped with centrifugal pumps using a gas oil seal flush. When a pump is running, the normal seal flush requirement is 2 to 4 gpm of seal flush per inch in the pump's shaft diameter. An expensive mechanical seal in good condition would correspond to the 2 gpm value. A cheap mechanical seal in poor condition would correspond to the 4 gpm value.

But that's not my point. If the pump is not running, it will still consume about 65% of these seal flush rates. For a very large pump with a 3-inch diameter shaft, which has both inboard and outboard mechanical seals, that's about 300 BSD of gas oil recirculated back to their fluid coker reactor. In the reactor, about 15 wt% of the gas oil is destroyed. All that needs to be done to avoid this waste is to block in the seal flush to an idled pump about 20 minutes after it is shut down. Of course, the operators must remember to open the seal flush flow before restarting the idle pump.

I've estimated Syncrude destroys about 0.1 vol% of their production by flushing idled pumps and using outmoded and inefficient mechanical seal designs (see chapter 11), on their black oil pumps.

Optimizing Pump Impeller Sizing

The sun radiated heat waves from the face of the granite mountain. But in obedience to God's commandment, my labor of atonement continued. I chiseled the story about centrifugal pumps (see chapter 9).

"I recall P-308. This was the 300 horsepower motor-driven gas-oil circulation pump at the Texaco Vacuum Tower in Eagle Point, New Jersey. The flow control valve on the pump discharge was always barely open. Perhaps 20%. The pump discharge pressure was 150 psig. About 90–100 psi was lost across the control valve. I could hear the throttling sound of the gas oil rushing through the control valve. I could hear the loud whining sound of the fully loaded motor driver calling out to me."

"Norm, Norm, help me!"

When you see a control valve in the field running less than 50% open at normal flow rates, or if you see operators throttling back on a gate valve to keep a control valve in a reasonable controllable position, check the curves for that pump. Assume that the required control valve ΔP is 20% to 30% of the frictional loss through the piping and heat exchangers. Pick out the reduced impeller size from the pump curves that will satisfy the required control valve ΔP.

At Eagle Point, I reduced the gas-oil circulation pump's impeller by 10%. Rotational work is proportional to:

$$(\text{Impeller diameter})^3 = \text{Amperage change.}$$

Thus reducing the impeller from 10 inches to 9 inches saved about 26% of the motor amperage (see chapter 11 for details), or about 58 kilowatts. My home in New Orleans consumes an average of 3 kilowatts, so the 58-kilowatt savings could power 20 homes like mine. Also, the unloaded motor ran cooler which preserved the mechanical integrity of the motor's windings. Erosion to the downstream control valve internal trim was also reduced.

The work required the time of two machinists and a pipe fitter for a full day. Power savings paid for their time in one week. The existing impeller was trimmed in the machine shop, so material costs were zero.

Variable Speed Centrifugal Pumps

My hammer became heavier with each swing. But I labored on. I inscribed the tale of the Chevron, El Segundo coker fractionator heavy gas-oil pumparound pump. This was a giant pump with a variable speed motor. An electronic device varied the frequency of the electric power to the motor to control the flow which totally eliminated the parasitic energy loss across the downstream flow control valve because there was no control valve.

Of course, variable speed electric motors are beyond the scope of my message from God. It's ordinary steam turbine-driven pumps that the Divine Creator has charged me to review. The speed of these pumps should not be controlled to hold a set speed. Rather, the pump and turbine speed should be directly controlled via the governor (see figure 36–1) which controls the motive steam flow to the turbine horsepower outlet.

This idea will permit the operators to open the existing control valve and bypass 100%. As work from the turbine varies with speed cubed:

$$\text{Work proportional (Speed)}^3$$

Reducing the turbine speed from 3,600 rpm to 3,200 rpm will save 20% of the motive steam to the turbine. Again, this totally eliminates the parasitic energy losses associated with the control valve.

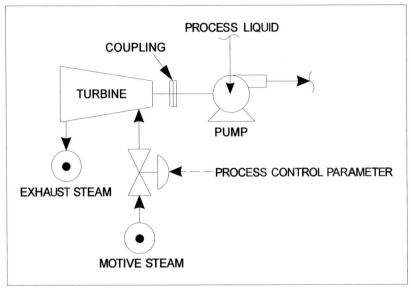

Fig. 36–1. Direct speed control by process parameter.

This concept I have seen in use at an old Gulf Refinery in Port Arthur, Texas, and at the Texaco Alky Unit in Eagle Point, New Jersey. Varying the flow of motive steam to a driver to control:

- Vessel Level
- Pressure
- Flow

is a very old practice, predating the use of control valves. Isn't this how the cruise control on your car works?

Turbine Hand Valves

In 1980, I met a girl during the endless strike in Texas City. I thought to impress her with my engineering knowledge by showing her how to save steam on a turbine. I was working as an operator on the Sulfur Complex where we had three 500-horsepower air blowers.

The governor speed motive steam supply control valves were mostly closed. The large pressure losses through these valves reduced the ability of the steam to do work. This is called an *iso-enthalpic expansion*. Or, in accordance with the Second Law of Thermodynamics, the entropy of the steam was increased.

To increase the amount of work that could be extracted from each pound of steam, I closed a hand or port valve on the steam chest. This caused the motive steam inlet control valve (see figure 36–1) to open. Now I could extract more work from each pound of steam and thus needed less steam to drive the turbine and pump.

Your turbine comes with curves that will tell you how much steam you'll save for each hand valve closed. It's about 5% to 10% per hand valve. If the governor is 100% open, try to reduce its speed by 50 to 100 rpm.

The young lady was impressed with my steam-saving ability. But it could be that relationship is one of my sins that landed me on this granite mountain with a hammer and a chisel.

Flare Sweep Gas

My former friend, Jerry called me last week. Former, because I died last night. He wanted to calculate the amount of purge gas required to exclude air from a refinery flare header.

It's true that we should maintain a slight positive pressure on the flare system. It's also true that sweep or purge gas from the refinery fuel gas system is commonly used for this function. But it is all wasteful and unnecessary. To maintain a positive pressure in the flare header without wastefully flaring fuel gas, refer to figure 36–2. The depth of the water seal determines the pressure in the flare system. No external source of purge gas is needed.

But be careful. If the water level drops too low, then air can be sucked into the flare system. If the water level gets too high, then back pressure from the flare system will retard venting vessels to the flare. I worked on this problem in a refinery in Madras, India (now known as Chennai). A seal depth of about 10 inches of water in the flare is about right.

We have all noticed flares have a tendency to puff when flaring rates are low. That's normal. It's just that the water seal is periodically blown by the gas entering and pressuring up the flare header. The water seal will reestablish itself once the flare header pressure drops. If you are bothered by the intensity of the puffs, then reduce the depth of the water seal in the bottom of your flare.

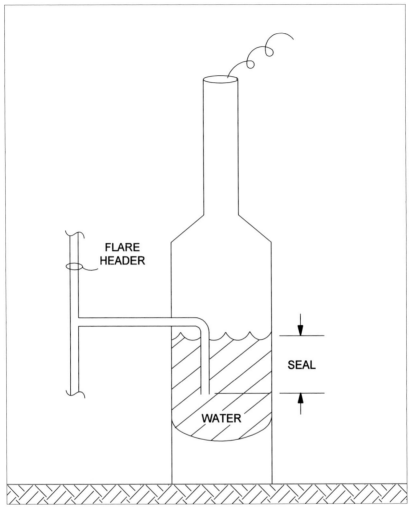

Fig. 36–2. Water seal maintains flare positive pressure.

Wasting Hydrogen

"Thou shalt not burn hydrogen in refinery fuel gas," I chiseled.

Don't read this if the hydrogen in your plant fuel is less than 10%. But, if you work in a typical refinery with an FCCU, distillate hydrotreating, a cat feed hydrotreater, delay coker unit, naphtha reformer, and hydrogen plant, then the hydrogen content of plant fuel gas should never be more than 20 mol%. Typically, each 3 moles of hydrogen generated, liberates a mole of CO_2.

When I worked in Aruba, the H_2 content of fuel gas was variable. Sometimes it was 14%; sometimes 32%. When it jumped above 20%, I would complain to the plant manager. He would ask Ray, the operations manager, if hydrogen was being vented to fuel. Ray would deny that anything was ever amiss. Regardless, next morning the lab would report H_2 in fuel was 12%.

Generating hydrogen in a steam-hydrocarbon reformer furnace consumes lots of energy in addition to venting the CO_2 absorbed from the syn gas. Think about this before you vent recycle gas to fuel from your hydro-desulferizer. Do you really need to maximize the H_2 partial pressure in your reactor all the time?

Insulation Integrity

My bare feet burned on the hot, cruel granite. It reminded me of the times I've crawled across a poorly insulated furnace transfer line. If one wanted to calculate such ambient heat losses, I suggest the following formula:

$$Q = UA(T_S - T_A)$$

where:

$\quad$ Q = Heat loss, BTU/Hr.

$\quad$ A = Area, square feet

$\quad$ T_S = Skin temperature of hot surface, °F

$\quad$ T_A = Ambient temperature, °F

$\quad$ U = Heat transfer coefficient, BTU/Hr./Ft.^{2}/°F

For a warm surface between 120°F to 200°F, when the wind is less than 10 mph, I suggest a U = 2. When the surface is hot, over 500°F, and the wind is strong, over 20 mph, I suggest a U = 4.

I check surface temperatures with my infrared temperature gun. This instrument should not be used on shiny, reflective surfaces such as stainless steel piping. The method works on both bare vessels and insulated pipe.

For large properly insulated process units, ambient heat losses may represent 1% to 2% of the total heat inputs. For smaller process units, poorly insulated, I have observed that such ambient heat losses might be over 30% of heat inputs.

When I was still alive, I was always annoyed to see removable heat exchanger channel head covers lying on the ground where they had been dropped during last year's turnaround.

Sampling Residuals and Asphalt

I am amazed at the difficulties my former clients have had in obtaining samples of hot asphalt and vacuum resid. Complex sample coolers using flushing oil were constructed. Large quantities of flushing oil were lost to the sewer. Or else, hot resid was poured directly into an open can. If spilled, the sample would auto-ignite. Once in the can the sample would smoke and the lighter hydrocarbon components would evaporate as evolved white vapors escaping from the hot can. It was all totally unnecessary.

The correct way to obtain such samples (and really almost any sample) without any hydrocarbons being lost to the environment is as follows:

1. Make a steel bottle out of a 1½-inch piece of pipe, about 12 inches long.
2. Weld a ¾-inch valve and threaded nipple to one end.
3. Connect this bottle to a bleeder on the discharge of a pump from which the sample is to be drawn.
4. Open the ¾-inch valve for a few seconds.
5. Close the valve.
6. Wait until the bottle is cool enough to handle. This may take half an hour.
7. Unscrew the entire bottle and take it to the lab.

Convective Section Air Leaks

My right shoulder ached from swinging the hammer. I had rather hoped to rest during the night. But the sun had not yet moved in the sky. So I had to continue. I chiseled the following:

A modern complex refinery consumes up to 8% of crude input as fuel and losses, if one includes:

- Heating value of imported natural gas
- Catalytic coke burn

- Imported and self-produced electric power
- Flaring
- Evaporative losses
- Plant fuel transportation requirements

The majority of this 8% is consumed in fired heaters. Typically, such heaters have leaking natural draft convective sections. These convective sections operate at a negative pressure of about 0.5 inches of water. Tramp air drawn into these leaks cool off the convective flue gas. This reduces heat recovery from the convective section as follows:

$$\Delta F = \frac{(\Delta T)(\Delta O_2)}{500}$$

where

ΔT = Stack temperature minus ambient temperature, °F

ΔO_2 = Percent oxygen difference between the flue gas in the stack and the radiant section

ΔF = Percent of furnace fuel wasted due to the cold air leaks

When the heater is out of service, close all dampers and air registers. Throw red smoke bombs into the firebox and identify the air leaks which may then be repaired with a roll of aluminum duct tape. A typical result of such an effort would reduce fuel consumption by 2% or more.

Air-Fuel Mixing Efficiency

"Thou shall repair all secondary air registers," I inscribed in stone, *"so that they may be closed when a burner is taken out of service."* You see, the purpose of a burner is to mix the combustion air and fuel. Poor air-fuel mixing efficiency requires more excess O_2 to maximize heat release from fuel. If air enters the firebox through:

- Broken sight ports
- Open air registers for burners not in service
- Pilot light ignition ports
- Miscellaneous firebox leaks

Such tramp air leaks degrade air-fuel mixing efficiency. A typical energy savings for correcting this problem might be several percent of the heater's fuel supply.

Actually, the best way to improve air-fuel mixing is to use low NO_x burners, which mix the air and fuel in two or three stages. I had been working with Murphy Oil outside of New Orleans the week before my sudden demise on their fired crude heater. They had maximized heat release from their fuel gas with an excess oxygen of 1.8%. Murphy used staged low NO_x burners with no deservable tramp air leaks. Visually, the burner air-fuel mixing efficiency looked great. This 1.8% excess oxygen is about the best I've ever seen (unless the heater is air deficient). The 1.8% excess O_2 corresponds to about 9% excess air. NO_x are also agents of global warming. Thus, another good reason to use low NO_x burners.

Air Preheaters

Preheating cold combustion air with hot flue gas is fine if the fuel is natural gas with 5 ppm of H_2S. For refinery fuel gas which has 100 to 10,000 ppm of H_2S, or fuel oil which can have 50,000 ppm (wt%) of sulfur (i.e., the Coastal Refinery in Aruba), ordinary air preheaters are not practical. The problem is the precipitation of sulfuric acid. The acid is formed by the oxidation of sulfur to SO_3.

The H_2SO_4 will precipitate at 250°F for 100 ppm H_2S fuel gas or 450°F for 5 wt% sulfur in fuel oil. Cold combustion air always causes localized H_2SO_4 corrosion and tube leaks in air preheaters. The bigger problem is that the hot flue gas side of the air preheater plugs with $FeSO_4$ deposits which restricts the flow of combustion air into the heater. The reduced air flow promotes the formation of aldehyles, ketones, light alcohols and CO. That is, atmospheric solar dimming agents.

My experience has been that the combustion air entering the air preheater needs to be kept above 120°F to retard localized H_2SO_4 precipitation. For an existing air preheater, an inline burner can be used to preheat the forced draft fan discharge from 40°F to 120°F. About 2% of the furnace fuel will be consumed.

If you have occasional peaks of sulfur in your fuel on cold days, a leaking and partially plugged air preheater can definitely be anticipated. Most air preheaters I've seen have suffered this fate. Alternately, you can wait until global warming corrects the problem for us in another 50 years.

Optimizing Excess Oxygen

I gasped for breath. It was becoming difficult to breathe. I prayed for help. God answered my prayers.

"Norman," said the Lord. "Current CO_2 concentrations are 1,385 ppm. That creates a sense of shortness of breath."

"Oh God, how could this be? It was only 390 ppm when I died."

"Mankind failed to heed my warnings!" thundered the Lord. "Global warming has heated the oceans, which have then released their dissolved CO_2 to add to the CO_2 released from fossil fuel oxidation and the calcining of limestone to make cement."

Fearfully, full of remorse, I then inscribed the following.

"CO_2 emissions from a fired heater should always be minimized by running at maximum efficiency. If the heater is running on TRC (automatic temperature control), maximum efficiency is defined as that combustion air rate that minimizes the required fuel to reach the required temperature set point. No analyzers are needed to determine this optimum point. Neither oxygen or carbon monoxide or combustible analyzers. Simply manipulate the combustion air rate to minimize fuel gas consumption.

"If the combustion air source is a forced draft blower that supplies air to the individual burner registers, then manipulating the blower inlet guide vanes will permit the operator to optimize heater efficiency, guided only by the fuel gas meter."

I recalled that at Murphy Oil, the combustion air rate was limited by a draft limitation. Then the operator would adjust the air flow to maximize the crude heater outlet temperatures.

Incidentally, analyzers located above the convective tubes are not accurate. They are strongly influenced by air leaks in the convective section, which vary with draft. But analyzers are not needed anyway to optimize the flow of combustion air.

Air flow can and should be adjusted by closing the stack damper provided that a positive pressure does not develop below the bottom row of convective tubes or below the bottom run of the radiant section shock tubes.

Calcining Cement

Over 5% of evolved CO_2 is produced from making cement and lime. Most of the CO_2 on our planet is stored as mineralized carbonates. CO_2 was extracted from seawater by shellfish. Their crushed shells eventually turned into chalk, limestone, and marble.

Producing cement in a calciner reverses the process. Also, calcining petroleum coke, used as a fuel, liberates CO_2. Any project you initiate that requires new process equipment will involve new foundations, pipe stanchions, and paved surfaces. Try to reuse existing equipment in place when expanding process units. A few of many such options are:

- **Condensers** Use low fin tube replacement bundles for clean services.
- **Reboilers.** Use high flux tubing replacement bundles.
- **Pumps.** Maximize impeller sizes and control valves ports.
- **Turbines.** Increase nozzle block port sizes.
- **Reciprocating Compressors.** Increase diameter of the plate-type pulsation dampeners.
- **Heat Exchangers.** Re-tube with stainless steel tubes to resist fouling deposit accumulation on the smooth, shiny tube surface.
- **Refrigeration Systems.** Increase the molecular weight of the refrigerant, if limited by the compressor speed.

Exchanger On-Line Spalling

"By the sweat of thy brow, thou shalt . . ." I guess I had that in common with Adam. The salty sweat ran into my eyes as I chiseled on.

"Heat exchanger efficiency in fouling services can be improved by starting and stopping the process flow. My personal experience on this subject is extensive, but limited to cold crude preheat with the crude on the tube side. Here's my procedure:

1. Bypass the crude around the exchanger. Leave the shell-side flow going.
2. Block in the upstream crude flow to the exchanger.
3. Wait fifteen minutes. While waiting, if possible, pump a few barrels of heavy aromatics into the crude side, such as

reformate splitter bottoms or light cat cracker cycle oil, to dissolve stubborn fouling deposits.

4. Open the valve you closed in step 2 and block in the tube side bypass.

The change in the tube temperature and the consequent thermal expansion and contraction of the tubes spalls off or melts off fouling deposits. I noticed this effect as a child. Pouring water over a hot rock in a campfire could split the rock that had lain peacefully for years on the forest floor.

Distillation Tower Pressure

The granite face of the mountain was covered by my inscriptions. Ages had passed. My arms were burned black by the unrelenting sun. I cried out to God, but the heavens were silent. Thus I continued.

The main consumer of energy in process plants is the reboiler and feed preheater. Most of this heat is used to generate reflux for fractionation. To minimize the tower heat input, we need to minimize the reflux rate. This can be done by optimizing the tower pressure. Too low a tower pressure will cause the tower to flood. Too high a pressure will cause the fractionation trays to weep due to low vapor velocities. Also, higher pressures reduces the relative volatility of the key components. To optimize the tower pressure, proceed as follows:

1. Place the reflux control valve on manual at a fixed flow.
2. Observe the difference between the top and bottom temperatures.
3. Reduce the tower operating pressure slowly. By slowly I mean about 1% of the absolute operating pressure (i.e., psia) every 20 minutes.
4. If the ΔT becomes larger, or remains the same, continue cutting back the tower pressure.
5. At some point the ΔT will shrink. The tower pressure that maximizes this ΔT is the optimum.
6. Cut the reflux rate until the ΔT is reduced back to that observed in step 1.
7. Cycle back to step 3 and continue on.

Note that even if ΔT remains the same, a lower tower pressure will still save energy because the column now operates cooler.

But, if the lower tower pressure causes loss of tray efficiency due to entrainment (i.e., jet flooding), you have dropped below the optimum tower operating pressure.

Methane Emissions at the Wellhead

Methane concentrations in the atmosphere have increased by about 1.1 ppm in the last 200 years. But methane is 23 times per mole more powerful a greenhouse gas than CO_2, which has increased 108 ppm in the same period. Thus, methane is contributing about 20% to global warming.

I recall that during my days on earth, the many pleasant times I spent in the desert outside of Laredo, Texas, troubleshooting natural gas wells.

But I also recall that all the instruments (level and pressure) were operated not by instrument air pressure but by natural gas, which was vented to the atmosphere. As I had 200 to 300 low production wells to care for, methane emissions must have been substantial. A month before my untimely demise, I was hiking in the swamps near my home. I noticed that Exxon had retrofitted all their gas condensate wells with bottled air for instrumentation purposes. One small step for mankind!

Centrifugal Compressors Using Gas Turbine or Turbine Drives

Cleaning a fixed-speed motor-driven centrifugal compressor rotor increases the amperage load on the driver. For variable speed drivers, cleaning the rotor, results in diminished driver speed. This reduces the driver work to the cube of the reduced speed.

For fouling services in a refinery, or in natural gas operations, I have found that injecting about 1 wt% of a higher boiling point liquid into the suction of the rotor will keep the wheels from drying out. The trick is to use a spray nozzle that creates a finely divided mist that keeps the vapor saturated as it absorbs the heat of compression.

I've seen rotors disassembled. They foul where the gas dries out. The spray pushes the dry-out point out of the rotor. I've seen this used to advantage at the Tenneco Refinery near New Orleans, and at the Amoco Plant in Texas City.

Slug Washing Exchangers

The fiery sun had drawn the moisture out of my body as I continued chiseling—no water; no shade; no evening coolness. This is rather like the day I was slug-washing the overhead condensers at the Chevron Coker in Los Angeles. Improved overhead condenser heat transfer reduced the delayed coker fractionator pressure. This reduced coke drum. Each 9 psi reduction in drum pressure increased liquid yields by 1 vol% (on fresh feed) at the expense of coke. This could permit Chevron to produce more gasoline and jet fuel and diesel oil from the same amount of crude, which could reduce crude oil production somewhere in the world by 800 BSD. I estimated my slug-washing procedure might reduce the air cooled condensers pressure drop by 3 psi. The current condenser pressure drop was 7 psi. My procedure was to water wash the worst twelve of the 24 bundles as follows:

1. The worst bundle will have the coolest air outlet temperature and the coolest process outlet temperature. The air outlet is cool because the process-side tubes are badly plugged with ammonia chloride, ammonia bisulfide, and ammonia sulfide salts. The ammonia salts are the product of the thermal cracking in the coke drums. The salts restrict the tube-side flow which reduces the process outlet temperature.

2. The salts have sublimed out of the vapor phase. They are insoluble in hydrocarbons, but very soluble in water. Add as much water as possible to one bundle at a time. Twenty minutes per bundle is sufficient. Caution! Watch the water boot level in the overhead receiver.

3. As you successfully clean a condenser, both the air outlet temperature from the condenser and the process outlet temperature will increase. The hotter the better, because the process side flow has increased. The overhead receiver temperature will gradually drop as you dissolve off the salts.

It is not necessary to diminish the process flow when slug-washing condenser tubes. For shell and tube exchangers with the process fluid on the shell side, the procedure is the same except that the operator is . . .

Suddenly a shadow past over me. A black cloud crept between my parched body and the golden red sun. A cool wind stirred the air. Then rain—glorious, blessed rain—washed my salt encrusted lips. I lifted my eyes to heaven and saw the Lord's promise to man arching above the granite mountain—a rainbow.

And then I woke. It was all just a dream. Or maybe a warning. Or maybe both.

Troubleshooting Checklist for Environmentally Friendly Process Concepts

Reducing pump seal flush
Optimizing pump impeller size
Variable speed centrifugal pump
Eliminate pump discharge
   control valve
Steam turbine hand valves
Eliminate flare sweep gas
Conserving hydrogen supplies
Measuring insulation integrity
Sampling very heavy hydrocarbons
Heater convective section air leaks
Fired heater burner efficiency

Furnace air preheaters
Optimizing excess oxygen in
   natural draft-fired process heaters
Heat exchanger online spalling
Optimize distillation tower pressure
Reduce methane emissions at
   natural gas wells
Centrifugal compressors
   turbine drivers
Gas-fired turbines
Slug-washing exchangers
   with water

The People Problem

The process engineer, when tackling a troubleshooting job in a refinery unit, faces a gamut of challenges-not the least of which is his interaction with people. The problem is complicated because of dealings with people at three levels: refinery management, operating supervision, and shift operators. The methods of communication and motivation used will be different for each level because they are driven by different goals and objectives. Each has a different view of the problem being investigated.

Dealing with Shift Operators

The secret to success in troubleshooting is gathering as much information as possible about a process operation. The largest source of this information is the daily observations made by the shift operators. Establishing a line of communications with these men and women is a high priority.

However, shift operators often resent engineers. The shift operators suspect (often correctly) that the new engineer knows much less than they do about their process unit. Here are a few nontechnical pointers useful in establishing contact with shift operators:

- Study the unit process flowsheet prior to visiting the unit. Learn the equipment numbering system.
- Wear work clothes. This shows a willingness to work with the operators at their level.
- Ask to speak to the chief operator upon entering the control room. Introduce yourself to the chief and ask permission to observe the unit's operation.
- Do not put on management airs. It is best to portray yourself simply as a technical person.
- Do not get right down to business. Drink a cup of coffee and chat awhile.

- Do not show up at shift change time. Operators use the last half hour of a shift to wash up and change clothes.
- If possible, come out to the control room on the 4 P.M. to 12 A.M. shift. The operators are not usually busy on this shift because they do not have to support the day shift maintenance effort.
- Never ask for or expect much cooperation from the graveyard shift. Everyone is simply too tired to do anything but respond to emergencies.

Try for involvement

Asking an operator's opinion and listening attentively is a simple and sure way to gain cooperation. Very often, operators know exactly what has transpired but do not understand the origin of the problem. Listen patiently to their theories because eventually they will describe personal observations.

Having gained a degree of rapport with the shift operators, get down to the business at hand. In addition to collecting the operators' experiences, you will want to observe how the unit reacts to a variety of process changes.

A chief operator will object to changes that upset the smooth functioning of the unit. Mainly, he or she is concerned that the process will become unstable and that it will be difficult to get the unit back under control. Also, the efforts being requested may seem pointless.

Clearly explain the objective. Avoid technical terms and draw pictures.

Before proposing a change in an operating parameter, you should have thought through the consequences on the process operation. Avoid making the impression that the idea to be tried is a flash-in-the-pan proposal.

The troubleshooter should be able to differentiate between a legitimate concern for the unit's integrity and operator inertia. Either factor may be a reason for shift operators to refuse to carry through a process change.

Opening valves

Regardless of your position or experience level, do not adjust valves that can affect the process operations without an operator's permission. This is true whether or not you are in a unionized plant. Ordinarily, if you know how to do the job safely you can, without specific approval of an operator, but with his or her general permission:

- Draw samples from small connections.
- Connect pressure gauges.
- Remove or install dial thermometers in thermowells.
- Blow down gauge glasses (being careful not to block in the level indicating level-trol while you do so).
- Open valves at utility stations.
- Check low-point bleeders or high-point vents.

Of course, you still need to give the chief operator a general idea of where you will be working and what you will be doing. He may have issued a hot-work permit (i.e., for welding) in the area in which you plan to catch a light-hydrocarbon sample. The chief operator is responsible for your safety when you are on the unit.

Reporting to Refinery Management

The troubleshooter will be expected to recount experiences and conclusions to management at the end of the assignment. Hopefully, you can report that the problem has been resolved. Often, you will have to recommend that the facility be shut down to repair the damage.

The sensitive aspect of such a report will be the implication that operating personnel have been doing an inadequate job. Since a percentage of process unit failures are related to operator errors, this can be a difficult problem. The objectives to achieve are:

- Get the facts of the incident across to management.
- Avoid poisoning your future relations with the operating supervisors.
- Accept credit for resolving the problem where it is deserved.

Unlike many other professions, operating process units requires a strict discipline. Misconceptions and falsehoods are put to the test quickly. If something does not work in the field, it is wrong. The people you are talking with have adjusted to this environment and expect that your report to them will conform to brevity, simplicity, and accuracy. A few techniques can help you to achieve these objectives:

1. Start out by describing the background of the problem. Don't assume the manager knows all of it.

2. Avoid jargon. Say, "We sent the debutanizer reflux pump to the shop for repairs," not "We pulled P-3B."

3. Find something nice to say about the unit operation. An operating superintendent has a piece of his ego tied up in the unit.

4. Avoid *you* and *I*; concentrate on using the pronouns *we* and *our*. Say, "We have upset the trays in the fractionator," not "You have a problem, as I have determined that the trays are upset."

5. Distribute a process flowsheet marked up with the operating data on which you will be basing your conclusions. A tabulation of field observations and supporting data should be handed out.

6. If operating blunders have transpired, ask the unit superintendent to relate them to management. You should have briefed the superintendent beforehand. It will be easier for the unit superintendent to say it than to listen to you tell it.

Theft

Part of our job as refinery process engineers is to close a plant's material balance. On several occasions I have been unable to accomplish this critical task because some of the products were stolen. A few examples might help:

- Tank field operators loading barges of jet fuel would also load jet fuel onto private boats that pulled up to their jetty on off-hours. (Yorktown, Virginia)

- When a plant was built, some foresighted person ran a ¾-inch line out into the surrounding desert from a diesel storage tank. (Middle East)

- A portion of the sulfur production from a refinery was diverted to private sale by a manager in the product sales department. (Norco, Louisiana)

- The entire delayed coke production from one refinery was periodically diverted for private sale by the manager of petroleum coke sales. (Yorktown, Virginia)

- Tank car products (rail and road) were under-weighted. This large theft was obscured by the underreporting of crude oil receipts to the refinery from Siberia. I had always wondered why the crude unit mass meters had been removed in a previous turnaround. (Baltic Republic)

- Extracted mercaptan hydrocarbons were given away to a waste processing facility. (Whiting, Indiana)

- Natural gas condensate was deliberately dumped into waste water from a two-phase separator. The waste-water tank was then drained into a tank truck for disposal. (Laredo, Texas)

- Sea water pumped out of a crude tanker typically contained a few tenths of a percent of the crude oil in the tanker. (Aruba)

- During the long 1980 strike at the Amoco Refinery in Texas City, the refinery mass balance improved by about 0.4 wt% without any apparent explanation, after taking into account flaring and evaporative losses from slop oils. I presented this anomaly to the Vice President, Dr. Horner. He reacted in a rather angry manner, as he didn't like me and often referred to me as "That troublemaker from the Process Department."

Working with Operating Superintendents

Next to a kamikaze pilot, a career as an operating superintendent is about the toughest way for a person to earn a living. The staff engineer troubleshooting a refinery problem should recognize the stress that the superintendent works under. With this in mind, here are a few helpful pointers:

1. Do not write a long, formal report documenting your recommendations with copies to every department in the company. If you have something to tell the operating superintendent, do so in person. Then follow up the conversation with a handwritten note with a single copy to his boss.

2. The superintendent will identify with both his unit and the people reporting to him. Do not bruise his ego by unnecessarily disparaging either.

3. Do not assume the superintendent knows everything about her unit. If she did, you wouldn't be there.

4. Don't surprise the superintendent in front of his boss. When you find an operating deficiency on his unit, tell him about it first.

5. Show the superintendent that you want to help her, not just proffer advice. Making a suggestion over the phone is advice; showing up at 8 A.M. in the control room to try out an idea is help.

In general, the troubleshooter should keep in mind that the operating superintendent must bear the brunt of everyone else's mistakes. Treat him kindly! Someday, you may be in his shoes.

In summary, you must get the cooperation of the people in the field and avoid establishing an adversarial position. At the same time, the ultimate objective of solving the problem and implementing a solution must be achieved. This is the real challenge in troubleshooting refinery processes. The operating engineer functions at the interface between fallible man and remorseless technology. She must understand both to be effective.

Role of the Process Consultant

The position of the outside expert, whether a company employee or an actual outside expert, is often quite difficult. From 1977 to 1980, I occupied such a position for Amoco Oil. Since 1983, I have been an independent process consultant for refinery processes. The difficulties I have encountered can be classified as:

- **Stealing ideas.** A concept that I have proposed may have been previously suggested by another company employee. Not infrequently, I have first been alerted to a process opportunity by that employee.

- **Safety issues.** My proposal is deemed to be in violation of standard refinery safety procedures. Often there are real hazards involved, but sometimes it's more a matter of a new HAZOP review meeting.

- **Creates additional work load.** Any new techniques are initially going to involve more work for someone in the plant's organization.

- **Not-invented-here problem.** If the new process concept is any good, why didn't the unit personnel think of it first, before the outside expert.

- **Current practices are wrong.** The flat-out suggestion that unit personnel have been doing something wrong for many years often is met with hostility.

- **We've tried it before.** The previous attempt to implement the new concept was tried under similar circumstances and it didn't work then.

- **The problem is not our fault.** People often try to formulate the explanation for a problem in the context that does not reflect negatively on themselves.

- **Nothing can be done anyway.** The problem is outside the scope of human intervention.

On a political level, I have never been able to contend with these sorts of challenges. I would rather rely on my concept of engineering ethics. Based on field observations, calculations, and experience, I present my conclusions to plant management. Not infrequently, my frank assessment has cost me a lucrative contract. It's a decision we each have to make for ourselves.

Norman Lieberman
New Orleans
October 2008
1-504-887-7714
norm@lieberman-eng.com

Appendix

Appendix to Chapter 1

Water Dew-Point Calculation

Step 1: Calculate the following values:

MR = Moles of top reflux

MN = Moles of light naphtha product

MG = Moles of wet gas product

MW = Moles of water (includes stripping steam and moisture from desalter)

Step 2:

$$PW = (PT) \cdot \left(\frac{MW}{(MR + MN + MG + MW)} \right)$$

where:

PT = Tower top pressure, psia

PW = Water-vapor partial pressure at top of tower, psia

Step 3:

On a steam table, look up the temperature (saturated steam) that corresponds to PW. This is the water dew-point temperature.

Appendix to Chapter 8

Calculating Bubble Points

Bubble point = condenser outlet temperature

$$P_T = \sum_1^i ViXi$$

Calculating Dew Points

Dew point = evaporator outlet temperature

$$\frac{1.0}{P_T} = \sum_{1}^{i} \frac{Yi}{Vi}$$

where:

P_T = Total absolute pressure

$\sum_{1}^{i}$ = Summation sign

Vi = Vapor pressure of ith component at condenser or evaporator outlet temperature

$\dfrac{Yi}{Xi}$ = Mole fraction of ith component in circulating refrigerant

(**Note:** Composition of the vapor leaving the evaporator and the liquid leaving the condenser is identical in a closed-loop refrigerant system.)

Appendix to Chapter 17

Rough-Tray Pressure-Drop Calculation Technique

$$\Delta P = 0.7(V_H)^2 \cdot \frac{\rho_v}{\rho_L} + 0.4(W_L)^{0.67} + \frac{H_w}{4}$$

where:

V_H = Velocity of vapor through the holes in the tray deck, fUsec

ρ_v = Density of vapor

ρ_L = Density of liquid

W_L = Weir loading, gpm per in. of weir length

H_W = Weir height, in.

ΔP = Total tray pressure drop, in. of liquid

To convert from inches of liquid (hydrocarbon) in the tower to psi multiply by:

$$\frac{\rho_L(lb/ft^3)}{1,730}$$

Note: For many types of valve trays, the following source is useful to more exactly calculate tray ΔP: *Ballast Tray Design Manual*, F. W. Glitsch and Sons Inc., P.O. Box 6227, Dallas, Texas 75222.

Index